T0327405

TACTICAL WIRELESS COMMUNICATIONS AND NETWORKS

TACTICAL WIRELESS COMMUNICATIONS AND NETWORKS

DESIGN CONCEPTS AND CHALLENGES

George F. Elmasry
XPRT Solutions, Inc., A DSCI Company, NJ, USA

WILEY

A John Wiley & Sons, Ltd., Publication

Library of Congress Cataloging-in-Publication Data

Elmasry, George F.
  Tactical wireless communications and networks: design concepts and challenges / George F. Elmasry.
      p. cm.
  Includes bibliographical references and index.
  ISBN 978-1-119-95176-6 (cloth)
  1. Communications, Military. 2. Wireless communication systems. I. Title.
  UA940.E56 2012
  623.7′34–dc23
                                                                                    2012023990

A catalogue record for this book is available from the British Library.

Print ISBN: 9781119951766

Set in 10/12pt Times by Laserwords Private Limited, Chennai, India

Contents

About the Author

Dr. George F. Elmasry was born in Egypt and received a Bachelor of Science in Electrical Engineering from Alexandria University, Egypt in 1985. He then continued on to receive a Master of Science and a Ph.D. in Electrical Engineering from New Jersey Institute of Technology (NJIT) in 1993 and 1999 respectively. He has 20 years of industrial experience in commercial and defense telecommunications and is currently leading XPRT Solutions – a DSCI Company which specializes in research and development of communications systems with an emphasis on Department of Defense Cyber, Command, Control, Communications, Computers, and Combat-support (C6) space.

Dr. Elmasry has an interdisciplinary background in electrical and computer engineering and computer science. He has much experience in many areas of these fields, including research, patenting, publication, and grant proposal activities. He has an in-depth knowledge of commercial and tactical telecommunication systems, experience with technical task leads, and team building for middle and upper management. Dr. Elmasry has over 50 publications and patents, which pertain to network management, network operations, quality of service and resource management, network and transport layer coding, joint source and channel coding, source coding, modeling and simulation of large-scale networks, security and information assurance, cross layer signaling, topology management, multidimensional interleaving, and spread spectrum communications.

In addition to his publications and patents, Dr. Elmasry has been a member of the technical committee for the annual Military Communication Conference (Milcom) since 2003, where he has led session organization, paper reviews, and session chairing. Dr. Elmasry is also a senior member of the Institute of Electrical and Electronics Engineers (IEEE), a member of the Armed Forces Communications and Electronics Association (AFCEA) International, a member of Sigma Xi, and a member of Alpha Epsilon Lambda – NJIT graduate-student honor society. Dr. Elmasry holds countless awards, including the prestigious Hashimoto Award for achievement and academic excellence in electrical and computer engineering.

Foreword

This textbook authored by Dr. George Elmasry, is the definitive text on the subject of tactical wireless networks. Dr. Elmasry, General Manager of XPRT Solutions, Inc., a subsidiary of DSCI, is an author of numerous peer reviewed papers, inventor, and patent holder, and recognized authority regarding the subject matter. The text provides a comprehensive view and unique insight in its treatment of tactical wireless networks by bridging theory with practical examples. In creating this work, Dr. Elmasry has succeeded in creating a thorough and reliable reference source for network engineers and scientists.

Steven DeChiaro
CEO and Chairman of the Board, DSCI

Preface

Before I can present design concepts or tactical wireless communications and network challenges, I feel the need to mention the challenges of writing for a field where some information is not available for public domain and cannot be included in this book's context. Another challenge is the use of military jargon and the extensive number of abbreviations (and abbreviations of abbreviations!) in the field. Engineering books are naturally dry, and I have attempted to make it light by presenting the concepts in layman's terms before diving into the technical details. I am structuring this book in such a way as to make it useful for a specialized graduate course in tactical communications and networking, or as a reference book in the field.

My own experience is similar to that of many engineers and scientists who moved from the commercial wireless communications and networks field, during the Internet boom in the 1990s, to the tactical wireless communications and networks field during the boom at the turn of the millennium. My unique experience is a result of the diverse projects I have been involved with. They have enabled me to see the larger picture of tactical networks, and have positioned me in such a place that I can take the challenge of writing this book.

The book, in many ways, draws a correlation between tactical wireless networks and commercial wireless networks, in order to make the reader aware of their different requirements, expectations, needs, and constraints of information assurance, and so on. I have always been influenced by how, at the turn of the millennium, commercial wireless networks have made more technological leaps than tactical wireless networks. Historically, in the USA, defense-based technologies were decades ahead of commercial technologies. The Internet began as a DARPA project, and defense networks were connected decades before academia and commercial entities had such technology. With the technological leaps in commercial wireless networks, expectations of tactical wireless networks increased and have yet to be fully met. If a soldier can use a lightweight powerful smart phone in his day-to-day life, the same soldier would question the necessity of carrying a few pounds of tactical radio, with diminished capabilities and features. In today's tactical wireless networks, security features (such as intrusion detection, anti-jamming, etc.) are certainly ahead of commercial networks. However, applications in tactical wireless radios are no match for the commercial capabilities that came with 3G and 4G wireless devices. Also, the scalability and price cuts in commercial wireless services seem to be decades ahead. One cannot expect the price of tactical communication equipment to be comparable to commercial wireless, given the limited market size. However, one hopes that the success of commercial wireless can influence the model used in tactical networks, in order to achieve cost saving. Reduced cost leads to leaps in technology, since funding can be channeled towards developing more advanced capabilities.

This book presents tactical wireless communications and networks in a breakdown model. The ultimate goal is to be of help to different types of readers, such as graduate students specializing in communications, or simply newcomers to the field. This book introduces tactical wireless communications and networks in terms of protocol stack layers (just like commercial wireless communications and networks), and points to the differences and similarities between tactical and commercial models. This book should also help engineers to get a better understanding of how to face the challenges in this field, through the use of concepts and design approaches introduced here, such as the open architecture model, and cross layer signaling. Hopefully this text can help engineers and scientists gear tactical network designs to a more manageable model, by breaking the general problem of a tactical wireless network node into smaller entities of what needs to be done at each stack layer. I have long been an advocate of open architecture in tactical wireless communications networking. For the most part, I am presenting tactical wireless networks with an open architecture framework. Tactical wireless networks are not the same as commercial wireless networks by any means. However, some engineering concepts that have proven successful in the commercial model, especially the open architecture concept, can be adapted by tactical wireless networks.

This book is presented in three parts as follows:

The first part is composed of four chapters, which focuses on theoretical basis. Chapter 1 is an introductory chapter. Chapter 2 discusses the physical layer as it relates to tactical radios. Chapter 3 discusses the Data Link Layer (DLL) layer and information theory in the context of tactical networks. Chapter 4 covers the Medium Access Control (MAC) and network layers as they apply to tactical networks, with some focus on Queuing Theory.

The second part is composed of three chapters and focuses on the evolution of tactical radios. In Chapter 5, legacy radios are presented with Link-16, as an example of its generation of spread-spectrum frequency-hopping radios. The Tactical Internet with IP touch points is introduced as well. Chapter 6 covers the IP-based tactical radio, with a focus on the JTRS WNW waveform as the representative of its generation of radios. The Global Information Grid vision is introduced as the natural evolution of the Tactical Internet. Chapter 7 covers cognitive radios as the futuristic generation of tactical radios.

The third part is composed of four chapters dedicated to the open architecture model of tactical networks and the seamless interoperability between tactical and commercial technologies. Chapter 8 introduces an open architecture model for tactical networks. Chapter 9 details some of this open architecture model's interface control documents (ICDs) and the deviations from the commercial model that require add-on functions to the different protocol stack layers. Chapter 10 is dedicated to the integration of commercial cellular technology into the tactical theater. Chapter 11 discusses the challenges of network management and network operations in tactical networks.

I certainly hope that this book helps the different readers to get a better grasp of the very complex problems of tactical wireless communications and networks. I also hope that it will contribute to the advancement of the field.

George F. Elmasry

List of Acronyms

2G	second generation
3G	third generation
3GPP	third generation partnership project
4G	fourth generation
ACK	acknowledgment
ADC	analog-to-digital converter
ADD	application deployment descriptor
ADNS	automated digital network system
AFH	adaptive frequency hopping
AI	artificial intelligence
AIM	advanced InfoSec machine
AMPS	advanced mobile phone system
AN	airborne network
AoC	area of coverage
API	application programming interface
ARQ	automatic repeat request
ASCII	American standards code for information interchange
ASIP	advanced system improvement program
AWGN	additive white Gaussian noise
BACN	battlefield airborne communications node
BER	bit error rate
BFSK	binary frequency shift keying
BGP	border gateway protocol
BPSK	binary phase shift keying
BSC	binary symmetric channel
C2	command and control
CAC	call admission control
CCSK	cyclic code shift keying
CDMA	code division multiple access
CDS	connected dominating set
CLS	cross layer signaling
CNR	combat net radio

ComSec	communications security
ConOps	concepts for operations
CONUS	continental United States
CORBA	common object request broker architecture
CoS	class of service
COTS	commercial-off-the-shelf
CPFSK	continuous phase frequency-shift keying
CPU	central processing unit
CR	cognitive radio
CRC	cyclic redundancy check
CRS	cognitive radio settings
CS	connecting set
CSDF	cyclic spectral density function
CSMA	carrier sense multiple access
DARPA	Defense Advanced Research Projects Agency
DAC	digital-to-analog
dB	decibel
DCD	device configuration descriptor
DFA	dynamic frequency access
DGSN	distributed GPRS support node
DiffServ	differentiated services
DISN	Defense Information Systems Network
DLL	data link layer
DNS	domain name service
DoD	Department of Defense
DPD	device package descriptor
DRSN	Defense Red Switch Network
DS	direct sequence
DS	dominating set
DSCP	differentiated services code point
DSL	digital subscriber line
DSN	dominating set neighbor
DSP	digital signal processors
DTV	digital television
DVS	DISN video services
EACK	explicit acknowledgment
ECE	explicit congestion experienced
ECN	explicit congestion notification
ECT	ECN capable transport
EDCA	enhanced distributed channel access
EF	expedited forwarding
EGP	exterior gateway protocol
EPLRS	enhanced position location reporting system
ESP	encapsulating security payload
FCC	Federal Communications Commission
FCI	framework control interfaces

FDM	frequency division multiplexing
FDMA	frequency division multiple access
FEC	forward error correction
FH	frequency hopping
FIFO	first-in-first-out
FPGA	field-programmable gate arrays
FSI	framework services interfaces
FSK	frequency shift keying
GDC	generic discovery client
GDS	generic discovery server
GGSN	gateway GPRS support node
GIG	global information grid
GN	gateway node
GNU	GNU's not Unix
GPP	general purpose processors
GPRS	general packet radio service
GPS	global positioning system
GRE	generic routing encapsulation
GSM	global system for mobile communications
GW	gateway
HAIPE	high assurance internet protocol encryption
HDP	HAIPE discovery protocol
HF	high frequency
HMS	handheld manpack and small-form-fit
HNW	high-band networking waveform
HQ	headquarter
Hz	hertz
I/O	input/output
IA	information assurance
ICD	interface control document
IDL	interface definition language
IdM	identity management
IEEE	Institute of Electrical and Electronics Engineering
IGMP	internet group management protocol
IMTS	improved mobile telephone service
INC	internet network controller
InfoSec	information security
IntServ	integrated services
IP	Internet Protocol
IP v4	Internet Protocol version 4
IP v6	Internet Protocol version 6
IPSec	Internet Protocol Security
ISDN	integrated services digital network
JDIICS	Joint Defense Information Infrastructure Control System
JGN	joint gateway node
JNMS	joint network management system

JTIDS	joint tactical information distribution system
JTRS	joint tactical radio systems
kHz	kilohertz
LAN	local area network
LDAP	lightweight directory access protocol
LPD	low probability of detection
LPI	low probability of interception
LSA	link state advertisement
LSD	link state database
LSU	link state update
LTE	long term evolution
MAC	medium access control
MANET	mobile ad hoc networking
M-ary	M-array
MBAC	measurement based admission control
MBRM	measurement based resource management
MCDS	minimum connected dominating set
MDL	mobile data link
MDS	maximum distance separable
MDS	minimum dominating set
MI	mobile Internet
ML	maximum likelihood
MPEG	Moving Picture Expert Group
MSEC	message security
MSE	mobile subscriber equipment
MSK	minimum shift keying
MTS	mobile telephone service
MTU	maximum transmission unit
NACK	negative acknowledgment
NASA	National Aeronautics and Space Administration
NCS	network control station
NCS-E	EPLRS network control center
NCW	network centric waveform
NetOps	network operations
NIPRNET	Non-secure Internet Protocol Router Network
NM	network management
NMS	network management system
NP	non-deterministically polynomial
NSA	National Security Agent
ODMA	orthogonal domain multiple access
OE	operating environment
OFDMA	orthogonal frequency division multiple access
OLSR	optimized link state routing protocol
OQPSK	offset QPSK

OS	operating system
OSI	open system interface
OSPF	open shortest path first
OTM	on-the-move
P2DP	pack-2 double pulse
P2SP	pack-2 single pulse
P4SP	pack-4 single pulse
PBNM	policy based network management
pdf	probability distribution function
PIM	protocol-independent multicast
PN	pseudo-noise
PoP	points-of-presence
POSIX	portable operating system interface
POT	plain old telephony
PPPoE	point-to-point protocol over Ethernet
PSK	phase shift keying
PUE	primary user emulation
QAM	quadrature amplitude modulation
QoS	quality of service
QPSK	quadrature phase shift keying
R&D	research and development
RBCI	radio based combat ID
RCC	radio common carrier
RF	radio frequency
RFC	request for comments
ROCs	receiver operating characteristics
ROSPF	radio OSPF
RS	Reed–Solomon
RSVP	resource reservation protocol
RSVP-AGG	RSVP aggregate
RTCP	real-time transport control
RTP	real-time protocol
RTT	round-trip timing
SAD	software assembly descriptor
SBU	sensitive but unclassified
SCA	software communications architecture
SCD	software component descriptor
SCI	sensitive compartmented information
SDR	software defined radio
SDU	secure data unit
SER	symbol error rate
SHF	super high frequency
SINCGARS	single channel ground and airborne radio system
SIP	system improvement program

SIP	session initiation protocol
SIPRNET	secret Internet Protocol router network
SiS	signal-in-space
SLA	service level agreement
SMS	short message service
SNMP	simple network management protocol
SNR	signal-to-noise-ratio
SOA	service oriented architecture
SOAP	simple object access protocol
SoS	speed of service
SPD	software package descriptor
SPI	service provider interface
SPR	software programmable radio
SRW	soldier radio waveform
STD-DP	standard-double-pulse
TCG	tactical cellular gateway
TCP	transmission control protocol
TDMA	time division multiple access
TH	time hopping
TI	tactical internet
TMG	tactical multinet gateway
TOC	tactical operational command
ToS	type of service
TransSec	transmission security
TS	top secret
TSEC	transmission security
UCDS	unifying connected dominating set
UDP	user datagram protocol
UHF	ultra high frequency
U-MAC	universal media access control
UML	unified modeling language
UMTS	universal mobile telephone system
USAP	unifying slot assignment protocol
USB	universal serial bus
USRP	universal software radio peripheral
UWB	ultra wideband
VHF	very high frequency
VoIP	voice over IP
WAN	wide area network
WIN-T	warfighter information network – tactical
WNW	wideband networking waveform
WRAN	wireless regional area network
WSGA	wireless system genetic algorithm
XML	extensible markup language

Part One

Theoretical Basis

1

Introduction

This book assumes that the reader has basic knowledge of wireless communications, including different types of wireless links, an understanding of the physical layer concepts, familiarity with medium access control (MAC) layer role, and so on. The reader is also expected to have a grasp of computer networks protocol stack layers, the Open System Interface (OSI) model, and some detailed knowledge of the network layer and its evolution to Internet Protocol (IP) with protocols such as Open Shortest Path First (OSPF) in addition to Transmission Control Protocol (TCP) and User Datagram Protocol (UDP) over IP. Also, some knowledge of topics such as queuing theory is assumed. In the first part of this book, we will review some of these topics in the context of the tactical wireless communications and networking field before we cover the specifics of the field.

One can divide engineers and scientist in the field of tactical wireless communications and networking into two main groups. One group emphasizes the physical layer and dives into topics such as modulation techniques, error control coding at the data link layer (DLL), and the air interface resource management, and so on. The second group emphasizes networking protocols diving into the network layer, the transport layer, and applications. The first group, primarily composed of electrical engineers, likes to build radios assuming that everything above the DLL that has to interface with the radio is commercial-off-the-shelf (COTS). The second group, mostly composed of computer scientists, assumes that everything below the network layer is just a medium for communications. This book will explain tactical wireless communications and networking in a balanced manner, covering all protocol stack layers. This will provide the reader with a complete overarching view of both the challenges and the design concepts pertaining to tactical wireless communications and networking.

The evolution of tactical wireless communications and networks followed a significantly different path from that of commercial wireless communications and networks. A major milestone of tactical wireless communications development occurred in the 1970s, with the move from the old push-to-talk radios to the first spread spectrum and frequency hopping radios (with anti-jamming capabilities). Since World War II, commanders and soldiers on the ground have effectively communicated with radios forming voice broadcast subnets. These subnets, with their small area of coverage, functioned independent of a core network. Over time, the core networks were created to link tactical command nodes to the command-and-control (C2) nodes and then to headquarters. Commanders on the ground carried push-to-talk

Tactical Wireless Communications and Networks: Design Concepts and Challenges, First Edition. George F. Elmasry.
© 2012 John Wiley & Sons, Ltd. Published 2012 by John Wiley & Sons, Ltd.

radios to communicate with their soldiers and relied on communications vehicles to link them to their superiors through a circuit switched network with microwave or satellite links. Although enhanced versions of these technologies are still deployed today, this book will consider them as legacy architecture and "a thing of the past." This text focuses on the Global Information Grid (GIG) vision where the tactical theater is full of IP-based subnets that communicate seamlessly to each other with network management policies that enforce the military hierarchy.

1.1 The OSI Model

Every computer networking book, in one way or another, emphasizes the OSI model. The OSI model has its roots in the IBM definitions of networking computers from the early days. Defining such interfaces, while the science of computer networking was a new field, was an effective approach to accelerate the development of networking protocols. With the OSI model, there are different protocol stack layers, with each stack layer performing some predefined functions. These protocol stack layers are separated and utilize standard upward and downward interfaces. These layers work as separate entities and are peered with their corresponding layers in a remote node. Figure 1.1 demonstrates a conventional OSI with seven layers: the application layer, the session layer, the transport layer, the network layer, the DLL, the MAC layer, and the physical layer. These layers communicate to their peer layers (A and B are peer nodes) as shown by the horizontal arrows. Traffic flows up and down the stack, based on well-defined interfaces as indicated by the vertical arrows. Note that some text books may have different variations of these layers. For example, some textbooks may present a presentation layer under the application layer and before the session layer to perform data compression or encryption. Other models (especially for point-to-point links) omit the MAC layer, but we are especially interested in the MAC layer in this book

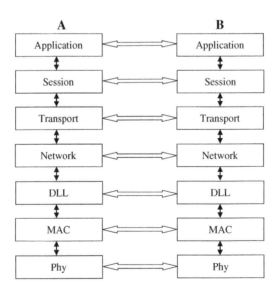

Figure 1.1 Conventional OSI protocol stack layers.

since it is an important part of multiple-access tactical radios. Some models also refer to the IP layer as a network sub-layer. Here, we consider the IP layer as the network layer using IP. Regardless of these variations of the OSI model, the wide use of IP today has created a standard network layer with standard interfaces below it (IP ports to radios, point-to-point links, multiple access wired subnets such as Ethernet, optical links, etc.) and IP ports above clients, servers, voice over IP (VoIP), video over IP, and so on.

Let us summarize the OSI model before we jump into the tactical wireless communications and networking open architecture model. You can refer to other computer networking books to read more about the OSI model details.

The *application layer* is simply a software (SW) process that performs its intended application. Your e-mail is a simple example of an application process that runs on your PC or phone (which is essentially a network node).

The *session layer* has roots in the plain old telephony (POT) networks where call connection information is given to the transport layer. Today, the session layer could be used for authentication, access rights, and so on.

The *transport layer* can be understood as two peer processors (in nodes A and B in Figure 1.1) that perform the following necessary functions:

- Break messages down into packets when transmitting, and reassemble when receiving.
- If the layers below are not reliable (packet loss is encountered), a reliable end-to-end protocol may be utilized by this layer.
- Perform end-to-end flow control. Please refer to TCP flow control as an example of this function.
- Session multiplexing (if there are many low rate sessions between the same node pair), and session splitting if there are high rate sessions going between the same node pair.

The *network layer* in the OSI model is ideally a single process per node. As you will see later, this convention changes with the tactical model. This layer performs many tasks including packet-based flow control and routing. A network packet has a payload and a header. The header contains the information needed for this process to perform its intended functions (packet flow control, routing, etc.). Notice that with the layer independence of the OSI model, the header information of the network layer is independent of the header information of the DLL layer. With the OSI model, each layer generates its own header information. The network layer packet (including the header) is treated by the DLL as a set of information that needs to be delivered reliably to the peer processor. Each layer generates its own header based on the information it has, or parameters passed from the adjacent layers. The network layer also generates its own control packets (e.g., Link State Updates–LSUs, route discovery packets, etc.). For the remainder of this book, we will adhere to the convention of referring to the IP layer as the network layer.

The *data link layer*, also known in some textbooks as the *data link control layer*, is a peer processor that ensures the reliability of the underlying bit pipe. The network layer above can send packets reliably, based on the DLL protocols. The DLL treats the network layer packets as a stream of information bits that need to be transmitted reliably. The DLL adds a header and trailer to the packet received from the network layer, forming a DLL frame. Note that this frame length is not constant since the network layer packets are of variable length. The DLL overhead size (headers and trailers) depends on the error control coding protocol used. Certain DLL protocols can stop transmission in the presence of errors above a specific threshold. Other DLL protocols may not have error correction capabilities and depend on the reliability of the transport layer. Other DLL protocols may have error detection and/or error correction capabilities. How much reliability should be at the DLL; how much reliability should be at the higher layers of the protocol stack (on a hop-by-hop basis); and how much reliability should be left for the transport layer (for an end-to-end path over the network) is a matter of debate. Network coding is an interesting area of research that you should look at to understand this much-debated topic. Cross layer signaling can also be used to optimize the performance of the protocol stack where the tradeoff between DLL overhead and transport layer overhead needed for reliability can be optimized.

The *medium access control layer*, or MAC layer, plays a major role in multiple access waveforms. Contrary to point-to-point links, with multiple access waveforms, this layer is responsible for managing the multiple access media. Protocols for collision avoidance (making sure two nodes do not transmit on multiple access media at the same time) are at the heart of the MAC layer. In this book, you will see examples of tactical IP radios where different types of medium control are implemented.

The *physical layer* transmits a sequence of bits over the physical channel. There are many physical media used today ranging from your cable/digital subscriber line (DSL) modem at home to your cellular phone, Ethernet cable, and optical media. Each medium has its modulation/demodulation technique that maps a bit or a sequence of bits to a signal. Notice that the bit stream from a MAC frame has to be transmitted in a synchronous manner since each bit needs a specific time duration t to be transmitted. The medium's speed or rate in bits per second is $1/t$. The MAC layer emits bits to the physical layer at this rate. Intermit periods between MAC frames (when the MAC layer has no information to send) means that the channel is idle for these periods. Before IP became the standard for the network layer, each physical media needed a MAC or DLL layer that could interface to it, and standardization was a nightmare. Now, we have IP-based modems that can interface to an IP port. IP is the gold standard and should stay with us for a long time. As you will see in the rest of this book, we approach the open architecture for tactical wireless communications and networking as IP-based and thus can bring a wide variety of technologies to the war theater, making them communicate seamlessly.

The OSI concept allowed the science of computer networking to evolve quickly, since the separated entities meant that engineers could focus on developing the layers of their specialties without having to worry about defining interfaces to upward or downward layers. Computer networking relies on this model with some changes.

1.2 From Network Layer to IP Layer

With the IP model, the community has established a five-layer model as shown in Figure 1.2, with the application layer, transport layer, IP layer, DLL/MAC layer, and the physical layer. In addition to the reduction of the OSI layers from seven to five layers, this model refers to the IP as "layer 3" and the DLL/MAC "layer 2" while the physical layer is layer 1.

Notice that the relationship between the network layer and the DLL/MAC is one-to-many. The network layer can send and receive packets from multiple DLL layers (links). This concept evolved with Internet Protocol to become an IP port. An IP router can also have ports dedicated to multiple workstations, servers, and so on, each with its own transport

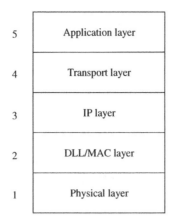

Figure 1.2 The IP as the network layer with a five-layer OSI model.

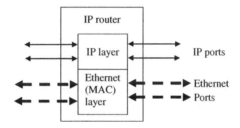

Figure 1.3 IP routers with IP and Ethernet ports.

and session layers. As we dive into tactical networks learning about the tactical edge, we will debate the benefits of relying on the transport layer for reliability versus relying on the tactical edge IP layer, where aggregation of traffic allows us to exploit the benefits of statistical multiplexing in achieving reliability. You will also learn about addressing reliability at the transport layer, tactical edge, DLL, and/or through network coding. You will also be introduced to cross layer signaling through this five-layer model. You will see why IP-based tactical radios can have some layer 3 capabilities with rich cross layer signaling to layer 2.

Figure 1.3 shows a conceptual view of an IP router with both IP ports and Ethernet (MAC) ports. With this naming convention, a workstation carrying a client or a server can be Ethernet-based and can connect to an Ethernet port in a router or a switch. IP-based tactical radios can connect to an IP port. Although the peering of the protocol stack layers still applied to the IP-based model in Figure 1.2, the IP layer is the focal point of this protocol stack model and IP route discovery can be of a client/server, a wired subnet, a wireless subnet, and so on.

1.3 Pitfall of the OSI Model

The literature is full of criticism of the OSI model and any of its deviations including the IP model. The fact of the matter is that IP technology is now the dominating technology

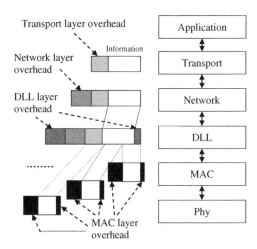

Figure 1.4   Overhead with the OSI stack layers.

and it is with us to stay. Techniques such as cross layer signaling, merging of protocol stack layers (especially layers 2 and 3), tradeoffs between network coding, and transport layer reliability are the path to more optimal performance of both commercial and tactical wireless communications based on the IP model. The attempts to develop a new technology or a super-layer concept are facing many challenges and could take a very long time to materialize, given the dominance of IP. One of the known problems with the OSI model is the amount of overhead bits transmitted over the physical media in comparison to the information bits. Since each layer works independently with its own headers, the ratio of overhead to information contents can be very high. Figure 1.4 demonstrates this problem where the information content in the packet (at the transport layer) is expressed by the white rectangle. The transport layer adds its own header(s) shown in light gray (think of UDP and RTP–real-time protocol–headers or the TCP header). The network layer adds its own header, shown in the medium gray (consider the IP header). The DLL treats the entire packet (information and headers) as a bit stream and adds its own overhead that may contain redundancy bits for error correction, as well as trailers, as expressed in the dark gray portion. The DLL also breaks down the bit stream corresponding to the packet (payload and headers) into small segments (to create MAC frames) with the MAC layer adding its own MAC header and trailers shown in black, and so on. Especially with small payload packets (such as VoIP), the ratio of information bits to the actual bits modulated over the air could be very small.

If you consider the accumulative amount of overhead bits compared to the information bits (for each packet that is carrying application-layer information) and that network protocols at the different stack layers generate their own overhead packets (control packets), you will see how inefficient the OSI model is. Control traffic comes from protocols such as session initiation and session maintenance, negative and positive acknowledgment, and so on. The IP layer introduces a large amount of overhead from routing protocols. Consider Hello packets, links state database (LSD) packets, LSU packets, and so on. In tactical mobile ad hoc networking (MANET), as the number of nodes per subnet increases, this control traffic volume increases, and considering that some links and radios have limited bandwidth, the ratio of control traffic to user traffic can get extremely high and further increase the inefficiency of the OSI model. Moreover, because of the unreliable nature of tactical wireless links, redundancy packets from error control coding such as network coding can further decrease the available bandwidth resources for user traffic.

1.4 Tactical Networks Layers

The OSI model does not apply directly to the tactical networks for many reasons to include security and information assurance requirements. It is the US National Security Agent (NSA) that defines the Communications Security (ComSec) standards for IP-based networking. NSA defines the High Assurance Internet Protocol Encryption (HAIPE) as the standard for IP-based encryption. Notice that HAIPE differs from commercial Internet Protocol Security (IPSec) in many ways including hardware separation of the plain text IP layer and the cipher text IP layer. HAIPE is the GIG ComSec encryption. Notice also that coalition forces may be required to adhere to a form of ComSec similar to HAIPE. The introduction of ComSec in tactical networks creates two network layers at each node. One is the plain text network layer (sometimes referred to as the red IP layer) and the other is the cipher text network layer (sometimes referred to as the black IP layer) separated by the encryption layer. As will become clear later in this book, the plain text networks are separated from the cipher text core network, creating two independent networking layers. If we take the OSI model in Figure 1.1 and try to create a tactical networks equivalent, we would need to introduce some modification to include ComSec as a layer by itself since HAIPE standards require the plain text and cipher text IP layers to work independently, as separate entities. Also, in a tactical wireless networking protocol stack model, one would need to consider a form of cross layer signaling where control signaling between the stack layers is allowed. Cross layer signaling is covered later in this book where we show how ComSec introduces more complexity to the cross layer approach of tactical networks.

With IP-based tactical wireless communications and networks, as you will see in subsequent chapters, the physical, MAC, and DLL layers form what we will refer to as the radio. The network layer plays a major role on top of the radio with ComSec encryption creating two independent IP layers. Figure 1.5 shows the mapping of the OSI model to the tactical networks model. Notice that the radio implementation of the DLL and MAC layer can have a different mutation specific for each tactical or commercial radio. Also notice the presence of plain text IP and cipher text IP as two independent layers separated by ComSec encryption.

With the model shown in Figure 1.5, COTS-based IP routers can interface to a tactical radio that has IP ports and has an IP layer that peers to the IP layer of the router. COTS-based IP routers can also interface to tactical links that have a simple IP modem. In either case, we will refer to the layers below the IP layer as the radio, to maintain consistency and to present tactical wireless networking in an open architecture theme.

The remainder of the first part of this book will cover some of the important theoretical bases for each of the tactical networks stack layers modeled in Figure 1.5. These theoretical

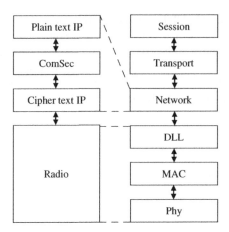

Figure 1.5 Mapping of the OSI to the tactical networks stack layers.

bases are presented in the context of tactical MANET, explaining the challenges and design concepts of tactical wireless communications and networks.

1.5 Historical Perspective

One can argue that computer networking has its roots in the T-carrier system developed by Bell Labs in the 1960s for digital telephony. The term integrated services digital network (ISDN) was coined when a worldwide telephone network was developed. Following IBM's introduction of the OSI concept, many proprietary networking techniques started to appear. The DARPA NET project was a major breakthrough in linking computers, not only the telephone sets. This project then turned into the Internet, and IP began taking root. Now, the tactical communications and networking field is in the path to be completely IP-based. The vision of the global information grid reaching the tactical theater requires that a network of networks should offer seamless communications to the warfighter anywhere in the world, and IP was selected to be the way forward for this vision. Today, the war theater could potentially be full of sensors, robots, unmanned vehicles, and so on. Also, there is a new generation of warfighters who are from the information age and are used to texting and sending images and short video clips. In addition, there is an explosion of potential applications that can aid the warfighter in his mission. All these factors created the need for greater capacity in the tactical theater and the need for seamless communications between the different subnets, leading to a new generation of tactical radio such as the Joint Tactical Radio System (JTRS) radios and the research in cognitive radios, as well as exploring the use of commercial wireless technologies in the tactical theater. This opens the door for tactical wireless communications and networking to utilize an open architecture approach. One can expect, as time goes by, that the boundaries between tactical and commercial wireless research topics to gradually lessen to where both communities are interested in cognitive radios, cross layer signaling, layer merging, the role of network coding, and so on.

Bibliography

1. Bertsekas, D. and Gallager, R. (1994) *Data Networks*, 2nd edn, Prentice Hall.
2. Schwartz, M. (1987) *Telecommunication Networks Protocols, Modeling and Analysis*, Addison-Wesley.
3. Tanenbaum, A. (1996) *Computer Networks*, 3rd edn, Prentice Hall.
4. Elmasry, G. (2010) A comparative review of commercial vs. tactical wireless networks. *IEEE Communications Magazine*, **48**, 54–59.
5. Lee, J., Elmasry, G., and Jain, M. Effect of security architecture on cross layer signaling in network centric systems. Proceedings of Milcom 2008, NC9-3.
6. Elmasry, G. and D'Amour, C. Abstract simulation for the GIG by extending the IP cloud concept. Proceedings of Milcom 2005, U503.

2

The Physical Layer

While the previous chapter presented an overview of the protocol stack layers and how they apply to tactical networks, this chapter presents integral details to the physical layer, such as modulation technique, signal detection in the presence of noise, signal attenuation, fading models, the role of spread spectrum, and frequency hopping (FH) in tactical radios. The material in this chapter is intended to present the physical layer details as they apply to tactical wireless radios.[1]

2.1 Modulation

As a junior high-school student, I was introduced to the concept of modulation when my science teacher asked me if I could get to my grandmother's house, which was in a different town, on foot. I responded with "I would not be able to do it. It is far away and we usually take the train." The teacher told me "Aha! The train is your carrier; you get *modulated* on the train, which can go faster and when you arrive to your grandmother's town, you are *demodulated* from the train." As I later studied modulation in depth, this example struck me as being inaccurate. Nevertheless, even to this day, I believe that the comparison of the train to the carrier signal, and the comparison of myself to the information signal still rings true. A simple example of modulation is cell phone use. Your voice is at a frequency ranging between 300 Hz and 6 kHz and the frequency is modulated over the radio frequency signal between the cell phone and the base station. This carrier signal remains at a much higher frequency and can propagate for long distances with much less attenuation than your actual voice signal.

Before we continue, let us review some of the important aspects of digital modulation over a point-to-point link or a multiple access medium. A stream of bits is modulated at the sending node, transmitted over the medium and demodulated at the receiving end. A symbol is a sequence of bits in the stream that allows us to modulate one or more bits at a time. Modulation techniques are designed around the number of symbols, M, needed to be transmitted over the channel. For example, we can have $M = 2$ symbols (0 and 1) and we

---

[1] The tactical communications and networking community commonly use the term "radio" to indicate a point-to-point or multiple access wireless communications technology. In this book, we assume that the protocol stack layers below the IP layer are part of the radio.

Tactical Wireless Communications and Networks: Design Concepts and Challenges, First Edition. George F. Elmasry.
© 2012 John Wiley & Sons, Ltd. Published 2012 by John Wiley & Sons, Ltd.

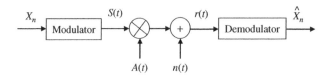

Figure 2.1 Generic channel model of modulation of digital signals.

modulate one bit at a time; or we can have $M = 4$ symbols (00, 01, 10, and 11) where we modulate two bits at a time, and so on. An M-ary modulator takes as an input one of the M symbols and produces one of M analog waveforms that can be transmitted over the physical media. There are many functions needed to be incorporated in the physical layer in order to make modulation work effectively to include frequency conversion, signal amplification, and transducers.

Figure 2.1 presents a generic channel model which focuses on modulation and demodulation of a digital signal. In this model, we start with M symbols and have a signaling rate of T_S. That is, every T_S seconds, the modulator gets an input $S(T)$ which represents one of the M symbols. At any given time t, the output of the modulator $S(t)$ is one of M waveforms. $S(t)$ can be expressed as $S(t) = S_{X_n}(t)$. The waveform $S(t)$ can be a set of voltages or currents (over a wire), a set of optical signals (over an optical medium), a set of electric field intensities (over the air), and so on. In many cases, $S(t)$ is limited to a time interval bounded by T_S seconds.[2] Simply, the modulator emits signals at a rate of $R_S = 1/T_S$ symbols per second. Note: M is a power of two in most cases.[3] If the symbol rate of this channel model is R_S, then the bit rate is mR_S or m/T_S bps.

In the model presented in Figure 2.1, the modulated signal is distorted by attenuation, fading, noise, and other interference. $A(t)$ expresses attenuation and fading while $n(t)$ expresses added noise. The received signal, in most cases, is very different from the transmitted signal. Communications references use models for attenuation and fading such as Rayleigh, Rician, and log-normal models. One of the most common noise models implemented in the field is the additive white Gaussian noise (AWGN) model. Reference [1] is known as a classic textbook example of noise models. Although these models help engineers design excellent demodulators, there are some caveats that accompany them. For starters, some of these models are linear, while channels are not. For example, signal amplifiers are non-linear; this is especially true for high power signals such as satellite communications where the channel non-linearity characteristics can considerably deviate from the models used in the design. Models could be time-invariant, while in reality channel gain is time variant. Another issue in channel modeling is the difficulty in accounting for frequency dispersion in a channel model. The AWGN model commonly used is often far from reality. With tactical radios, we need to consider the fact that the enemy could use sophisticated jammers that introduce noise patterns as far away from AWGN as possible; in such a case, the AWGN model seems

---

[2] In some waveforms, spectrum management techniques call for pulse overlap – detection and estimation of modulated signals is an interesting field of study that you can reference. Simply remember that a pulse is not an even-shaped pulse and the leading and tail ends of it can have low spectrum density. The tradeoff between pulse overlap and pulse energy at the receiver is an interesting area of study.

[3] For one bit per symbol, M = 2 (0 or 1). For two bits per symbol, M = 4 (00, 01, 10, 11). For three bits per symbol, M = 8 (000, 001, 010, 011, 100, 101, 110, 111), and so on. The $T1$ example presented in this section shows an unusual case where we have two digital symbols and three modulated signals.

too simple to use. Another issue to consider for tactical radios is the possibility that the same radio can be deployed in various environments (flat terrain, mountain areas, or urban environments) where each of these deployments introduces different channel characteristics. Physical and environmental effects (rain, fog, etc.) are another reality of tactical radios. The warfighter needs to have reliable communications under these adverse conditions. Demodulators provide many techniques to enhance signal detection by increasing the probability of detecting the symbol corresponding to the transmitted signal. These techniques will be reviewed later in this chapter.

Channel characteristics are the starting point of deciding which signal to use as a carrier. Noise, interference, and distortion are just examples of channel characteristics. Bandwidth is a key factor in considering modulation techniques. When bandwidth is limited, spectrum shaping techniques are used to reduce signal distortion and simplify signal decoding. The tactical radios carried by soldiers or used with unmanned sensors need to be very compact, so battery size needs to be reduced as much as possible. Reducing power consumption becomes an important factor in designing modulation techniques. Consequently, signal decoding at low signal-to-noise-ratio (SNR) becomes a driving factor. Tactical radios can trade SNR for greater bandwidth occupancy. The design and use of error control coding techniques are considered in conjunction with the design of the modulation techniques (the relationship between modulation and error control coding is absolutely crucial in tactical radio design) to provide optimal radios. Another set of common channel characteristics in tactical radios comes from mobile platforms (e.g., jet fighters moving at high speed) which cause a Doppler shift. Jamming also plays a major role in designing tactical radios. These characteristics result in a distorted carrier signal (which is sinusoidal with a phase reference angle). The demodulator can sometimes fail to find the carrier signal phase reference angle. With these radios, engineers design a demodulator that is capable of processing the received signal without the need for the carrier signal phase reference; these are "non-coherent" demodulators.

Let us review a case of modulation that exemplifies how the channel characteristics affect the modulation technique selected. This case is used in the North American $T1$ digital transmission system, for which the physical medium is a coaxial cable or twisted wire pair that is used for long distances. This medium does not allow for long consecutive positive or negative polarity (the signal is presented as current over the wire and has to alternative for transmission over long distances to reduce attenuation). In the $T1$ case, binary symbols are used (0 and 1). *Zero* is transmitted as no current and *one* is a current pulse. Current pulses use alternating polarity. Figure 2.2 demonstrates this example with a sequence of bits 001101101. In Figure 2.2, T is the transmission time per bit. The $T1$ modulation technique can be considered an unusual type of M-ary modulation where the analog signals have $M = 3$ (M analog waveforms), but the digital signals have two symbols.

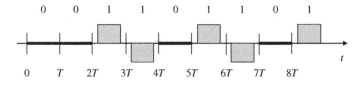

Figure 2.2   Alternate polarity in $T1$ transmission.

2.1.1 Signal-in-Space (SiS)

Signal-in-space (SiS) is an important part of tactical radios design; this section gives a review of its basic concepts.

Let us begin this discussion from the principles of harmonics. The classical Fourier series is one example of expressing a signal in terms of harmonics or frequencies as series. Mapping the coefficients of the series to coordinates in a space can create a geometric representation of the signal in that space. Demodulation techniques utilize this concept of SiS, and turn the waveform detection problem into a decision theory problem where estimation of which symbol the received signal represents turns into a decision of which of the original signals is closer to a random variable (representing the received signal). References [2–4] are classic books in this area.

Let us review this concept starting from defining the unit of energy assuming that the signal is a voltage signal across a one ohm resistor. The integration of the square value of the signal (energy) over a specific time period (T_i, T_f) is *one*. This signal is expressed as $\phi_j(t)$.

2.1.1.1 Orthonormal Signals Sets

Let us start from a set of waveforms expressed as $\{\phi_m(t), m = 0, 1, 2, \ldots\}$. This set is defined over the specific time period (T_i, T_f). Notice that m and $T = T_f - T_i$ can be finite or infinite (imagine infinite dimensions). This set of waveforms is *orthonormal* if,

$$\int_{T_i}^{T_f} \phi_i(t)\phi_j(t)dt = \delta_{ij} = \begin{cases} 1, & i = j, \\ 0, & i \neq j, \end{cases} \tag{2.1^4}$$

Let us discuss two different examples of orthonormal sets. The first example is the set of sinusoids which have frequencies that are multiples of $1/T$. This set can be expressed as:

$$\{\phi_m(t)\} = \left\{ \left(\frac{1}{T}\right)^{1/2}, \left(\frac{1}{T}\right)^{1/2} \cos m\omega_0 t, \left(\frac{1}{T}\right)^{1/2} \sin m\omega_0 t, \ldots \right\}, m = 1, 2, \ldots, \tag{2.2}$$

where $\omega_0 = 2\pi/T$.

Any pair within Equation 2.2 satisfies Equation 2.1, even if the set was infinite.[5] Figure 2.3 shows a set of n sinusoidal (Fourier) orthonormal frequencies during a time interval $T = T_f - T_i$.

The second example is shown in Figure 2.4 as a set of four pulses that are not overlapping. A finite set of N non-overlapping pulses can be expressed as:

$$\phi_m(t) = \left(\frac{1}{T'}\right)^{1/2} rect\left[\frac{1 - mT'}{T}\right], \quad m = 0, 1, 2, \ldots, N - 1, \tag{2.3}$$

where $T = (T_f, -T_i)/N$ and the pulse unit (height at $t = 0$) is $rect\left[t/T\right]$.

$\phi_m(t)$ is the basis for creating SiS. This can be understood in two steps. The first step is to imaging a set of signals (M-signals). The second step is to express each signal in this set.

---

[4] In Equation 2.1, δ is the impulse function, which has a value of 1 at a specific instant and 0 elsewhere.
[5] It is hard for a human to imagine an infinite set of orthonormal signals because we live in a 3-D space perception. Mathematicians, however, have expressed formulas in N dimensions, and even infinite dimensions, since the 1700s.

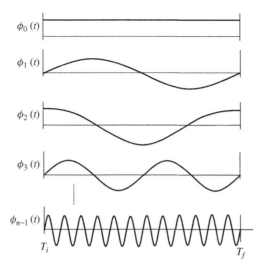

Figure 2.3   Set of n sinusoidal orthonormal frequencies.

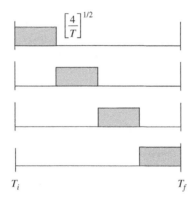

Figure 2.4   Set of non-overlapping four pulses forming an orthonormal basis.

A signal in this set (the ith signal) is expressed in the N-dimensional space as an N-term series using N-basis functions as follows:

$$S_i^{(N)}(t) = \sum_{m=0}^{N-1} S_{im}\phi_m(t). \tag{2.4}$$

This signal has a finite energy that is expressed by the series coefficients as follows:

$$S_{ik} = \int_{T_i}^{T_f} S_i(t)\phi_k(t)dt. \tag{2.5}$$

Notice that Equation 2.5 is obtained by taking Equation 2.4, multiplying both sides by $\phi_k(t)$ and integrating over the time interval (T_i, T_f). Recall that $\phi_k(t)$ is one of the basis functions. In other words, if we project the ith signal onto the kth basis function we get Equation 2.5. Due to the condition expressed in Equation 2.1, we can now conceive N-dimensional orthogonal projection and utilize the geometric characteristics of the signal to detect and estimate which symbols the received signal represents, while reducing the probability of erroneously decoding a signal.

Notice that N is not fixed. Not every set of signals has the same dimension. However, theoretically, as N grows, the integral square error diminishes. That is, for a given set of signals $\phi_m(t)$, convergence is not affected by adding more dimensions to the signal. The set is then called a complete set or said to have a complete orthonormal basis. The Fourier set (the example shown in Figure 2.3 is just one example) is known to have a complete orthonormal basis since, for a finite period of time and with signals with finite energy, the integral square error diminishes as N grows.

So, given that we can use sinusoidal waveforms (each sinusoidal waveform is a harmonic) as the basis to construct signals, and given that we can create a signal of different harmonics with different coefficients expressing the energy of each harmonic, and given that we can express this signal as an N-dimensional series, and given that we can construct a different signal (with a different variation of the power of each harmonic), we can imagine having an N-dimensional space where the different signals can be projected over this N-dimensional Euclidean space. Now, electromagnetic waves (as one example) can be constructed as such, and transmitted over a medium. The receiver will receive the transmitted signal plus noise, attenuation, and so on. The receiver can project the received signal over the N-dimensional space and find out which of the original signals is closer to the received signal. Now with this understanding, you are ready to move on to the next subsection where we define signal constellations and orthogonal signals. Many tactical radios are designed based on orthogonal signals.

2.1.1.2 Signal Constellations and Orthogonal Signals

Let us visualize a set of signals: each signal $S_i(t)$ is mapped to a point in an N-dimensional Euclidean space (signal space where N is the size of the basis set). This point representation in the signal space is $S_i = (S_{i0}, S_{i1}, S_{i2}, \ldots, S_{i(N-1)})$. Now let us show how a signal is constructed, mapped to the signal space, and how it is recovered.

Figure 2.5 provides an N-dimensional signal with base functions $\phi(t)$ used to construct the ith signal in a signal set (of size M). Figure 2.5 simply correlates to Equation 2.4.

Now, imagine this signal as one of M signals in an N-dimensional space. If this signal is received, one would want to construct it. That is, one would want to recover the coefficients $S_i = (S_{i0}, S_{i1}, S_{i2}, \ldots, S_{i(N-1)})$, since the base functions are known. Ideally, in the absence of noise or attenuation, we would receive the exact signal $S_i = (t)$ and project it over the N-dimensional space to find the coefficients. Equation 2.5 demonstrates how this is accomplished by projecting the received signal over the kth basis function to obtain the kth coefficient. Figure 2.6 is the application of Equation 2.5 for all N-coefficients.

The science of modulation and coding defines a *signal constellation* as a set of M points in N-space. A given set of information symbols of size M or less is mapped onto a signal constellation of size M, which is projected onto the N-dimensional space. The geometric characteristics of this definition are utilized to design modulation and coding techniques and in detection/estimation of modulated signals. Let us define the signal energy using the defined geometric characteristics. To do this, we must start by conceptualizing a two-dimensional

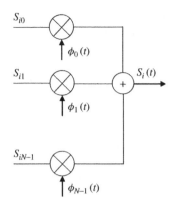

Figure 2.5 Constructing a signal.

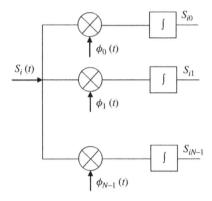

Figure 2.6 Recovering signal coefficients.

space, and envisioning a given point in this space. The squared distance from the origin to this point is simply the energy of the signal corresponding to this point. The squared distance representing signal energy applies to any N-dimensional space. Now, envision two points in the two-dimensional space; each point corresponds to a signal with its own energy. The squared distance between these two points (the squared intrasignal distance) is an important factor as well. Now, if we use the signal definitions in the equations above, we can express the squared distance of the ith signal in the constellation in the N-dimensional space as $\sum_{m=0}^{N-1} S_{im}^2$ and the squared intrasignal distance between the ith and the jth signals as $d_{ij}^2 = \sum_{m=0}^{N-1} (S_{im} - S_{jm})^2$. A specialized book on modulation may show you, in detail, how the squared distance of the ith signal is actually the signal energy driven from the integration of the square of the ith signal over the time interval (T_i, T_f) and that the squared intrasignal distance d_{ij}^2 between the ith and the jth signals is actually the integration of $[S_i(t) - S_j(t)]^2$ over the time interval (T_i, T_f).

There are some rules that should be followed when designing modulation techniques. First, N should be minimized; there exists a direct relationship between increasing N and requiring higher bandwidth. Also, increasing N can complicate modulation techniques. Choosing the basis functions can simplify or complicate modulation. The basis functions and the signal set chosen need to have some degree of correlation. The best way to get a sense of picking basis functions with a signal set is through practice.

Example 2.1: We have a binary message (0, 1), and we have an antipodal signal Antipodal refers to one signal being the negative of the other, regardless of how the signal is shaped. We can express this case as $S_1(t) = -S_0(t)$. One can create a single space for this set where $\phi_0(t) = cS_0(t)$. Note that c is a constant (needed to normalize the signal energy). The constellation for this signal is simple as shown in Figure 2.7.

Example 2.2: We have a binary message (0, 1) – no antipodal pulse signal We can elect the binary signal set (S_0, S_1) over T_S as shown in Figure 2.8a. The base functions (ϕ_0, ϕ_1)

Figure 2.7   Binary message and antipodal signal constellation.

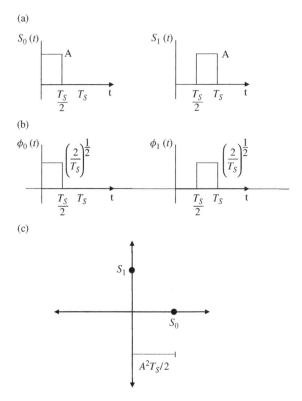

Figure 2.8   (a) binary signal set; (b) basis functions (scaled version of the signals); (c) signal constellation.

are chosen as shown in Figure 2.8b, such that they correspond to the signals (they are scaled appropriately to the signals). Notice that both (S_0, S_1) and the base functions are orthogonal. The result of this selection is the two-dimensional signal space shown in Figure 2.8c. Notice that with this selection, each signal has a squared distance from the origin equal to $(A^2 T_S/2)$. If you calculate the energy of each signal in the set (S_0, S_1) from Figure 2.8a, you will find that it is $(A^2 T_S/2)$ as well. In this case, the signal energy is $E_S = A^2 T_S/2$. Notice how the squared intrasignal distance (the squared distance between S_0, and S_1,) is $A^2 T_S$.

One can think of different variations of this example. For example, if we make the number of pulses equals to four (still with no overlapping) instead of two, we can create a four-dimensional space. However, this space is not minimal.

Example 2.3: Binary phase shift keying (BPSK) and M-ary phase shift keying (M-ary PSK) Now let us think of another approach to modulating the binary message of symbols (0, 1). We can choose a base function set that is sinusoidal (driven from a carrier signal with frequency $\omega_c t$). Let us think of creating the base functions through phase shifting of the carrier signal, using a phase shift in multiples of $2\pi/M$ radians. Let us modulate the signal for a period of T_S, while using a reference signal energy equal to E_S. Using trigonometry, one can create a set of two orthonormal functions as follows:

$$\phi_0(t) = \left(\frac{2}{T_S}\right)^{1/2} \cos \omega_c t, \tag{2.6a}$$

$$\phi_1(t) = \left(\frac{2}{T_S}\right)^{1/2} \sin \omega_c t. \tag{2.6b}$$

One can then find that these base functions can create a two-dimensional constellation for any given set of M information messages. Any signal in this constellation is expressed by:

$$S_i(t) = \left(\frac{2E_S}{T_S}\right)^{1/2} \cos \left(\omega_c t + \frac{2\pi i}{M}\right). \tag{2.7}$$

In the case of the binary message $(M = 2)$ we are interested in showing that we can have:

$$S_0(t) = \left(\frac{2E_S}{T_S}\right)^{1/2} \cos \left(\omega_c t\right) \tag{2.8a}$$

$$S_1(t) = \left(\frac{2E_S}{T_S}\right)^{1/2} \cos \left(\omega_c t + \pi\right) = -\left(\frac{2E_S}{T_S}\right)^{1/2} \cos \left(\omega_c t\right), \tag{2.8b}$$

which yields a *single*-dimension space.

The relevance of creating this binary phase shift keying (BPSK) of a sinusoidal waveform, as opposed to the pulsed choice above, should now be apparent. Not only does the BPSK use a single dimension, any M size array can use a two-dimensional signal space. This is referred to as the M-ary phase shift keying (M-ary PSK). The 4-ary PSK can be expressed as follows:

$$S_i(t) = \left(\frac{2E_S}{T_S}\right)^{1/2} \cos \left(\omega_c t + i\frac{\pi}{2}\right), \quad i = 0, 1, 2, 3, \quad 0 \le t \le T_S. \tag{2.9}$$

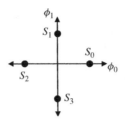

Figure 2.9   Cartesian representation of 4-ary phase shift keying.

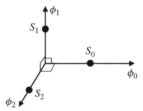

Figure 2.10   Cartesian representation of a three-dimensional orthogonal signal set.

Figure 2.9 presents the 4-ary PSK signal set in Cartesian representation. This is referred to as the quadriphase shift keying or quadrature phase shift keying (QPSK), or 4-PSK. Note how this can also be considered a bi-orthogonal set.

Example 2.4: General orthogonal signals A signal set is orthogonal if all the signals in the given set have equal energy E_S and

$$\int_{T_i}^{T_f} S_i(t)S_j(t)dt = \begin{cases} E_S, & i = j \\ 0, & i \neq j \end{cases}. \tag{2.10}$$

One would expect that $N = M$ with orthogonal signal sets, and that the base functions would be a scaled version of the signals. That is, $\phi_i(t) = cS_i(t)$, where c is a constant. In this N-dimensional signal constellation, each signal is a distance $E_S^{1/2}$ from the origin along the corresponding coordinate axis. Figure 2.10 demonstrates the Cartesian representation of 3-ary signal constellation.

2.2   Signal Detection

Although this book does not specialize in types of noise, fading models, or the details of signal detection, we will review some important cases that are relevant to tactical radio SiS. We will start with the simplest case: detecting a binary antipodal signal with equal energy, as shown in Figure 2.7, in the presence of AWGN. The AWGN causes the received signal to be in a different location of the single-dimensional Cartesian space as shown in Figure 2.7. Intuitively, one can decide that the received signal corresponds to S_0 if it is positive ($S_i > 0$) and corresponds to S_1 if it is negative ($S_i < 0$). Practically, a binary correlation receiver is

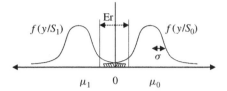

Figure 2.11   Binary antipodal signal detection in the presence of AWGN.

implemented (with foreknowledge of the existing of AWGN). Here, we will focus essentially on the detection technique without diving into the mathematical representations. The AWGN creates a Gaussian distribution of the noise random variable as shown in Figure 2.11. Think of this as the projection or superimposing of the AWGN represented by the bell curve (the probability distribution function or pdf of the noise random variable) onto the signal which has energy difference $(S_0(t) - S_1(t))$ of E_d. The curve on the right-hand side represents the conditional pdf of the received signal y, given that the transmitted signal is S_0, while the curve on the left-hand side represents the conditional pdf of the received signal y, given that the transmitted signal is S_1. The mean of y is μ_0, given that the transmitted signal is S_0; whereas μ_1 is the mean of y given that the transmitted signal is S_1. Then σ^2 is the variance of the decision variable (and is not dependent on which signal is transmitted, but rather depends on the noise energy and the signal distance). For the case of the binary antipodal signal, the optimal threshold midway between the two averages is zero. Notice that as the noise energy increases, σ increases, causing the bell curve to widen and the shaded areas under the curve to increase, which increases the probability of detecting the wrong signal (bit error rate). One can see the importance of optimizing the signal energy difference such that $(\mu_0 - \mu_1)$ is maximized. The shaded area under the curve is referred to as the Gaussian tail integral or the probability of error detection $p(\varepsilon)$. In the case of the binary antipodal signal it is given by:

$$p(\varepsilon) = \int_0^\infty \frac{1}{\left(2\pi N_0/2\right)^{1/2}} e^{-n^2/N_0} dn = Q\left[\left(\frac{2E_S}{N_0}\right)^{1/2}\right], \qquad (2.11)$$

where E_S is the signal energy or the squared signal space distance.

Equation 2.11 emphasizes the importance of increasing the signal distance to minimize the probability of error detection (since we have no control over noise). The term "signal-to-noise ratio" is used widely in literature to indicate the health of the received signal. You can find tables in other communications references of the Q-function in Equation 2.11 which express the direct relationship between SNR and the probability of detecting the wrong signal. The probability of detecting the wrong signal translates into the probability of bit error rate as explained below.

The probability of bit error rate is referred to as "hard-decision coding," where a bit is determined to be either zero or one. soft-decision coding can be best conceptualized from Figure 2.11, but will be properly introduced later in this book. With soft-decision decoding, a signal detected at the far right is decoded as S_0, a signal detected at the far left is decoded as S_1, and a signal detected around the cut-off region (middle) is decoded as an erasure. As you will see, some error control coding techniques are more powerful when coupled with erasure.

Note that marking a symbol as an erasure requires defining an erasure threshold. As the erasure threshold approaches the cut-off value (Er approaches zero in Figure 2.11), we approach hard-decision demodulation. On the other hand, the wider the erasure area (larger Er), the more erasure symbols are decoded, making the error control code used as mostly erasure code. On one hand, the increase in erasure symbol frequency can make the error control coding more powerful, but on the other hand, it can cause the loss of valuable information since a symbol marked as an erasure might have been decoded correctly. There are tradeoff studies for the relationship between soft-decision demodulation and erasure coding that attempt to find the optimized cut-off for symbol erasure that maximizes the benefits from soft-decision demodulation and erasure-based error control coding.

2.2.1   Signal Detection in Two-Dimensional Space

It is easier to see the AWGN in the single dimension of the binary antipodal signal. As we move to two-dimensional space, the signal points can exist anywhere within this space. We need to then get an enhanced perception of how the midway points are selected (or how we define the decision zone for each signal in this two-dimensional space). Figure 2.12 attempts to help us conceptualize the existence of AWGN in the two-dimensional space and how decisions zones are established. First, imagine the pdfs of the AWGN as a mountain or a church bell (instead of a bell curve in the single space). Imagine this mountain expressed as contours over the two-dimensional space. The inner circles of these contours are higher probability density and the outer circles are lower probability density. Second, instead of a midway point, imagine the need for a dividing line or a perpendicular bisector that defines each signal decision zone. In Figure 2.12, we have S_0 and S_1 placed randomly in the two-dimensional space and we define r as the received signal plus noise.

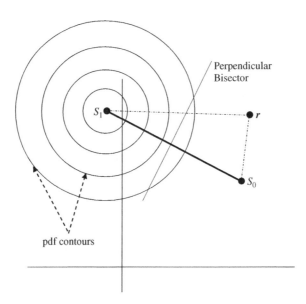

Figure 2.12   Conceptualizing two-dimensional signal space and decision zones.

The geometry of the two-dimensional signal space shows that if the distance between r and S_0 is less than the distance between r and S_1, r is in the S_0 region and the detected signal will be decoded as the symbol corresponding to S_0. On the other hand, if the distance between r and S_1 is less than the distance between r and S_0, r is in the S_1 region and the detected signal will be decoded as the symbol corresponding to S_1. Notice that the only component of the noise vector that can cause a decision error is that in the direction of the thick line connecting S_0 to S_1. Notice also that this noise component has a zero mean and its variance is expressed as $N_0/2$. Similar to the binary antipodal signal, if this noise variable is more than half of the distance between S_0 and S_1, a decision error occurs. The equivalent to Equation 2.11 showing the probability of bit error is,

$$p(\varepsilon) = \int_{d/2}^{\infty} \frac{1}{(2\pi N_0/2)^{1/2}} e^{-n^2/N_0} dn = Q\left[\frac{d}{(2N_0)^{1/2}}\right]. \tag{2.12}$$

Both Equations 2.11 and 2.12 emphasize the fact that the only factor to minimize the probability of error detection is the squared signal-space distance. So when we design modulation techniques, we simply need to maximize the squared signal-space distance given the boundaries of the signal energy that we can use. In general, looking at Figure 2.12, the distance between the two signals is bounded by $2E_S^{1/2}$. That is, the distance, d, becomes $d \leq 2E_S^{1/2}$. If you consider the case of the binary antipodal signal with energy E_S, the distance, d, is actually $d = 2E_S^{1/2}$. You will then find that the optimal binary signal design is the antipodal design that yields the least probability of error detection.

Now let us cover the case of detecting M-ary PSK signal, which is commonly used with tactical radios and satellite modems. Equation 2.7 defines the signal in the general M-ary case. The basis functions are defined in Equation 2.6. If $M = 2$, we get the BPSK case, or PSK, which yields the antipodal set shown in Figure 2.7. The $M = 4$ case is shown in Figure 2.9. In general, an M-ary signal constellation has M points equally spaced on a circle with radius $E_S^{1/2}$. Figure 2.13 demonstrates the case of $M = 8$ and how the decision zones based on AWGN can be easily established.

The M-ary PSK is a practical modulation scheme for many reasons. First, simple phase shifter technologies of standard sinusoidal carriers are used. Second, signal amplification

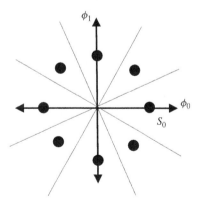

Figure 2.13 Decision zones for 8-ary phase shift keying.

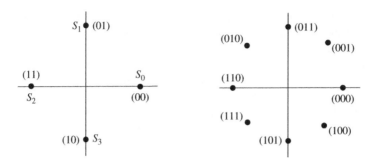

Figure 2.14   Gray code labeling for 4-ary PSK and 8-ary PSK.

does not cause much distortion since the carrier amplitude is constant.[6] Third, we can increase M and make the channel absorb more bits per second at higher signal-to-noise ratio. Tactical radios can have different modes of PSK based on the distance requirements, terrain, and so on. Each mode can use different signal power, and a different error control coding approach can be used to ensure that the bit-error rate is bounded by a required level. A specialized book on digital communications can explain the tradeoffs between increasing the symbols per second at the channel and adding more redundancy bits to tolerate increased bit error rate (when signal power cannot be increased). Such references cover techniques like soft-decision coding, trellis coded modulation, and so on, which can further increase tolerance to lower SNR. We will cover the Gray-code for bit labeling to give you a sense of how modulation and error control coding are related.[7]

With M-ary PSK, certain errors are more likely than others. The nearest neighbor errors are certainly the most likely to occur. Gray-code labeling makes sure that adjacent signals differ in the fewest bits possible. M-ary PSK can assure that adjacent signals differ only in a single bit. Figure 2.14 explains the Gray-code labeling for the cases of $M = 4$ and $M = 8$.

Understanding the probability of bit error in M-ary PSK helps to explain why it is so common. A good approximation would suggest that symbol error will occur only with adjacent symbols.[8] By using Gray code, only one bit error will occur in $\log_2 M$ bits. As a result, the probability of bit error P_b is the probability of signal error P_S divided by $\log_2 M$, that is,

$$P_b = \frac{P_S}{\log_2 M}. \tag{2.13}$$

Now let us apply Equation 2.13 to the case of 4-ary PSK. This means $P_b = P_S/2$. If we consider that the energy per bit is half of the energy per signal, this means that $E_S = 2E_b$.

---

[6] Signal amplification is necessary and these devices tend to introduce non-linear amplification. Phase shift or frequency modulation can protect the modulated signal against the non-linear amplification.

[7] In the case of orthogonal signaling, where the number of signals and the number of dimensions are the same, bit labeling is not needed since added noise will shift the received signal vector toward the other coordinates equally.

[8] Considering that the pdf of AWGN has lower density as distance increases, error probability for non-adjacent symbols is negligible in comparison to that with adjacent symbols. As the SNR increases, this approximation becomes more valid.

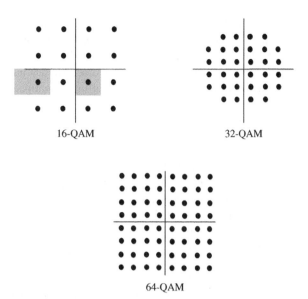

Figure 2.15 Quadrature amplitude modulation (QAM) signal constellation for 16, 32, and 64 symbols.

Now let us go to Equation 2.12 knowing that, with the 4-ary PSK, the distance between adjacent symbols is given by $d = (2E_S)^{1/2}$. This will yield,

$$p_b = Q\left[\left(\frac{2E_b}{N_0}\right)^{1/2}\right]. \tag{2.14}$$

Equation 2.14 shows that the probability of bit error rate with the 4-ary PSK with Gray code labeling is the same as the binary antipodal shown in Equation 2.11. The gain we get from 4-ary PSK over binary antipodal comes from the fact that we modulate two bits at a time instead of one bit at a time. In other words, 4-ary PSK with Gray code labeling can occupy half the spectrum of that of binary PSK for a given bit rate. Note that the orthogonal bases $\cos \omega_c t$ and $\sin \omega_c t$ occupy the same spectrum region. As an exercise, try to find the probability of bit error rate for the case of 8-ary PSK with Gray code labeling.

Now let us move to discussing quadrature amplitude modulation (QAM) starting from the 4-ary PSK, which can be considered the simplest form of QAM. In QAM, we modulate the orthogonal bases $\cos \omega_c t$ and $\sin \omega_c t$ by equally spaced distances (signal amplitude) and create simple constellations. Figure 2.15 provides examples of 16 QAM, 32 QAM, and 64 QAM. Can you think of the Gray code labeling for each case? Start with the 16 QAM. Can you see how the inner signals have four neighbors while the outer signals (edge and corner) have three or two neighbors? You can see how the decision zones are easily defined. The figure shows some decision zones for the 16 QAM case. However, the probability of bit error rate can be calculated as an upper bound in QAM given that it is not the same for all points in the constellation.[9]

---

[9] The study of probability of bit error rates with QAM can be found in digital communications references and is beyond the scope of this book.

As you have seen, a 4-ary PSK is superior to binary antipodal because it can double the bit rate for the same SNR channel. However, QAM is superior to 4-ary PSK. With the 16 QAM case, we quadruple the bit rate; nevertheless, we would need to increase the SNR in order to obtain the same bit error rate for the AWGN case. Studies have shown that we need to increase SNR by 6 dB with 16 QAM over that of binary antipodal in order to stay at the same bit error rate.

In tactical wireless communications, nodes that are stationary or on the quick halt can use microwave links in the range of 1–30 GHz. Microwave links offer high capacity wireless links and have an acceptable bit error rate. Naturally, they need hours to set up a mast and aim the dish. Tactical operation centers (TOCs) tend to use microwave links for reach-back to their command and control (C2) nodes. For example, one common microwave link in the 4 GHz range with channel bandwidth allocation of 20 MHz uses 64 QAM. Now that we are modulating sinusoidal signals, theoretically we can have up to one modulated symbol per hertz.[10] Practically, this type of link can transmit 15 Msps (mega symbol per second). With 64 QAM, we transmit six bits per modulated symbol giving us 960 Mbps. If Gray code labeling is used, a symbol error will result in one bit error (as a good approximation). At SNR of about 26 dB, the bit error rate of such a link can be in the range of 10^{-5}. The vendor of such a link will supply specifications about the relationship between the bit error rate and SNR. During bad weather conditions one can increase the signal power (microwave links have the advantage of a narrow beam allowing increasing power without interfering with other signals), in order to accommodate higher noise and still have the link operate at the same bit error rate.

More powerful microwave links that use up to 1024 QAM constellation offer a higher bit rate over the same spectrum and require a higher SNR.

2.2.2 Multidimensional Constellations for AWGN

So far, we have seen the one- and two-dimensional signal constellations for AWGN. One can conceptualize these worlds easily, and with little extra effort can conceptualize the three-dimensional world. With sinusoidal waves, we can obtain the orthogonal bases $\cos \omega_c t$ and $\sin \omega_c t$. The challenge we have been addressing for one- and two-dimensional signal sets is how (for a given M) to minimize the average signal power, and maximize inter-signal distance (to the nearest neighbor). The one- and two-dimensional vectors we have been using can be generalized to multidimensional vectors in multidimensional spaces. The mathematical representation of these multidimensional signals is very interesting to study. For example, if we compare a two-dimensional constellation to a four-dimensional constellation while keeping the number of bits per signal space unchanged, a 1.5 dB gain can be achieved. Another 1.5 dB gain can be achieved from eight-dimensional space over four-dimensional space. As modulators and demodulators have to deal with multidimensional spaces, they rely on look-up tables and use maximum likelihood detection, based on lookup tables. We transmit a point in signal space based on a correspondence of this point to a given signal. We then receive a different point expressing the signal and noise, and then use a lookup table to find the maximum likelihood signal that we can map the received point to.

---

[10] Understanding how to select the amplitude modulation pulse shape and the nature of the modulated signal power spectrum as well as the available technologies that can achieve the 1 symbol per hertz limit is beyond the scope of this book.

2.3 Non-Coherent Demodulation

If you operate a high-end oscillator at 200 MHz, it can shift by 1 Hz (per second, which is very small theoretically) leading to a phase error of 2π radians. Techniques for phase locking have been developed where the demodulator estimates the phase of the carrier signal. However, these techniques are at times ineffective and may need frequent phase locking periods (with channels that transmit in bursts such as time division multiple access – TDMA – this can make the channel very inefficient).

What makes non-coherent demodulation more important in tactical wireless communications is node mobility, which causes the phase of the carrier signal to be shifted; this is known as the Doppler shift. Doppler shift occurs in addition to the oscillator issues mentioned above.

So far, we have been addressing signal detection based on known references of the orthogonal bases functions ($\cos \omega_c t$ and $\sin \omega_c t$ in the case of PSK and QAM). This ideal case assumes that the demodulator is phase-synchronized to the received signal phase while given the amplitudes of the signals to be detected (known as coherent detection). When the received signal reference phase is distorted (imagine that the reference axis of the received signal is randomly shifted) the demodulation problem becomes more complex. The demodulator now has to use *non-coherent demodulation*, which assumes that the phase angle is a random variable with a mean around the transmitted phase (known phase).

The key objective of non-coherent demodulation is to reduce the frequency's uncertainty (due to oscillator and Doppler shift) to a fraction of the symbol rate. The error in the signal phase is assumed to be uniform over $[0, 2\pi]$, and constant over a period of time (which is achieved by properly choosing the oscillator given the transmission rate of the channel). The use of signals with equal energies (as in orthogonal sets) is needed to make non-coherent detection easy to implement. The non-coherent detection process is similar to that of coherent detection in the sense that we express the received signal as a vector and implement maximum likelihood detection. With non-coherent demodulation, we express the received signal as a vector in terms of the orthonormal functions (original base set) used to describe the signal and in terms of the orthonormal functions that are in phase quadrature with the original base set. The maximum likelihood decoder works with available data from $2N$-vector. Non-coherent demodulation requires more complex circuitry than coherent demodulation. Understanding non-coherent demodulation in detail requires diving into a few topics which are beyond the scope of this book, such as expressing the signal in complex envelope notation, and expressing the signal in phase quadrature. If you specialize in designing tactical radios, you will certainly need to delve into these topics.

2.4 Signal Fading

Now that you have some fundamental information regarding non-coherent demodulation and know that the received signal has uncertainty in its phase (in addition to the uncertainty created by AWGN), let us visit another uncertainty factor derived from deep amplitude fading of the signal. So far, you have seen how the probability of detecting the signal correctly depends on SNR. This dependency weakens with channel fading. Error control coding techniques developed to remedy the consequences of fading are visited in the next chapter. In this section, we will look at the effect of fading without factoring in error control coding.

The channel gain $A(t)$ is essentially a Rayleigh random process (which means that the signal power varies with time). As a result, in channels where fading is expected, one should shy away from QAM because there are multiple signal energies and the error in estimating the signal amplitude will have greater impact. PSK and FSK (frequency shift keying) are commonly used with fading channels. Also, as the signal amplitude fades, phase shifting can occur as well. This would make non-coherent detection a must for use with channels with deep fading.

Let us try to define the problem, in the sense that we need to know the average error probability calculated over a fading period. In other words, we need to know how much signal detection error we will encounter in the presence of fading. This will allow us to revisit our SNR requirements in order to compensate for the error due to fading. For example, if we require a certain SNR to achieve 10^{-5} bit error rate in the presence of AWGN, we may increase SNR to make this error rate less than 10^{-5} so that the presence of fading gives us the reliability we want to achieve. This approach will demonstrate how big of a penalty we can pay for fading, if our design is not flexible.

If the channel had a fixed gain a, then our probability of signal detection error can be expressed as $P(\varepsilon/a)$. In the presence of fading, which is expressed as a random process $f_A(a)$, we can calculate the average error probability as:

$$P_S = \int_0^\infty P(\varepsilon/a) f_A(a) da. \tag{2.15}$$

Equation 2.15 applies to other slow fading models such as Rician and log-normal. The pdf of the Rayleigh fading can be expressed as $f_A(a) = 2ae^{-a^2}$, where $a \geq 0$.

Now let us visit the probability of error detection which we estimated earlier by adding the presence of fading. Let us start from Equation 2.11 for the binary PSK antipodal signaling and add the presence of Rayleigh fading. Equation 2.11 can be expressed in terms of bit energy (bit energy and signal energy are the same for binary antipodal signaling) as $p_b = Q[(2E/N_0)^{1/2}]$. In the presence of Rayleigh fading, assuming the channel amplitude is a, it can be expressed as the conditional probability:

$$P(\varepsilon/a) = Q\left[\left(\frac{a^2 2E_b}{N_0}\right)^{1/2}\right]. \tag{2.16}$$

Notice that $a^2 E_b/N_0$ is the ratio between bit energy and bit noise density at a given instant.

Now let us find the unconditional average error probability from Equation 2.15.

$$P_b = \int_0^\infty (2ae^{-a^2}) Q[(a^2 2E_b/N_0)^{1/2}] da. \tag{2.17}$$

Processing this integration is beyond the scope of this book but can be found in specialized references. The result of this integration is:

$$P_b = \frac{1}{2}\left(1 - \left[\frac{E_b/N_0}{1 + (E_b/N_0)}\right]^{1/2}\right). \tag{2.18}$$

For $E_b/N_0 \geq 20$ or $13\,\text{dB}$, Equation 2.18 can be approximated as:

$$P_b \approx \frac{1}{4E_b/N_0}. \tag{2.19}$$

Equation 2.19 suggests that to obtain a probability of bit error rate of 10^{-5} we need E_b/N_0 to be $44\,\text{dB}$. When fading is not present, the binary antipodal signal requires E_b/N_0 to be $9.6\,\text{dB}$. Notice that $44\,\text{dB}$ is almost 2500-fold of $9.6\,\text{dB}$ (dB is a logarithmic scale of SNR). If we loosen our requirements to suggest that we need bit error rate to be 10^{-5} in the absence of fading and 10^{-4} in the presence of fading, we require an increase of SNR of only by about $10\,\text{dB}$. Yet we can loosen our constraints even further and require, for example, that 90% of time the probability of bit error is less than 10^{-5}. Given that fading periods can be short, there will be no need to increase SNR. As will be explained in the next chapter, error control coding techniques that are suited for fading channels can be used instead.

Performing similar analysis on coherent orthogonal signals with Rayleigh fading, the equivalent to Equation 2.19 is:

$$P_b \approx \frac{1}{2E_b/N_0}, \tag{2.20}$$

which yields a $3\,\text{dB}$ loss comparing to antipodal signaling (this is expected since envelope fading has no impact on phase).

For non-coherent orthogonal signals with Rayleigh fading, the equivalent to Equation 2.20 is:

$$P_b \approx \frac{1}{2 + (E_b/N_0)}, \tag{2.21}$$

which gives poorer performance than that of coherent orthogonal signals (4–$6\,\text{dB}$ factor).

2.5 Power Spectrum

The power spectral density (power spectrum) of the modulated signal is important to understand and is related to the signal energy efficiency studied so far. Note that the power spectrum is measured as an averaged property of the signal and is in the order of seconds (which can hide certain properties of the signal). Thus power spectrum should not be the only factor in choosing a signal for a given channel. However, selecting the signal format and studying the power spectrum leads to good design of communication media. A channel can have a certain frequency response that will affect the signal power spectrum. The channel frequency response can differ, based on the modulation technique. For example, if we have a channel that has poor frequency response at low frequencies, then we should select a signal that has low power density at the low frequency range.

Regulation also plays a major role in selecting a signal with specific power spectrum characteristics. For example, microwave links must have a certain spectrum pattern to satisfy regulations. At $10\,\text{MHz}$ away from the microwave center (carrier) frequency, the power spectrum measured over $1\,\text{kHz}$ bandwidth should be $60\,\text{dB}$ below the signal total power.

In the commercial world, telecommunications authorities such as the Federal Communications Commission (FCC) in the USA regulate spectrum patterns to ensure best use of the available spectrum. Even with military communications, there are spectrum regulations, and frequency bands are assigned to military communications systems with regulations that control spectrum efficiency.[11] In frequency division multiplexing (FDM), studying the power spectrum is essential to ensure that adjacent channels' spectral overlap is accounted for in designing these systems.

Theoretically, a signal is not bounded by a limited bandwidth, since its Fourier transform has infinite extent in frequency. Practically, a signal has a range of frequencies where most of its power spectrum is located. One can say that a signal has 90 or 99% of its spectrum concentrated in a given frequency range (bandwidth). The spectrum outside of this range is referred to as the sidelobes. One needs to be aware of the tradeoffs pertaining to spectral design. A signal that has low spectral density at the sidelobes, which reduces its interference to other signals, may have achieved this at the expense of making the main lobe wider. A wide main lobe can suffer more distortion during channel filtering.

The power spectral density describes the modulated signal in probabilistic terms that express the power distribution over frequency. This description is assumed to apply regardless of what information is carried by the signal.

The detailed mathematical description of power spectrum is beyond our scope. Let us work through an example that has historical significance and was published in 1928 by Nyquist [5]. This example should give you an idea of how power spectrum can be modeled in a way that is close to reality. In this problem, the pulse is not time limited and has important time domain and frequency domain characteristics. In the frequency domain (Fourier transform), it is described as a raised cosine pulse as shown in Equation 2.22.

$$
\Phi_0(f) = \begin{cases} T_S^{1/2}, & 0 \leq |f| < \dfrac{1-\beta}{2T_S}, \\[2ex] T_S^{1/2} \cos^2\left(\dfrac{\pi T_S}{2\beta}\left[|f| - \dfrac{1-\beta}{2T_S}\right]\right), & \dfrac{1-\beta}{2T_S} \leq |f| < \dfrac{1+\beta}{2T_S}, \\[2ex] 0, & \dfrac{1+\beta}{2T_S} < |f|, \end{cases} \tag{2.22}
$$

where $0 < \beta \leq 1$.

Figure 2.16 illustrates this raised cosine function. Notice that the power spectrum density is shown for one side (positive side). The transition from the passband (frequency band where the signal spectrum is expected) to the stopband (frequency band where we should not expect to find spectrum) is not a sharp transition but follows raised cosine function characteristics. The $1 + \beta$ is known as the excess bandwidth factor. The significance of what Nyquist demonstrated in his work was that the ideal situation with no inter-pulse interference only requires a minimal bandwidth of $\frac{1}{2T_S}$. As β increases, inter-pulse interference increases.

---

[11] It is worth noting that on continental USA, frequency bands assigned to military communications can be different from that assigned to US forces overseas (depending on the host countries). Even in war times, frequency bands are selected so as not to interfere with the local commercial communication systems, and sometimes differ from the enemy frequency bands such that jammers can be used on enemy communication systems freely. The latest tactical radios can operate on a wider frequency band to meet these challenges.

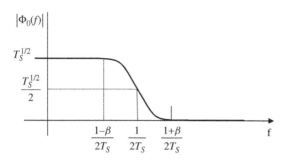

Figure 2.16   Raised cosine Nyquist pulse in the frequency domain.

However, given the raised cosine function characteristics, where $0 < \beta \leq 1$, no signal spectrum will exist outside of $\frac{1}{T_S}$ (twice the bandwidth). This approximation of the signal power spectrum was considered a breakthrough in the 1920s and allowed engineers to design communications systems based on sound power spectrum models which reduced inter-pulse interference. Simply, Nyquist started his analysis from a frequency domain convolution of a rectangular spectrum and a half cycle of a cosine shape spectrum.

The time domain representation of the Nyquist pulse is obtained by multiplying the inverse Fourier transform of the rectangular spectrum by the inverse Fourier transform of the half cycle of the cosine shape spectrum,[12] which is expressed as follows:

$$\phi_0(t) = \frac{1}{T_S^{1/2}} \frac{\sin(\pi t/T_S)}{(\pi t/T_S)} \left[\frac{\cos(\beta \pi t/T_S)}{1 - 4\beta^2 t^2/T_S^2} \right]. \tag{2.23}$$

Figure 2.17 exemplifies a Nyquist pulse as a function of t/T_S for a given β. Notice how, theoretically, this signal model has infinite time duration (because the Fourier transform is defined over a limited band). In practice, engineers truncate the time domain representation to have a better approximation of the signal.

The Nyquist model was considered a breakthrough since the time domain Nyquist pulse has many characteristics which complement how circuitry is built. Before Nyquist, the ideal time domain pulse model that has finite energy at the exact sampling time did not work well in practice (think of the way capacitors charge and discharge). The Nyquist pulse offers a good time domain model (where the energy of the pulses is close to reality and the frequency domain shows some excess bandwidth).

Today, we have better models of the signal spectrum and better understanding of the relationship between harmonics and spectrum interference. Figure 2.18 illustrates a typical power spectra model (showing the positive frequency portion) with a carrier f_c, bandwidth of W, and multiple harmonics. Comparing this to Figure 2.17, you can see how Nyquist's idea of having an excess bandwidth factor was proper, and since the second harmonic carries more energy than the subsequent harmonics, Nyquist's approximation with stopband was not far off at all.

---

[12] Convolution in the frequency domain is equivalent to multiplication in the time domain.

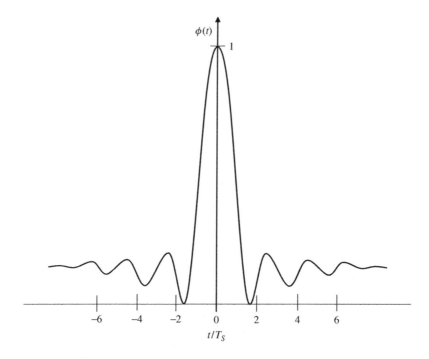

Figure 2.17 A Nyquist pulse carrying unit energy (not an exact plot of Equation 2.23).

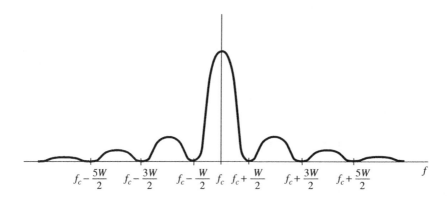

Figure 2.18 A typical signal power spectrum – only positive frequency shown.

2.6 Spread Spectrum Modulation

Spread spectrum modulation is an important area of interest when it comes to tactical wireless radios. This modulation technique was first used in tactical radios and later made available for commercial use. Generally, modulation techniques tend to minimize the bandwidth occupied by the modulated signal since the spectrum is probably the most important commodity in wireless communications. There are cases where a much wider spectrum is used (wider in comparison to the signal rate in symbol per second and can be in the order of hundreds or thousand time wider). With the use of a wider spectrum and proper modulation techniques, we can achieve a spectrum capacity close to that of orthogonal modulation. Multiple users share the spectrum in a given frequency range with no mutual interference using what is known as code division multiple access (CDMA). Spread spectrum is used for tactical radio and commercial wireless (3G air interface) for different reasons.

In tactical wireless communications, it has the following advantages:

1. Spreading the transmitter power means lower power spectral density and makes the signal less detectable by the enemy.
2. Spreading the signal makes it less vulnerable to enemy jamming.
3. A spread-spectrum signal is less sensitive to multipath interference.

Commercial wireless uses spread spectrum for reasons that include:

1. Lower power spectral density means less interference to narrow band signals using the same frequency range. With the spectrum becoming more and more crowded with users, this is a very important feature.
2. It creates the concept of shared resources, since not all the users are using the spectrum at once. More users can be given access to the spectrum. A cellular base station can have end users coming and going, utilizing the spectrum only for the call duration, and the spectrum is recycled.
3. Range determination and position location can be estimated easily with wideband signals. Cell phone locations can always be tracked accurately.
4. Resistance to multipath interference means reliable use in urban environments. You owe your good reception on 3G and 4G phones in crowded cities, where buildings reflect the signal in many directions, to the spread spectrum modulation technique.

Different types of spread spectrum modulation are implemented. Direct sequence (DS) spread spectrum and FH spread spectrum are very common. Some tactical radios use a combination of time and FH techniques.

2.6.1 Direct Sequence Spread Spectrum

To understand direct sequence spread spectrum, imagine an information bit sequence at a rate that is much slower than the channel capacity (this could be your encoded voice at 12 kbps rate). A code generator generates another binary stream at high rate (the channel rate at mega bps rate), which is referred to as the chip rate or chip code or chip sequence. Chip rate is a pseudorandom. The two binary streams are modulo-2 added to generate a high speed random sequence. Receiving this random sequence and knowing the chip code means that the information sequence can be recovered.

Figure 2.19 shows a DS spread spectrum modulator where the information bit stream is modulo-2 with the chip code producing the random sequence m_n, which is modulated with a carrier with frequency ω_c to produce the spread spectrum signal $s(t)$.

The chip code looks like a random binary sequence. However, the fact that both the transmitter and the receiver should know about the chip code, makes it deterministic. In your

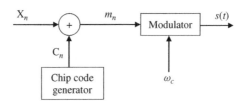

Figure 2.19   Direct-sequence spread spectrum modulation.

cellular phone, both the end user terminal and the base station know the chip code that your phone is using (it is actually assigned to the phone by the base station controller). In tactical applications, more elaborate and evasive approaches are used to assign chip codes, making it harder for an enemy to make sense of the signal. The chip rate sequence should have as many ones as zeros as possible and must have autocorrelation properties. In cellular applications, different chip codes are assigned to the different users of the spectrum. To reduce mutual interference, the different chip codes assigned to the different users should have as small cross-correlations as possible. There are many references about how to generate binary sequences for optimal spread spectrum multiplexing. Reference [6] is a classic paper in this area that was published in 1967. As you will see in the subsequent chapters, tactical radios that used spread spectrum started to be deployed in the 1970s, and commercial applications of spread spectrum such as 3-G started in the 1990s.

There are many techniques in producing chip codes. Because chip codes in tactical radios are a form of transmission security (TransSec), they may be classified. However, in cellular applications, one can determine the way that chip codes are generated if one obtains the chip code given to two end user terminals using the same base station carrier frequency. Because commercial cellular technologies are standardized (e.g., universal mobile telephone system – UMTS – standardization) for a given base station carrier frequency, the set of chip codes that have small cross-correlation and that reduce mutual interference is generated according to standardization. Obtaining two chip codes and reverse engineering the standards can produce knowledge of all the chip codes of that carrier frequency at the specific base station. This creates security vulnerability with commercial cellular that tactical use of spread spectrum avoids.[13] In tactical applications, the story is very different. The techniques used to generate chip codes and the seeds for the generators are classified, and it can change dynamically.

One common technique in generating chip codes uses a linear feedback shift register in conjunction with a primitive polynomial, which ensures the desired properties mentioned above. For a given code length, according to finite field theory, there exist a few sequences that can be generated from known primitive polynomials. In the 1970s, circuitries were developed to implement these techniques. With the abundance of memory we have in newer technologies, computer programs can generate large size tables that can be stored and used as lookup tables. Generating these look-up tables uses techniques that start from two known sequences (generated from primitive polynomials), are modulo-2 added and some phase shift implemented on the result to create larger sets of chip codes for CDMA applications.

---

[13] To standardize or not to standardize is a dilemma in tactical communications. As you see here, standardization can create vulnerabilities. On the other hand, as this book makes the case in subsequent chapters, lack of standardization has severe paybacks. The question should be how to create standardizations that do not introduce vulnerabilities.

You may be asking why this chapter presented modulation technique moving from binary antipodal signals to orthogonal sets, and wondering if there are any non-binary chip code sequences. Reference [7] presents quadriphase chip codes, which can lower the cross-correlation further between utilized chip codes.

Now let us understand how the power spectra for spread and non-spread modulation differ. Let us start by trying to understand how the modulated signal based on Figure 2.19 looks like. X_n, C_n, and m_n all have binary values ± 1 (map logical 0 to -1 and logical 1 to $+1$ because we can be using a pulse waveform in the actual implementation that makes multiplication equivalent to modulo-2 addition). For m_n, the code sequence properties mentioned above makes 0 and 1 appear at the same rate as if they were binary independent random variables or as if they followed a coin flip. This makes the modulated signal $s(t)$ stochastically similar to conventional PSK. The modulated signal $s(t)$ can be expressed as:

$$s(t) = A \sin\left[\omega_c t + \theta + \frac{\pi}{2} m(t)\right] = -Am(t) \sin(\omega_c t + \theta). \tag{2.24}$$

Notice that since $m(t)$ can be ± 1, we can simplify the equation, where θ is the initial carrier phase and is necessary to express here because in a set of chip codes for a given carrier, a different initial *random* carrier phase is associated with each code, in order to minimize their cross-correlation. Notice also that the PSK modulated signal is now at the chip code rate $R_c = B R_b$, where B is the spreading factor.

Figure 2.20 shows the spectral expansion of a spread spectrum signal compared to a non-spread signal where the lowering of the spectral density and the spectrum spread are dependent on the spreading factor B.

The detection of the DS spread spectrum signal can use a correlation receiver which integrates the spectrum over one bit (multiple chips) and compares it to a threshold (much like antipodal signal). Aligning the carrier phase is a must to ensure that the local code generated at the correlation receiver aligns with the incoming signal. This is a known challenge in implementing DS spread spectrum.

You may think that DS spread spectrum signals are less efficient than binary antipodal signals when it comes to error probability since there is less energy involved. Nevertheless, since the correlation receiver integrates over multiple chips (to decode one bit), it results in

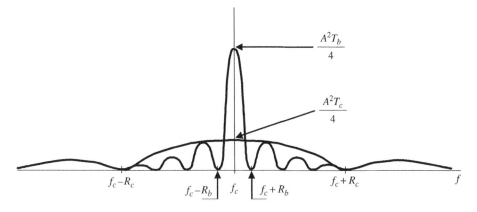

Figure 2.20 Comparing the spectra of spread and non-spread spectrum modulation – only positive side shown.

a probability of bit error rate equivalent to that of the binary antipodal signal expressed in terms of bit energy (not chip energy) to noise power density ratio. Increasing or decreasing the spread factor B has no effect on the probability of bit error when dealing with AWGN.

2.6.2 Frequency Hopping Spread Spectrum

Frequency hopping spread spectrum is an important technique in tactical wireless communications since it provides anti-jamming protection. In FH systems, we divide the available spectrum into slots. The carrier frequency hops among these slots at a rate known as the frequency hopping rate. If we have a hopping range of W Hz, we can then divide it into N slots where each slot size is Δf; this makes $W = N\Delta f$. At any given moment, the carrier frequency is decided by the frequency synthesizer that is driven by a pseudorandom generator which generates N unique input vectors to the frequency synthesizer. Figure 2.21 shows a pseudorandom generator that is clocked at the hopping rate, feeding the frequency synthesizer, which in turn generates the hopping carrier frequency for the modulator. Notice that the binary stream X_n is modulated over the carrier with frequency ω_i, which corresponds to the ith slot of the N slots available for hopping. The hopping rate is deciding by R_c.

In DS spread spectrum, we have one parameter of interest which is the spreading factor B. In FH spread spectrum, we need to define two parameters, which are the hopping rate (how fast we hop) and the hopping range.

The hopping rate is of great interest in tactical wireless radios. A system with *slow hopping* is one where the hop rate relative to the message rate is low (the carrier frequency stays at a given frequency long enough to modulate an entire message). With *fast hopping*, the carrier frequency changes multiple times during the modulation of one symbol. Fast hopping requires non-coherent detection. Recall that non-coherent detection is more complex than coherent detection, which can be used with slow hopping. Tactical radios use fast hopping for many reasons:

1. It is harder for the enemy to listen to the hopping pattern and follow the utilized frequency to decode the signal.
2. It is harder for jammers to follow the hopping pattern and disrupt the signal. Jammers have to spread their spectrum over the entire range weakening their effect.

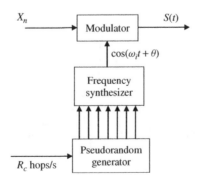

Figure 2.21 Frequency hopping (FH) spread spectrum (SS) signal modulation.

3. Fast hopping can mitigate the problems associated with fading channels. This is known as the frequency diversity benefit.

Figure 2.22 gives a simple example where we have $N = 8$ slots (the eight frequencies are shown in the vertical axis) and fast hopping results in three hops occurring during the modulation of one symbol (symbol time is T_S).

Figure 2.23 demonstrates a slow hopping case for the same $N = 8$ slots, and slow hopping results in two symbols being modulated during a single hop (symbol time is T_S).

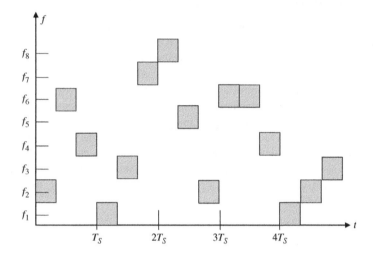

Figure 2.22 Fast hopping where three hops occur during the modulation of one symbol.

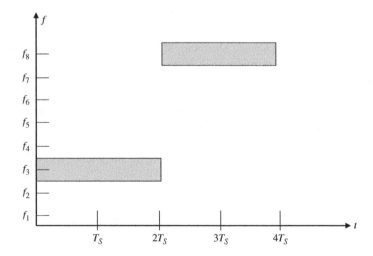

Figure 2.23 Slow hopping example with two symbols per hop.

Error probability in FH is not affected by the hopping and remains the same depending on the modulation technique as discussed earlier. Moving from slow to fast hopping neither gains nor loses performance of bit error detection in the presence of AWGN.

Note that tactical wireless radios face many challenges. FH spread spectrum offers the means to overcome some of these major challenges. You will see in the subsequent chapters some examples of tactical radios where FH is a core part of its design.

2.7 Concluding Remarks

2.7.1 What Happens Before Modulation and After Demodulation?

In the next chapter, we will discuss what happens before modulation and after demodulation over the tactical network node protocol stack layers. We will discuss some concepts of information theory where a set of information symbols are encoded before being modulated over the communications channel. We will discuss error control coding, source coding, the relationship between error control coding and modulation, and will review the channel capacity concepts and see how channel coding can help us use a simple modulation and demodulation technique. Channel coding techniques are usually implemented at the data link layer (DLL). An objective of this book is to help you understand how the protocol stack layers in tactical communications and networking relate to each other and how some techniques (such as error control coding) can be implemented at different layers of the protocol stack. Even if you specialize in a particular protocol stack layer, this book should help you choose design concepts that complement the other layers.

2.7.2 Historical Perspective

Today, with the advances in microprocessor technology and the low cost of memory, the ability of communications and networking to transmit information is bounded only by the physical layer's ability to transmit bits at a higher rate. There are physical limits on how many bits per second a medium can transmit. We have come a long way since Samuel Morse invented telegraphy in 1832. Many historians today refer to Morse as the father of information theory (rather than Claude Shannon). Morse code's ingenuity lay in the dot-dash approach. The letter "E," the most frequent letter in the English alphabet, was expressed as a dot (a symbol that takes the least time to transmit), whereas less frequent letters, such as "Q" or punctuation marks, were assigned longer patterns of dots and dashes. Many data compression techniques today follow the same concept as Morse code, where more frequent symbols in an alphabet are assigned a fewer number of bits and less frequent letters in the alphabet are assigned a larger numbers of bits. In the next chapter, you will review source coding and see how source and channel coding relate to the modulation techniques discussed in this chapter.

In the 1920s, Nyquist introduced the concept of optimizing transmission of digital signals over a given channel. This showed that inter-symbol interference can be reduced through sampling (this evolved to be sampling theory or Nyquist theory). Pulse code modulation appeared in the late 1930, opening the door for speech coding.

As the scientific community started to discuss information bits as probabilistic models, Claude Shannon published a breakthrough paper on the limits of throughput of information over a channel and introduced the term channel capacity. This prepared the way for modern information theory. His short paper in 1948 changed the scientific community, and began discussions on how to send information reliably and how to approach the optimum capacity of the communications channel. The next chapter will review some of these information theory concepts such that you can see how communication over a deterministic channel (with known and static characteristics) can be optimized.

Error control coding techniques started to take root in the 1950, where adding redundancy bits (parity check bits) demonstrated that better performance over the same channel can be achieved through proper coding techniques. By the 1980s, scientists and engineers were able to approach the Shannon bound or the theoretical limit or entropy of the communications channels through more complex modulation techniques, joining coding and modulation as in trellis coded modulation, soft decision techniques, and so on. The next chapter will review simple error control coding methods, which will emphasize their importance in tactical wireless communications and networking.

As mobile ad hoc networking (MANET) evolved, error control coding techniques at the different protocol stack layers started to emerge. In addition, the DLL error control coding, which is implemented on a bit stream before modulation, started to be used in network coding as it ensures hop-by-hop packet reliability through packet erasure techniques (which can be implemented at the IP layer) as well as transport layer packet erasure techniques. Although the Shannon model could not be expanded to the dynamic nature of MANET environments, a defense-in-depth approach can ensure reliable communications under the adverse conditions of tactical MANET. The next chapter will introduce you to these concepts.

Bibliography

1. Van der Ziel, A. (1986) *Noise*, Prentice Hall.
2. Wozencraft, J. and Jacobs, I.M. (1965) *Principles of Communications Engineering*, John Wiley & Sons, Inc., New York.
3. Franks, L. (1969) *Signal Theory*, Prentice Hall.
4. Weber, C. (1987) *Elements of Detection and Signal Design*, Springer-Verlag.
5. Nyquist, H. (1928) Certain topics in telegraphy transmission theory. *AIEE Transactions*, **47**, 617–644.
6. Gold, R. (1967) Optimal binary sequences for spread spectrum multiplexing. *IEEE Transactions on Information Theory*, **IT-13**, 619–621.
7. Botzas, S., Hammons, S., and Kumar, P.V. (1992) 4-phase sequences with near-optimum correlation properties. *IEEE Transactions on Information Theory*, **IT-38**, 1101–1113.
8. Popoulis, A. (1991) *Probability, Random Variables, and Stochastic Processes*, 3rd edn, McGraw-Hill.
9. Ross, S. (2009) *Introduction to Probability and Statistics for Engineers and Scientists*, 4th edn, Academic Press.
10. Leon-Garcia, A. (1994) *Probability and Random Processes for Electrical Engineering*, 2nd edn, Addison-Wesley.
11. Wilson, S. (1996) *Digital Modulation and Coding*, Prentice Hall.
12. Simon, M., Hinedi, S., and Lindsey, W. (1995) *Digital Communications Techniques: Signal Design and Detection*, Prentice-Hall.
13. Proakis, J. and Salehi, M. (1994) *Communications Systems Engineering*, Prentice-Hall.
14. Haykin, S. (1994) *Communications Systems*, John Wiley & Sons, Inc.

15. Ross, S.M. (2009) *Introduction to Probability and Statistics for Engineers and Scientists*, 4th edn, Academic Press.
16. Larson, H. and Shubert, B. (1979) *Probabilistic Models in Engineering Sciences*, vols. I and II, John Wiley & Sons, Inc.
17. Wu, W. (1984) *Digital Satellite Communications*, Computer Science Press.

3

The DLL and Information Theory in Tactical Networks

Chapter 2 covered some important aspects of the physical layer, including signals, modulation/demodulation, probability of symbol error rate, and spectral modeling. Now that you have seen the challenges of the physical layer in tactical wireless communications, you are ready to move up the protocol stack, while looking at the fundamental communications and networking issue of information transfer. This includes source coding, channel coding, and the challenges of error control coding at the different layers of the tactical wireless networking protocol stack. While signal representation and channel coding belong to the data link layer (DLL), not all the topics in this chapter do, such as source coding, transport layer coding, and network coding.

3.1 Information Theory and Channel Capacity

A fundamental aspect of communications theory is to consider a source of information symbols and discuss how we can represent these symbols as signals for reliable communication over a channel, while keeping physical limitations in mind. Information theory abstracts the signaling part (covered in Chapter 2) and focuses on the message information aspect. Information theory quantifies three basic concepts: the source information, the capacity of the channel to carry this information, and the *coding* (message representation) utilized in the channel. Information theory mathematically represents these three concepts for optimization of the information communication. If we have a source emitting symbols with an information rate that does not exceed the channel capacity, then there exists a *coding* technique that makes it possible to transmit the source information symbols over the channel with a small symbol error probability in the presence of channel noise. It will become more apparent that *coding*, here, means source coding, channel coding, and symbol to signal representation.

Let us refer back to Figure 2.1 which represents a generic channel model with the modulation of digital signals. Figure 2.1 is abstracted in Figure 3.1, as the noisy channel block. As mentioned above, channel encoding offers the means to communicate the source symbols to the destination almost error free. Source and channel coding are related in the sense that

Tactical Wireless Communications and Networks: Design Concepts and Challenges, First Edition. George F. Elmasry.
© 2012 John Wiley & Sons, Ltd. Published 2012 by John Wiley & Sons, Ltd.

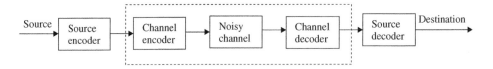

Figure 3.1 Source and channel coding in the context of an end-to-end communications system.

channel coding is selected to match the source coding in order to maximize the information communicated over the channel reliably.

Theoretically, source and channel coding are two distinct methods of coding. Practically, there are multiple joint uses for source and channel coding in commercial applications. When it comes to tactical wireless communications and networking, this issue becomes much more complex. There exists channel coding that can be implemented at the DLL, network coding that ensures hop-by-hop packet reliability, and forward error correction (FEC) that can be implemented at the transport layer. There is also error detection or hybrid error detection and error correction that can be implemented at the transport layer. There are error detection and/or error correction techniques that can be implemented at the tactical edge. The use of COTS (commercial-off-the shelf) voice and video codec over IP makes these issues exponentially compound since some codecs, such as MPEG, uses a joint source and channel coding approach. In this technique, the codec adds redundancy to the encoded information for error control coding. The complexity of error control coding in tactical communications and networking, and the existence of different protocol stack layers that work independently, create different approaches to error control coding. For example, in network coding, researchers are attempting to determine how reliable the MANET hops should be, how much redundancy should be invested in packet erasure correction (aside from the DLL error control coding which is applied to the bit stream), and how much reliability should be left to the transport layer. In the area of tactical edge, researchers promote leaving the transport layer untouched to ensure the use of COTS applications over the plain text subnets and instead are developing new proxy methods to ensure that COTS applications work seamlessly over an encrypted tactical core network.

The different error detection and error correction approaches in tactical networks are discussed in this chapter so that you can form a well-rounded picture of what information transmission means in tactical wireless communications and networking. To initially understand the information theory construct, we will focus on the model in Figure 3.1, regardless of how many stages of error detection and error correction exist between source coding and the noisy channel. For a single channel, the goal is to have information flow from the source to the destination and have error detection and error correction techniques yield the equivalent of a noise-free channel while approaching channel capacity. In tactical networks, the goal is the same except that we do not deal with a simple deterministic channel, but rather a multilayer stack and a complex network with many dynamic factors. The equivalent to maximizing channel capacity is maximizing information flow in the network (the ratio of application layer information bits to all other bits transmitted over the physical media). Yielding the equivalent to an error free network path that delivers information in a predefined speed (meeting QoS – quality of service – requirements) becomes a huge challenge. Unfortunately, today's tactical wireless networks are far from implementing an optimum end-to-end solution to the communications and networking problem due to the dynamic nature of many contributing factors and the existence of multiple independent layers.

Let us begin exploring the information theory construct from the model in Figure 3.1 where we saw a form of error detection and/or error correction existing between the source coding and the channel. Let us start by defining source coding.

3.1.1 Uncertainty and Information

If the destination in Figure 3.1 knows (with certainty) what information symbols are being emitted by the source, then no information actually needs to be delivered. Essentially, the information only needs to be delivered to eliminate uncertainty. The information measure is based on the probability of a message being delivered. If a source emits a single symbol at all times, then the probability that this symbol is known to the destination is 1. The destination is always certain about the information emitted by the source. Information theory is particularly interested in the probability of a message. If x_i expresses an arbitrary message and $P(x_i) = P_i$ is the probability of x_i being selected as the next symbol to be transmitted, then the information content associated with x_i is a function of P_i. Claude Shannon's definition of information measure uses a logarithmic function as follows:

$$I_i = -\log_b P_i = \log_b \frac{1}{P_i}. \tag{3.1}$$

The base logarithm b is related to the size of the source symbols ($b = 2$ for the binary case). I_i is the self-information of the message x_i, and its value depends only on P_i (the interpretation of the message is irrelevant).[1] The definition presented in Equation 3.1 is the basis for the following outcomes:

- For $0 \leq P_i \leq 1$, $I_i \geq 0$: the self-information, I_i, cannot be negative.
- As $P_i \rightarrow 1$, $I_i \rightarrow 0$: no uncertainty, no information.
- If $P_i < P_j$, $I_i > I_j$: information content increases with uncertainty.
- If the source emits two successive independent symbols x_i and x_j with joint probability $P(x_i x_j) = P_i P_j$, then,

$$I_{ij} = \log_b \frac{1}{P_i P_j} = \log_b \frac{1}{P_i} + \log_b \frac{1}{P_j} = I_i + I_j, \tag{3.2}$$

which means that the total information content of successive symbols equals the sum of the individual symbols' contribution.

To make these definitions easier to grasp, let us consider the binary case where we have a source that emits one of two equi-probable messages (this is the most elementary case). In this case, $I_i = \log_2 \frac{1}{P_i}$ bits. Because $P_0 = P_1 = 1/2$, $I_0 = I_1 = \log_2 2 = 1$ bit. That is, if we have a source that emits two equally likely symbols, the information content of each

---

[1] More than a century before Claude Shannon, Samuel Morse understood the information content concept. His code assigned the most frequently used letters in the English alphabet to dot-dash combinations that required the least transmission time, and assigned the less frequently used letters a combination which required longer transmission time.

message is one bit. Needless to say, if one of these two symbols is emitted with a higher frequency (occurs with higher probability) than the other, its information content is less than one bit and the information content of the less frequent symbol is more than one bit.

Example 3.1 A binary source emits two symbols with probabilities 1/10 and 9/10. The information content for each symbol is determined as follows:

$$I_0 = \log_2 10 = \frac{\log_{10} 10}{\log_{10} 2} = 3.32 \text{ bits.}$$

$$I_1 = \log_2 \frac{10}{9} = \frac{\ln 1111}{\ln 2} = 0.152 \text{ bits.}$$

The discussion above demonstrates how the symbol probability in a given source alphabet relates to its information content. It is important that you can differentiate between a binary source (emitting one of two symbols or having an alphabet of size two) and binary digits. The use of binary sources to explain information contents does not mean that we practically deal with such simple source coding. In Figure 3.1, the channel encoder can produce binary digits of equal probabilities. Source encoding, where information content matters, happens prior to channel encoding. A good way to conceptualize where source coding and channel coding happen in tactical networks is to envision a user compressing a file and attaching it to an e-mail. Compressing the file uses a form of source coding that is ASCII based (a source alphabet of size 256 or 128 ASCII symbols). The e-mail with the attached file turns into a TCP (transmission control protocol) session and the transport layer can introduce error detection (TCP retransmission) and error correction (you will see later in this text how the transport layer in tactical networks can have forward error correction). The tactical edge (aggregation point to the encrypted network core) may also have a form of error detection and/or error correction (e.g., a TCP proxy). At the network core, network coding can be enabled at some hops. Also, the DLL may introduce a form of error detection and/or error correction before the digital signal is transmitted over the tactical radio.

3.1.2 Entropy

Let us consider a source emitting symbols from an alphabet of size M denoted as $\{x_1, x_2, \ldots, x_M\}$, where each symbol x_i is emitted with a probability P_i. Naturally,

$$\sum_{i=1}^{M} P_i = 1. \tag{3.3}$$

For simplicity, let us assume that the source is stationary (which would mean that the probabilities are constant over time). Also, let's assume that the emitted symbols are statistically independent[2] and are generated at an average rate r symbols per second. Such a source, with stationary and independency properties, is referred to as a *discrete memoryless source*. Over an arbitrary symbol interval, this source will produce an amount of information

[2] Most sources emit symbols with some form of dependency. In the English language symbols q and u have dependency. Exploiting such dependencies in source coding can add further compression gain and more complexity.

that can be expressed as a discrete random variable having the values $I_1, I_2, \ldots I_M$. The expected value of information per symbol is the statistical average given by:

$$H(X) = \sum_{i=1}^{M} P_i I_i = \sum_{i=1}^{M} P_i \log \frac{1}{P_i} \text{ bits/symbol.} \tag{3.4}$$

$H(X)$ is referred to as the source entropy.[3]

Example 3.2 Find the entropy of the binary source with probabilities 1/10 and 9/10 discussed above.

$$H(X) = \frac{1}{10} \times 3.32 + \frac{9}{10} \times 0.152 = 0.332 + 0.136 = 0.468 \text{ bits/symbol.}$$

Example 3.3 Find the entropy of a binary source with probability of 1/2 and 1/2.

$$H(X) = \frac{1}{2} \times 1 + \frac{1}{2} \times 1 = 1 \text{ bits/symbol.}$$

These two examples demonstrate that greater deviation between the source symbol probabilities, results in a smaller source entropy. Therefore, more compression can be achieved with source symbols that have a larger variation between the symbol probabilities.

Now let us consider a source emitting a sequence of symbols (over a large interval of time) with an average rate r symbols per second. As the number of emitted symbols n becomes large, the total information to be transmitted approaches $nH(X)$ bits and the information transmits at a rate $rH(X)$ bits/second.

Information theory asserts that any discrete memoryless source can be encoded in binary digits (bits) and transmitted over a channel with *no noise* if the channel rate r_b is greater than or equals to $rH(X)$ bits/second.[4]

Entropy plays a major role in information theory. For a given source, $H(X)$ depends on the symbols' probabilities and the alphabet size M. $H(X)$ always satisfies:

$$0 \leq H(X) \leq \log_2 M. \tag{3.5}$$

The lower bound corresponds to no uncertainty (one symbol has a probability of 1 and the rest of the symbols have a probability of 0). The upper bound corresponds to maximum uncertainty, which occurs when all symbols have equal probability of $\frac{1}{M}$. That is $P_i = \frac{1}{M}$ for $i = 1, 2, \ldots, M$.

Let us visit $H(X)$ boundaries for the case of a binary source. In general, a binary source emits two symbols with probabilities $P_1 = p$ and $P_2 = 1 - p$. Using Equation 3.4, the entropy of the binary source is given by:

$$H(X) = \Omega(p) = p \log \frac{1}{p} + (1 - p) \log \frac{1}{1 - p}. \tag{3.6}$$

---

[3] Shannon borrowed the term entropy from statistical mechanics. You can find references such as [1] explaining how information theory entropy is related to thermodynamics entropy.
[4] Later in this chapter, we will discuss how information theory covers the case of noisy channel creating bounds for the transmission rate in bits/second based on the channel capacity.

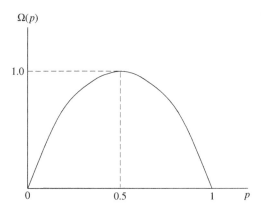

Figure 3.2   The entropy of a binary source as a function of the source symbols' probabilities.

The function $\Omega(p)$ expresses the entropy of a binary source as a function of p. Figure 3.2 plots $\Omega(p)$ as a function of p. The term $H(X)$ expresses the entropy of any source while the term $\Omega(p)$ will be reserved for the binary case with symbol probabilities of p and $1 - p$.

Figure 3.2 shows that the entropy of the binary source is maximized at $p = 1 - p = 1/2$ giving 1 bit/symbol. The entropy approaches 0 as the probability of one symbol approaches 1.

Proving the upper bound of Equation 3.5 for the non-binary source requires more detail, whereas the lower bound is easier to determine. By referring to Equation 3.4, we can note that as the source emits one symbol, whose probability approaches 1, the remaining symbols will have a probability which approaches 0. Therefore, symbols with probabilities approaching 0 will have summation terms which also approach 0. Knowing that $\log 1 = 0$ allows us to state that the lower bound must then equal 0. This is the simple proof for determining the lower bound. As for the upper bound, it can be conceptualized for the case of $p_i = \frac{1}{M}$ for all source symbols given that $H(X) \leq \sum_{i=1}^{M} \frac{1}{M} \log M = \log M \sum_{i=1}^{M} \frac{1}{M} = \log M$. The detailed mathematical proof is beyond the scope of this book.

3.1.3 Coding for a Discrete Memoryless Source

Let's look at the case of a discrete memoryless source which emits M symbols, all of which have an equal probability. Let's say that it emits these symbols at a rate of r symbols/s. Then, all the symbols have the same information content and the information transmission rate is $R = r \log M$ bits/s. One can then map the source alphabet M to an M-ary signaling where the signaling rate is equal to the symbol emission rate r. When the source symbols have different probabilities such that the information transmission rate $R = r H(X) < r \log M$, one needs to consider a coding process based on the fact that different symbols carry different amounts of information. If we have a binary encoder where symbols arrive at a rate of $R = r H(X)$, then the encoder will produce bits at a fixed rate of r_b. The relationship between the symbol rate r and the encoded bit rate r_b is of great interest. The ratio $r_b/r = N$

corresponds to the average number of encoded bits per source symbol. $\bar{N}$ can also be expressed as:

$$\bar{N} = \sum_{i=1}^{M} P_i N_i, \tag{3.7}$$

where N_i is the length of the codeword of the ith symbol.

Shannon's source coding theorem creates boundaries for $\bar{N}$ such that,

$$H(X) \leq \bar{N} < H(X) + \varepsilon, \tag{3.8}$$

where ε is positive.

Good source coding can produce a very small ε value. An optimum source coding mechanism would achieve the lower bound of $\bar{N} = H(X)$. A suboptimum source coding would have $\bar{N} > H(X)$. It is necessary for source coding to ensure that source information is not lost. It also needs to ensure that the decoder can uniquely decipher the received bits and then revert them back to the original symbols without ambiguity. This is referred to as lossless source coding or lossless compression.

Example 3.4 Let us now discuss the properties of an optimum code, a suboptimum code, and a code that is not uniquely decipherable. Consider a four-symbol source which emits symbols (A, B, C, and D) at probabilities (1/2, 1/4, 1/8, and 1/8) respectively. The entropy of this source is 1.75 bits/symbol. Let us consider the following codes that map A, B, C, and D consecutively as follows:

- Code I: (00, 01, 10, 11)
- Code II: (0, 1, 10, 11)
- Code III: (0, 01, 011, 0111)
- Code IV: (0, 10, 110, 111)

  – Code I is a simple mapping with fixed length and is uniquely decipherable, but with $\bar{N} = 2.0$, the code has an entropy to $\bar{N}$ ratio of 0.88.
  – Code II is not uniquely decipherable. Notice that $\bar{N} = 1.25$, which is less than the entropy of the source. This code destroys information and is unusable. For example, if the decoder receives the sequence 1011, it could be BABB, CD, or BAD, ...
  – Code III is uniquely decipherable (notice how each codeword starts with 0). With this code, $\bar{N} = 1.875$. This indicates that Code III is more efficient than Code I since $\bar{N}$ is closer to the entropy. This results in a entropy to $\bar{N}$ ratio of 0.925.
  – Code IV exemplifies an optimum code; each codeword is not a prefix to the other and is uniquely decipherable since $\bar{N}$ equals the entropy.

Example 3.5 The Shannon–Fano coding algorithm can be explained as follows:

- Divide the source symbols into two groups such that the sums of the symbol probabilities in each group are as close to each other as possible.
- Assign 0 to each symbol in the first group and 1 to each symbol in the second group.
- Divide each group into two subgroups and assign 0s and 1s with the same rule as above.
- Whenever a group has just one symbol, no further subdivision is required (codeword is complete).

- Successive implementation of these rules will eventually lead to all groups reduced to one symbol, and one can map each symbol to a unique codeword, reading from left to right.

Table 3.1 provides a visual for the implementation of the Shannon–Fano coding algorithm to an eight-symbol source. This example results in $\bar{N} = 2.18$, while the entropy for this source is 2.15. Keep in mind that for an eight-symbol source, one would have required three bits per symbols if a coding technique was not implemented. This coding technique approaches the source entropy.

We have covered simple examples of source coding, yet real world applications are more complex and implement a variety of techniques. Today, there are many source coding techniques, including Lempel–Ziv and arithmetic coding, with results approaching the source entropy. You are encouraged to see references in this area such as [2, 3]. In tactical wireless networks, COTS applications such as MPEG can use a form of source coding referred to as lossless compression.[5] In order to compress (zip) a file on your desktop, your computer uses a form of source coding. Some source coding techniques, such as arithmetic coding, do not require knowledge of the probability of each symbol. Instead, they collect this information on the fly, through frequency tables as encoding and decoding occur. Arithmetic coding works well for long sequences since the frequency tables yield source statistics that approach the probability of each symbol. This makes the number of encoded bits per symbol approach the entropy.

3.1.4 Mutual Information and Discrete Channels

So far we have covered the source coding aspect of information theory. Now, let us consider the existence of a transmission channel. Let us study the case where both the source and the channel are discrete. We are interested in defining the *channel capacity* with which appropriate coding can achieve error-free transmission over a noisy channel.

Table 3.1 Shannon–Fano coding example

x_i	P_i	Successive steps						Codeword
		I	II	III	IV	V	VI	
A	0.50	0						0
B	0.15	1	0	0				100
C	0.15	1	0	1				101
D	0.08	1	1	0				110
E	0.08	1	1	0	0			1110
F	0.02	1	1	1	1	0		11110
G	0.01	1	1	1	1	1	0	111110
H	0.01	1	1	1	1	1	1	111111
H(X) = 2.15								$\bar{N} = 2.18$

---

[5] Source coding in some references is referred to as lossless compression to distinguish it from lossy compression. Quantization (which turns analog information into digital information) is referred to as lossy compression since the information lost in quantization is not recoverable.

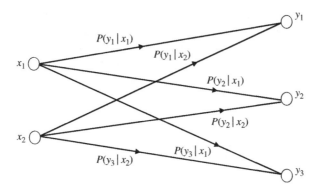

Figure 3.3 Forward transition probabilities of a discrete channel with two source symbols and three destination symbols.

Let us consider the case where a discrete source emits symbols from an alphabet X. If the channel is error free, the receiver will reproduce the same symbols emitted at the source. Channel impairment can result in a different symbol alphabet Y at the destination. Let us identify the information transferred in this case, beginning with some definitions:

- $P(x_j)$ is the probability that the source emits symbol x_i.
- $P(y_j)$ is the probability that symbol y_j is received at the destination.
- $P(x_i y_j)$ is the joint probability that x_i is transmitted and y_j is received.
- $P(x_i | y_j)$ is the conditional probability that x_i was transmitted given that y_j is received.
- $P(y_j | x_i)$ is the conditional probability that y_j is received given that x_i was transmitted.

Let us assume that the channel is memoryless and time-invariant, such that the conditional probabilities are independent of previous symbols and remain constant independently of time.

Of the five definitions above, $P(y_j | x_i)$ should have a special importance to us since it expresses the channel's "*forward transition probability.*" Figure 3.3 explains the importance of the forward transition probability. Here we have a noisy channel, given two source symbols and three destination symbols.[6]

For the channel in Figure 3.3, let us assume that the objective is to deliver y_1 when the transmitted symbol is x_1 and to deliver y_2 when the transmitted symbol is x_2. The symbol error probabilities are given by $P(y_j | x_i)$ for $j \neq i$. The importance of *forward transition probability* will become clear as we analyze information transfer in more detail.

Now, we introduce the term mutual information; this expresses the information transferred over a discrete memoryless channel as follows:

$$I(x_i; y_j) = \log \frac{P(x_i | y_j)}{p(x_i)} \text{ bits,} \tag{3.9}$$

where $I(x_i; y_j)$ measures the information transferred when x_i is transmitted and y_i is received.

---

[6] The receiver can use y_2 to decode an erasure symbol as discussed in Section 2.2.

If the channel is error free and every received symbol y_j can be uniquely mapped to a particular source symbol x_i, then $P(x_i|y_j) = 1$, and $I(x_i; y_j) = \log \frac{1}{P(x_i)}$. This would indicate that the transferred information is equal to the information content (self information) of x_i. On the other hand, if the channel is so noisy that there is no relationship between y_j and x_i, then $P(x_i|y_j) = P(x_i)$ and $I(x_i; y_j) = \log 1 = 0$. In this case, no information is transmitted. Most transmission channels are in between these two extremes (perfect transfer and zero transfer).

Now, let us define the average mutual information as:

$$I(X; Y) = \sum_{x,y} P(x_i y_j) I(x_i; y_j)$$

$$= \sum_{x,y} P(x_i y_j) \log \frac{P(x_i|y_j)}{P(x_i)} \text{ bits/symbol.} \qquad (3.10)$$

The summation subscript in the equation above means that the statistical average is taken over both the source and the receiver indices. $I(X; Y)$ is the average source information acquired per *received* symbol. In order to relate $I(X; Y)$ to *source* symbols, we need to express $I(X; Y)$ in terms of the source entropy $H(X)$. (Recall that $H(X)$ expresses the average information per source symbol.)

Let us use the following probability relationships to assist in this expression.

1. $P(x_i y_j) = P(x_i|y_j)P(y_j)$.
2. $P(x_i y_j) = P(y_j|x_i)P(x_i)$.
3. $P(x_i) = \sum_y P(x_i y_j)$.
4. $P(y_j) = \sum_x P(x_i y_j)$.

Now we can denote the mutual information as:

$$I(X; Y) = \sum_{x,y} P(x_i y_j) \log \frac{1}{P(x_i)} - \sum_{x,y} P(x_i y_j) \log \frac{1}{P(x_i|y_j)}.$$

The first term of this equation can also be expressed as:

$$\sum_{x,y} P(x_i y_j) \log \frac{1}{P(x_i)} = \sum_x \left[\sum_y P(x_i y_j) \right] \log \frac{1}{P(x_i)} = \sum_x P(x_i) \log \frac{1}{P(x_i)} = H(X).$$

The second term $\sum_{x,y} P(x_i y_j) \log \frac{1}{P(x_i|y_j)}$ has an interesting meaning since it conveys the information lost in the communications channel. In other words, the average mutual information is the source entropy minus the information lost in transmission. Thus one can express the mutual information as:

$$I(X; Y) = H(X) - H(X|Y). \qquad (3.11)$$

$H(X|Y)$ in Equation 3.11 is defined as:

$$H(X|Y) = \sum_{x,y} P(x_i y_j) \log \frac{1}{P(x_i|y_j)}. \qquad (3.12)$$

Interestingly, one can express the information transfer in a different manner than the destination entropy prospective, if we consider that,

$$I(X; Y) = I(Y; X). \tag{3.13}$$

This can yield:

$$I(X; Y) = H(Y) - H(Y|X), \tag{3.14}$$

where

$$H(Y) = \sum_y P(y_i) \log \frac{1}{P(y_i)}, \tag{3.15}$$

and

$$H(Y|X) = \sum_{x,y} P(x_i y_j) \log \frac{1}{P(y_i|x_j)}. \tag{3.16}$$

Equation 3.14 expresses the information transfer as the destination entropy $H(Y)$ minus the channel *noise entropy* $H(Y|X)$.

3.1.5   The Binary Symmetric Channel (BSC) Model

This model is shown in Figure 3.4, where we have two source symbols with probabilities $P(x_1) = p$ and $P(x_2) = 1 - p$. We also have two destination symbols with transition probabilities of $P(y_1|x_2) = P(y_2|x_1) = \alpha$ and $P(y_1|x_1) = P(y_2|x_2) = 1 - \alpha$. With this model, errors are statistically independent, and error probabilities are the same for both source symbols. This would suggest that the average error probability for this channel is $P_e = P(x_1)P(y_2|x_1) + P(x_2)P(y_1|x_2) = p\alpha + (1 - p)\alpha = \alpha$.

Now let us calculate the mutual information of the binary symmetric channel (BSC) in terms of p and α. As shown in Equation 3.14, $I(X; Y) = H(Y) - H(Y|X)$. We begin this calculation by using this formula, since $H(Y)$ can be easily calculated once we consider that the output of the channel is a binary source. This source emits symbols at probabilities $P(y_1)$ and $P(y_2) = 1 - P(y_i)$, with an entropy expressed as:

$$H(Y) = \Omega\left(P(y_1)\right), \tag{3.17}$$

where $\Omega()$ is the binary entropy function defined earlier in Equation 3.6.

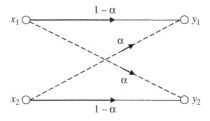

Figure 3.4   The binary symmetric channel (BSC) model.

Notice that $P(y_1)$ can be calculated with the BSC model (see the transition probabilities in Figure 3.4) as:

$$P(y_1) = P(y_1|x_1)P(x_1) + P(y_1|x_2)P(x_2) = p(1 - \alpha) + (1 - p)\alpha = \alpha + p - 2\alpha p.$$

Now, recall that we expressed the noise entropy $H(Y|X)$ above as $H(Y|X) = \sum_{x,y} P(x_i y_j) \log \frac{1}{P(y_i|x_j)}$. If we substitute $P(x_i y_j) = P(y_j|x_i)P(x_i)$, we can express the noise entropy as:

$$H(Y|X) = \sum_x P(x_i) \left[\sum_y p(y_j|x_i) \log \frac{1}{P(y_j|x_i)} \right]. \tag{3.18}$$

Substituting the BSC probabilities in the noise entropy equation, we get:

$$H(Y|X) = p \left[\alpha \log \frac{1}{\alpha} + (1 - \alpha) \log \frac{1}{1 - \alpha} \right] + (1 - p) \left[\alpha \log \frac{1}{\alpha} + (1 - \alpha) \log \frac{1}{1 - \alpha} \right]$$

$$= \alpha \log \frac{1}{\alpha} + (1 - \alpha) \log \frac{1}{1 - \alpha} = \Omega(\alpha). \tag{3.19}$$

Equation 3.19 implies that the channel's symmetry causes the noise entropy to be independent of p (the symbol probabilities). The entropy is only a function of the channel error probability α.

Now we can express the average mutual information as:

$$I(X; Y) = H(Y) - H(Y|X) = \Omega(\alpha + p - 2\alpha p) - \Omega(\alpha). \tag{3.20}$$

Equation 3.20 denotes the dependency of the mutual information of the BSC on both the source probability p and the channel transition probability α.

3.1.6 Capacity of a Discrete Channel

In the previous section, we demonstrated how a discrete memoryless channel transfers a definite amount of information $I(X; Y)$ that can be measured in the existence of channel error. If we consider that a given channel has fixed source and destination alphabets and known forward transition probabilities, then the only variable quantity in $I(X; Y)$ is $P(x_i)$. In order to maximize information transfer, one would need to base the technique on the source alphabet statistics $P(x_i)$. Source encoding, as discussed in previous sections, demonstrates how we can approach the source entropy. At this point, we can introduce the term channel capacity by beginning with the following relationship:

$$C_s = \max_{p(x_i)} I(X; Y) \text{ bits/symbol.} \tag{3.21}$$

C_s represents the maximum amount of information transferred per channel symbol. To express the channel capacity in terms of information rate (amount of information transferred per unit time), let us assume that s is the maximum symbol rate that the channel can transfer. Then channel capacity can be expressed as:

$$C = sC_s \text{ bits/s.} \tag{3.22}$$

The *channel capacity C* is the *maximum rate of information transfer*.

Shannon's fundamental theorem for a noisy channel states the following:

> If a channel has a capacity C and a source has information rate $R \leq C$, then there exists a coding system such that the output of the source can be transmitted over the channel with an arbitrarily small frequency of errors. Conversely, if $R > C$, then it is not possible to transmit the information errors.

Covering the proof of this theorem is beyond the scope of this book, but its plausibility can be explained by considering the following two cases.

Case I

We have an ideal noiseless channel and the source symbols are $\mu = 2^\gamma$. When $P(x_i) = 1/\mu$ for all symbols, the information transfer $I(X; Y) = H(X)$ is maximized (refer to the BSC example above). Thus,

$$C_s = \max_{p(x_i)} H(X) = \log \mu = \gamma \text{ bits/symbol.} \tag{3.23}$$

If s is the maximum symbol rate allowed by the channel, we get:

$$C = s\gamma \text{ bits/s.} \tag{3.24}$$

In order to transmit over the channel, one would need a coding scheme to match the source symbols to channels signals as shown in Figure 3.5. The source produces symbols at the rate $R \leq C$ symbols/s. Then the symbols go through a binary source encoder which generates bits at the rate of r_b bits/s. The binary flow is then converted into μ-ary channel symbols at the rate of $S = r_b / \log \mu = r_b / \gamma$ symbols/s.

Notice that if $R > C$, the binary encoder can either produce bits at the rate $r_b = s\gamma > C$, meaning some symbols will not be transmitted over the channel or at the rate $r_b = s\gamma = C$, where some source symbols will not be encoded (lossy compression). In either case, decoding error would occur even though the channel is noiseless. One can say that optimum source encoding will occur when $R = C$ (when we select the channel that can optimally transmit the source symbols).

Case II

We address the presence of noise referring to the BSC model discussed in Section 3.1.5. We found in Equation 3.20 that $I(X; Y) = \Omega(\alpha + p - 2\alpha p) - \Omega(\alpha)$. Recall that the channel capacity $C = sC_s = \max_{p(x_i)} I(X; Y)$. Notice that if $p = 1/2$, then $\Omega(\alpha + p - 2\alpha p) = \Omega(1/2)$. Referring to Figure 3.2, we know that $\Omega(1/2)$ equals to 1. Thus, $\Omega(\alpha + p - 2\alpha p) - \Omega(\alpha)$ is maximized at $p = 1/2$ and the capacity of the BSC is:

$$C_s = 1 - \Omega(\alpha). \tag{3.25}$$

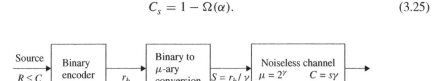

Figure 3.5 Encoding model for a noiseless discrete channel.

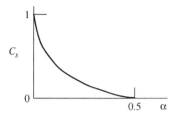

Figure 3.6 Capacity of the binary symmetric channel (BSC) as a function of the channel transition error probability α.

Figure 3.6 plots the capacity of this BSC versus α (since we know that the maximization of $I(X; Y)$ occurs at $p = 1/2$). By observing the curve, one can see that at $\alpha = 1/2$, $I(X; Y) = \Omega(\alpha) - \Omega(\alpha) = 0$. The curve illustrates how the channel capacity rapidly approaches zero as α approaches 0.5. Notice that with the BSC, if we replace α with $1 - \alpha$ the same curve properties apply.[7]

In order to achieve reliable transmission, both error control coding (channel coding) and source encoding are necessary. Channel coding allows the bit error probability of the bits produced by the source encoder to be much smaller than the channel bit error probability α. Figure 3.7 introduces the BSC encoder and decoder by demonstrating the transmitter and receiver steps. For example, a BSC error rate of approximately 10^{-3} can be enhanced to the order of 10^{-5} by utilizing a channel coding technique that has a coding rate of around $\frac{3}{4}$ (this introduces one extra bit for every three bits). This would lead the source decoder to see symbols with very low error rate probabilities. Shannon's theorem states that there exists a coding technique that yields virtually errorless transmission when $R \leq C$, where $C = s[1 - \Omega(\alpha)] \approx s$.

Figure 3.7 further clarifies the role of channel coding. The BSC has the capacity $C_s = 1 - \Omega(\alpha)$ and the source produces information at the rate R symbols/s. First, the source symbols enter a binary source encoder that produces binary bits (0 and 1 produced by the source encoder have equal probability). Second, we map blocks of the encoded bits to M-ary symbols (which are also equally probable). The alphabet size of the M-ary symbols is based on the number of bits (block size) mapped to symbols, where each symbol represents $\log M$ bits. Third, the BSC encoder represents each M-ary symbol by a channel codeword, consisting of N binary channel symbols. The average information per channel symbol is $\frac{\log M}{N}$. The channel symbols are generated at the rate s. Thus, the source information rate in relation to the channel symbol rate is:

$$R = s(\log M)/N. \tag{3.26}$$

In order to have reliable communications, we must have $R \leq C$ where $C = sC_s = s[1 - \Omega(\alpha)]$. From Equation 3.26, we find that Shannon's Theorem essentially requires that $(\log M)/N \leq C_s$. This means that M and N (the channel encoder parameters)[8] are

---

[7] If a channel is known to convert 0 to 1 and 1 to 0, the decoder can just toggle the bits and obtain the transmitted information correctly. In a BSC, error probability more than 0.5 can be turned to its complement value.

[8] Please do not confuse N here with $\bar{N}$, the average codeword length introduced in Section 3.1.3. Also, as we cover error control coding, we will use the naming conventions (k,n), where k is the length of the unencoded codeword (output of the M-ary) in Figure 3.7, and n is the codeword length (output of the channel encoder), and k/n is known as the coding rate.

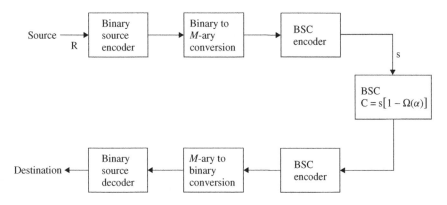

Figure 3.7 The role of channel coding in an encoding system for a BSC.

related as follows:

$$M = 2^{N(C_s - \varepsilon)}. \tag{3.27}$$

In Equation 3.27, $0 \le \varepsilon < C_s$, when ε is made arbitrarily small, $R \to C$. Choosing the proper channel coding means that the output of the channel decoder in Figure 3.7 (M-ary symbols) has a very low probability of symbol error. Notice that the longer the channel coding codeword length (N), the more we approach the channel capacity. One can prove that as $N \to \infty$, the Shannon bound is approached and we can achieve reliable communications. However, it is impractical to have $N \to \infty$. This would require infinite delay over the channel. Practical channel coding techniques have finite codeword lengths and introduce finite delay. Nevertheless, some of these codeword lengths do come pretty close to the Shannon bound. In the next section, we will review the basic concepts of channel coding.

3.2 Channel Coding, Error Detection, and Error Correction

In Chapter 2, we discussed communications over a channel, where a symbol (could be one or multiple bits) is transmitted and received over the channel, independent of the previous or consecutive symbols. We covered many modulation techniques, and showed the probability of detecting a symbol in error while in the presence of AWGN. We also illustrated different channel molding techniques and their relationship to symbol error detection. If there is no inter-symbol interference, the demodulator can apply an optimum detection technique to decode each symbol by itself.

In this chapter, we have moved into information theory concepts. We have demonstrated that with channel coding, the input symbols to the modulator are interrelated, which introduced a memory aspect into the signaling over the channel. Although we introduced the concept of channel coding as mapping a block of bits to an encoded codeword, there are many different categories for channel coding. Depending on the channel coding technique,[9]

---

[9] Choosing which channel coding to use depends on the channel characteristics. There are codes designed for fading channels, high bandwidth channels, and band-limited channels. In some cases, processing power is needed and its lack can limit the choices of certain channel coding techniques.

this interrelationship could either be addressed in a block-by-block manner or on a sliding window representation. Channel coding introduces redundancy in a controlled approach. The addition of channel coding can result in achieving reliable communications where the information symbol rate communicated over the channel can approach the physical limitation of the channel (channel capacity).

There are many ways to reach channel capacity, such as the use of orthogonal signal sets over AWGN, or increasing the signal set dimension M. However, the demodulator becomes exponentially complex as M increases. Yet building a complicated demodulator is impractical. Channel coding offers us the potential to approach channel capacity without the need to build a complex demodulator. Channel coding constructs symbol sequences in a large-dimensional space, while the modulation technique works on elementary symbols sets. For example, we can use a binary channel encoding technique to produce a code sequence where code symbols are communicated by means of binary antipodal signaling. However, the signal sequence does occupy high-dimensional space. The modulator and demodulator complexity is far less than that of using orthogonal signals. The spectral occupancy is an easier concept to grasp and model, while in the meantime we achieve equivalent performance in terms of channel capacity. As described above, channel coding maps the M-ary symbols in Figure 3.7 to the symbols entering the modulator.

There are different families of channel coding. Two important families are block codes and convolutional codes. Block codes are divided into linear block codes and non-linear block codes. Some linear block codes are considered as cyclic codes. Convolutional codes can be either linear or non-linear. Although channel coding is not within the scope of this book, we will cover some basic concepts and simple examples. Channel coding techniques used in today's commercial and tactical wireless systems are increasingly sophisticated and utilize ever-increasing processing power to implement error correction codes that were studied theoretically only decades ago.

In tactical wireless communications, minimizing the signal power is crucial (reducing spectrum footprint to prevent eavesdropping and jamming is one reason; increasing battery life is another). This desire to reduce power makes the need to use vigorous error control coding techniques ever more pressing.

3.2.1   Hamming Distance and Probability of Bit Error in Channel Coding

Figure 3.8 illustrates a simple Hamming distance example where error detection and/or error correction can be achieved. In general, an n-bit codeword can be represented in an n-dimensional space as a vector whose coordinates are the bits in the codeword. Figure 3.8a depicts the n-dimensional representation of two messages (zero and one) after being encoded, where channel coding represents these messages through simple repetition as (000, 111) codewords.[10] The solid dots represent the codewords in the three-dimensional space. In Figure 3.8b, we have four messages (00, 01, 10, 11) that are represented by codewords (000, 011, 101, 110) through adding a parity-check bit. *Hamming distance* is the mean for measuring the separation between the codewords. For example, in Figure 3.8a

---

[10] Repetition codes are very inefficient and they are used here for illustration only.

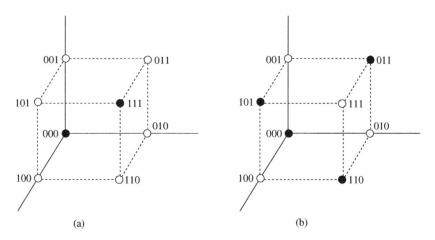

Figure 3.8 Three-bit codewords in vector representations. (a) Repetition code of one-bit messages and (b) parity-check code for two-bit messages.

the two codewords are separated by three bits (the minimum bit difference between the two codewords is three bits). These codewords have a minimum Hamming distance of $d_{min} = 3$. In Figure 3.8b, the minimum Hamming distance is two (the minimum difference between any two of the four codewords is two bits). Hamming distance determines the error detection and error correction capability of the code.

Error detection is possible when the number of bit errors introduced by the channel within the codeword is less than d_{min}. If the channel introduces bit errors within a codeword that is more than d_{min}, we could receive another valid codeword, and then error detection is not possible. It is obvious from Figure 3.8a that if the channel reverses all three bits (000 will be received as 111 and 111 will be received as 000), the resulting codewords would also be valid. However, if the channel introduces one or two bit errors, we will receive a vector that is not a valid codeword. In Figure 3.8b, if the channel introduces a single bit error, we will receive a vector that is not a valid codeword. If the channel introduces two bit errors, we will receive a valid codeword that is different from the transmitted one. Generally, if the number of error bits within a codeword exceed d_{min}, the erroneous codeword (the received vector) may actually correspond to *another* valid codeword and the error may not be detected.

In order to achieve the following error control capabilities, the respective code requires d_{min} as follows:

- To *detect* up to ℓ bit errors per codeword, $d_{min} \geq \ell + 1$.
- To *correct* up to t bit errors per codeword, $d_{min} \geq 2t + 1$.
- To *correct* up to t bit errors and *detect* up to $\ell > t$ bit errors per codeword, $d_{min} \geq t + \ell + 1$.

In Figure 3.8a $d_{min} = 3$ and $\ell \leq 3 - 1 = 2$ meaning that this code can be used to detect up to two bits per codeword. The same code can be used to correct $t \leq \frac{3-1}{2} = 1$ bit error per codeword. If we have a code with $d_{min} = 7$, we can use it to: (i) correct up to three bits per codeword; (ii) detect up to six bits per codeword; and (iii) or correct two-bit errors and detect four-bit errors.[11]

One would like to create codes where the highest d_{min} is achieved with the least number of redundancy or parity-check bits added to the message, while also making the codeword size as small as possible (for simplifying processing and reducing memory). In general, a block code turns k bits into n bits where $n > k$ and $n - k$ are the parity check bits or added redundancy bits. For a (k, n) code, the upper bound of the minimum Hamming distance is:

$$d_{min} \leq n - k + 1. \tag{3.28}$$

When selecting a channel coding for a specific channel, one needs to consider factors such as the channel characteristics and the error detection and correction capabilities of the code. One important factor is the coding rate, the ratio of information bits to the sum of redundancy, and information bits. In order to maximize information flow, we strive to maximize this ratio. Another means to measure the effectiveness of the code is the coding gain in terms of signal-to-noise-ratio (SNR) under certain constraints. This measurement expresses how much the SNR needs to be increased (if channel coding is not used) in order to obtain the same bit error rate that was achieved with channel coding.

3.2.2   Overview of Linear Block Codes

Linear block codes are based on the use of a *generation* matrices and *parity-check* matrices. As discussed above, the channel encoder is fed M-ary symbols. Each symbol consists of k bits where $M = 2^k$. Based on the specific channel encoding roles, the channel encoder transfers each symbol to a binary n-tuple vector where $n > k$. The binary n-tuple vector is also called a codeword. For each of the $M = 2^k$ symbols, the channel encoder creates a unique codeword. There is a one-to-one correspondence between the messages and the codewords; the codewords have unique characteristics. For example, a code is called a linear block code if the modulo-2 sum of any two codewords produces a codeword. A block code is expressed as an (n, k) code. The example shown in Table 3.2 is for the linear block code (7,4). You can see that this code has a minimum Hamming distance of three (any two codewords differ in three or more bits). The code can correct one bit or detect errors in two bits. Also the four rightmost bits of the codeword are the same as the four bits of the uncoded symbol (these types of codes are called systematic codes).

An (n, k) linear block code, C, creates a k-dimensional subspace of the vector space V_n from the binary n-tuples. The linear block encoder finds k linearly independent codewords, $g_0, g_1, \ldots, g_{k-1}$ in C, such that every codeword is a linear combination of these k codewords, that is,

$$v = u_0 g_0 + u_1 g_1 + \ldots + u_{k-1} g_{k-1}, \tag{3.29}$$

where $u_i = 0$ or 1 for $0 \leq i < k$.

---

[11] In channels with AWGN, the probability of having more than two bit errors per codeword is far less than the probability of having two or less bit errors per codewords. In most cases, the code is used for correction (called forward error correction). In the unlikely event of having more than two bit errors, error detection may mean requesting a retransmission of the codeword or discarding it.

Table 3.2 Linear block code (7, 4)

Symbol	Codeword
0 0 0 0	0 0 0 0 0 0 0
1 0 0 0	1 1 0 1 0 0 0
0 1 0 0	0 1 1 0 1 0 0
1 1 0 0	1 0 1 1 1 0 0
0 0 1 0	1 1 1 0 0 1 0
1 0 1 0	0 0 1 1 0 1 0
0 1 1 0	1 0 0 0 1 1 0
1 1 1 0	0 1 0 1 1 1 0
0 0 0 1	1 0 1 0 0 0 1
1 0 0 1	0 1 1 1 0 0 1
0 1 0 1	1 1 0 0 1 0 1
1 1 0 1	0 0 0 1 1 0 1
0 0 1 1	0 1 0 0 0 1 1
1 0 1 1	1 0 0 1 0 1 1
0 1 1 1	0 0 1 0 1 1 1
1 1 1 1	1 1 1 1 1 1 1

One can then create a $k \times n$ matrix where the k linearly independent codewords are the rows of this matrix as follows:

$$
G = \begin{bmatrix} g_0 \\ g_1 \\ \cdot \\ \cdot \\ \cdot \\ g_{k-1} \end{bmatrix} = \begin{bmatrix} g_{00} & g_{01} & g_{02} & \cdots & g_{0,n-1} \\ g_{10} & g_{11} & g_{12} & \cdots & g_{1,n-1} \\ \cdot & \cdot & \cdot & \cdots & \cdot \\ \cdot & \cdot & \cdot & \cdots & \cdot \\ \cdot & \cdot & \cdot & \cdots & \cdot \\ g_{k-1,0} & g_{k-1,1} & g_{k-1,2} & \cdots & g_{k-1,n-1} \end{bmatrix}, \tag{3.30}
$$

where $g_i = (g_{i0}, g_{i1}, \ldots, g_{i,n-1})$ for $0 \leq i < k$.

If the message to be encoded is $u = (u_0, u_1, \ldots, u_{k-1})$, then the corresponding codeword is:

$$
v = u \times G
$$

$$
= (u_0, u_1, \ldots, u_{k-1}). \begin{bmatrix} g_0 \\ g_1 \\ \cdot \\ \cdot \\ \cdot \\ g_{k-1} \end{bmatrix} \tag{3.31}
$$

$$
= u_0 g_0 + u_1 g_1 + \ldots + u_{k-1} g_{k-1}.
$$

The matrix G is the generator matrix[12] for the code C. Any k linearly independent codewords can be used to form a generator matrix for the (n, k) code. As a matter of fact, the code can be completely specified by the k rows of the generator matrix G. The encoder can utilize G to generate any codeword based on the input message.

---

[12] Because the rows of G span the (n,k) linear code C, it is called the generator matrix.

The (7,4) linear code in Table 3.2 has the following generator matrix:

$$G = \begin{bmatrix} g_0 \\ g_1 \\ g_2 \\ g_3 \end{bmatrix} = \begin{bmatrix} 1 & 1 & 0 & 1 & 0 & 0 & 0 \\ 0 & 1 & 1 & 0 & 1 & 0 & 0 \\ 1 & 1 & 1 & 0 & 0 & 1 & 0 \\ 1 & 0 & 1 & 0 & 0 & 0 & 1 \end{bmatrix}. \tag{3.32}$$

If the message to be encoded $u = (1\ 1\ 0\ 1)$, then the corresponding codeword is obtained as follows:

$$v = 1.g_0 + 1.g_1 + 0.g_2 + 1.g_3$$
$$= (1\ 1\ 0\ 1\ 0\ 0\ 0) + (0\ 1\ 1\ 0\ 1\ 0\ 0) + (1\ 0\ 1\ 0\ 0\ 0\ 1)$$
$$= (0\ 0\ 0\ 1\ 1\ 0\ 1).$$

A linear block code is systematic if the codeword is divided into two parts. The first part, consisting of k bits, is identical to the message. The second part, consisting of the $(n - k)$ parity-check bits, is the redundant checking part. The parity-check bits are linear sums of the information bits. Thus, for linear systematic block codes, each codeword follows the format shown in Figure 3.9.

Note that if we swap two entire rows or two entire columns of the generator matrix, we still have a unique one-to-one correspondence of messages to codewords. Therefore, in Figure 3.9 it is irrelevant if the message part is on the left or the right. Linear block codes have many interesting properties that are beyond the scope of this book.

Notice that what applies to a binary digit in this coding approach can also apply to non-binary symbols. We covered the case when the message to be encoded consists of binary symbols, since it is easier to understand. Error control coding techniques can be constructed over Galois field $GF(2^m)$ elements where $(m > 1)$. Here, the message to be encoded can consist of non-binary elements. In computer networking, it is common to use $GF(2^8)$, where each element of the message to be encoded is a byte (eight bits) and the elements of the $GF(2^8)$ are mapped to the ASCII 128 table. Packet erasure coding techniques that utilize this mapping are commonly used at the transport layer or in network coding as shown in Appendix 3.A. Please refer to error control coding references for more details.

3.2.3 Convolutional Codes

Convolutional codes differ from block codes in that their encoder contains memory. The n symbols (bits in the binary case) produced by the encoder at any given moment, depend on both the k input symbols at that time and on the m previous input blocks. A convolutional code is expressed as (n, k, m) with k input, n output, and m memory. Some convolutional

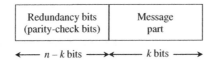

Figure 3.9 Format of systematic linear block code.

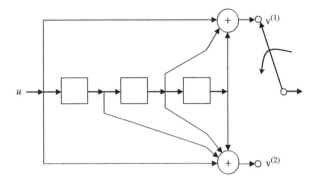

Figure 3.10   A simple example of convolutional codes.

codes use a small n and k with $k < n$, but have large memory m. A larger m enhances the error correction capability of the code.[13] When $k = 1$, the input information sequence does not consist of bit blocks; it is then treated as a bit steam that can be processed continuously. Convolutional codes were first utilized in the space program in the 1970s. Many of the latest commercial and tactical communications technologies use turbo codes which are a mutation of convolutional codes that are proven to have powerful error correction capabilities.

Figure 3.10 illustrates a simple example of building a binary convolutional encoder for the (2,1,3) code. There are three stages of shift registers ($m = 3$), and two modulo-2 adders ($n = 2$) and a multiplexer that is needed to make the encoder output serial. This encoder is simply a linear feed-forward shift register. All convolutional code encoders can be built in much the same way.

In Figure 3.10, we have an information sequence $u = (u_0, u_1, u_2, \ldots)$ entering the encoder (one bit at a time), and we have two output sequences $v^{(1)} = (v_0^{(1)}, v_1^{(1)}, v_2^{(1)}, \ldots)$ and $v^{(2)} = (v_0^{(2)}, v_1^{(2)}, v_2^{(2)}, \ldots)$. $v^{(1)}$ and $v^{(2)}$ are basically the convolution of u with the impulse responses of the encoder (notice that the encoder has two different impulse responses). In order to understand the impulse response of this encoder, we let $u = (100\ldots)$ and observe the two output sequences. For an encoder with m unit memory, the impulse response lasts for at most $m + 1$ time units.[14] The two impulse responses are expressed as:

$$g^{(1)} = (g_0^{(1)}, g_1^{(1)}, g_2^{(1)}, \ldots, g_m^{(1)}),$$

$$g^{(2)} = (g_0^{(2)}, g_1^{(2)}, g_2^{(2)}, \ldots, g_m^{(2)}).$$

For the encoder in Figure 3.10, the impulse responses are:

$$g^{(1)} = (1 \ 0 \ 1 \ 1),$$

$$g^{(2)} = (1 \ 1 \ 1 \ 1).$$

---

[13] In his work, Shannon mentioned the relationship between random errors and approaching channel capacity; larger m in convolutional codes assists in spreading bit error to make it random.

[14] First 1 enters the encoder (at the far left and not in a specific memory location), then 1 moves to the first memory element at the far left while 0 is entering the encoder, then 1 moves into the middle memory, while 0 is at the far left memory element. At this point, another 0 is entering the encoder, the 1 moves to the far right memory with 0s in the other two memory elements and a third 0 enters the encoder. As 1 leaves the encoder, all outputs become 0s.

These impulse responses are considered the generator sequences of the code, and one can write the encoding equations of a convolutional encoder as:

$$v^{(1)} = u * g^{(1)}, \tag{3.33a}$$

$$v^{(2)} = u * g^{(2)}, \tag{3.33b}$$

where $*$ expresses the discrete convolution with a modulo-2 operation.

The outcome of the convolution operation for all $l \geq 0$ is:

$$v_l^{(j)} = \sum_{i=0}^{m} u_{l-i} g_i^{(j)} = u_l g_0^{(j)} + u_{l-1} g_1^{(j)} + \ldots + u_{l-m} g_m^{(j)}, \quad j = 1, 2, \tag{3.34}$$

where for all $l < i$, $u_{l-i} = 0$.

For the encoder in Figure 3.10, the output sequences can be expressed as:

$$v_l^{(1)} = u_l + u_{l-2} + u_{l-3}, \quad v_l^{(2)} = u_l + u_{l-1} + u_{l-2} + u_{l-3}. \tag{3.35}$$

If you look at the figure, you will see how the top output sequence does not receive an input from the left memory element, corresponding to Equation 3.35. As the figure demonstrates, the two output sequences are multiplexed into what becomes the encoder output or codeword for transmission over the channel. The codeword can be expressed as:

$$v = v_0^{(1)} v_0^{(2)}, v_1^{(1)} v_1^{(2)}, v_2^{(1)} v_2^{(2)}, \ldots. \tag{3.36}$$

3.2.4 Concatenated Coding and Interleaving

Prior to the establishment of packet erasure coding for MANET networks, the error control coding community had relied upon Viterbi decoding of convolutional codes[15] in a concatenated fashion with Reed–Solomon (RS) codes. The reasoning behind this convention was that when convolutional codes fail, they tend to cause a burst of error and the decoder can detect a block of erroneous bits (which can be considered an erasure). RS codes are well suited to correcting erasures. When fading occurs in a channel, convolutional codes will fail and erasures will occur. A typical concatenated coding approach is shown in Figure 3.11, where the data sequence is first RS encoded, then interleaved at the symbol level (the RS code symbol) before it enters a convolutional encoder. At the receiving end, when the convolutional decoder fails, the error burst is spread (from the de-interleaving process). The nature of the spread is such that each block of bit-erasure is the size of the RS code symbol, and the bursts are divided among multiple RS codewords. The RS decoder is then faced with erasure symbols that are more or less random and is designed to be able to correct a certain level of erasure symbols.

In tactical networks, if convolutional codes or turbo codes are used at the DLL of the radio, erasure can occur and cause the loss of an entire IP packet. Network coding can use RS codes to recover these lost packets. One area in error control coding that applies to tactical networks is the use of RS codes for packet erasure. References such as [4–6]

---

[15] Viterbi introduced a maximum likelihood decoding technique which made the use of convolutional codes easier to implement.

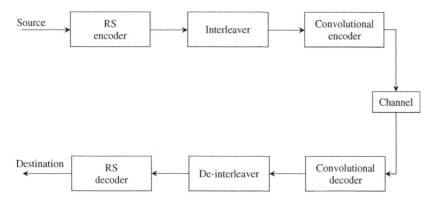

Figure 3.11   The use of concatenated coding with convolutional and RS codes.

can give you an idea of how RS codes can be used to overcome packet erasure in tactical networks. The next section briefly explains the areas of network coding and transport layer coding, both of which use a form of RS codes for packet erasure. Appendix 3.A presents a case study for the use of RS codes at the transport layer of tactical networks with an encrypted core.

3.2.5   Network Coding versus Transport Layer Packet Erasure Coding

Let us imagine a tactical MANET subnet where link reliability fluctuates in such a way that the bit error rate ranges between 10^{-6} and 10^{-2}. Over a network path, if the end-to-end path reliability (from all the hops) remains within the 10^{-6} range of bit error rate, the transport layer can function (i.e., requests for retransmission or packet drop are rare). On the other hand, when the end-to-end path reliability (from all the hops) is at the 10^{-2} level, the packet error rate will be high. The ratio of discarded packets then becomes higher, to the extent that the application will not be able to function (for real-time applications, packet are lost and the application layer at the receiving node is burdened by information loss; for non-real-time applications, packets can go through retransmission requests that can cause the transport layer to reduce its flow and come to a halt or abort the session). In tactical MANET, the link condition at the RF layer can fluctuate drastically. Therefore, one would need techniques that ensure some level of hop reliability such that the transport layer can function, and the end-to-end network path can give the user the expected QoS. Network coding attempts to create some form of stability for the transport layer by ensuring that as packets traverse the network hop-by-hop, each hop is reliable (that is, the probability of causing erroneous packets is bounded by some desired limit). Consequently, the end-to-end path reliability (of many MANET hops) stays within a desired limit that ensures a preferred level of QoS and the functionality of the transport layer. Network coding is especially crucial for multicast over MANET networks. The term hop-by-hop is used here loosely since, in MANET, some nodes can act as a relay between two different nodes that are out of reach (due to distance or terrain). Also, multiple access radios make a packet hop from one point to multiple points.

There are obvious tradeoffs presented here. How far can we burden the transport layer? How far should network coding go (how many redundancy packets can it generate)?

How about odd moments (like slow fading)? Should we introduce so much redundancy at the hop-by-hop level that we can ensure a high level of reliability at the cost of information flow? (Remember that more redundancy packets for error control coding means reducing the information packet flow.) With so many nuances in tactical MANET (nodes move fast and link conditions change fast), it is difficult to develop a crystal-clear information theory model similar to the Shannon model.

Instead of attempting to develop complex models (that would hardly hold due to the dynamics of MANET), one can regard the per-hop error control coding and the transport layer error control coding as the tools for developing a defense-in-depth mechanism. The per-hop error control coding should adapt to changes in the channel conditions (without drastically reducing information flow or being designed based on odd cases). This will create a form of path reliability (most of the time) such that the transport layer will only have to "worry" when the hop-based error control coding fails to create the desired per-hop, and consequently per-path, reliability.

Tactical MANET radios may be using error control coding at the DLL to ensure bit stream reliability; this differs from network coding. Network coding is packet based and can be implemented at the higher layers to ensure packet reliability on a hop-by-hop basis. At the receiving end, after the receiver has obtained enough encoded packets, it is capable of reconstructing the original data packets. With tactical MANET, if we can ensure a certain level of link reliability (at the DLL for the bit stream), a certain level of robustness at the packet level on a hop-by-hop basis, and develop a more robust technique at the transport layer, we can achieve the desired level of QoS under adverse network conditions. This does not require developing a network model similar to that of Shannon's channel model; rather, one relies on a defense-in-depth approach and uses coding approaches that attempt to maximize the information flow.

Network coding and transport-layer-packet-erasure coding share some common ideas. They both encode the original packets into an infinite[16] stream of encoded packets, and their respective receivers reconstruct the original data packets when they have collected a certain number of encoded packets. The main difference is that network coding happens at the intermediate node, where packets are encoded, while the transport layer packet erasure coding occurs only at the source encoded node. Thus, network coding can encode packets from different streams mixed together, while transport layer coding works on each session individually.[17]

Network coding specifically increases the reliability of multicast sessions in MANET networks. There are two approaches for network coding. One approach can be categorized as *intra-session* network coding, where only packets in the same session are encoded together. The second approach allows coding across different sessions or *inter-session* network coding. Here, overhead (encoded) packets are transmitted over the broadcast wireless media and are exploited at the different hops to further enable network coding. *Inter*-session network coding has been shown to further increase the throughput of multicast sessions over MANET.

---

[16] Here, infinite is in the sense that some packet erasure coding techniques are not limited to the number of redundancy packets they can produce. Some techniques can keep generating redundancy packets until an acknowledgment is received. Please refer to [5] for an example of infinite stream of encoded packets.

[17] The case study in Appendix 3.A makes the argument against using infinite stream of encoded packets with high importance traffic in tactical networks in order to meet the speed of service (SoS) requirements.

While designing a network coding technique that fits in the end-to-end context of MANET, the amount of overhead incurred by additional coding operations must be minimized and must be outweighed by the benefits provided by the network coding. There are many techniques that can maximize the benefits of network coding while minimizing the amount of overhead. With *inter*-session network coding, an increase in the number of nodes receiving an encoded packet can maximize the benefits of network coding. Also, understanding how network coding relates to channel fading is essential. In fading channels, the performance of the network is dictated by the channel gains of the different MANET users during the current transmission. However, the channel gains may not be known prior to transmission. In this case, the amount of information that is received varies based on the transmission quality of the channel. Curtailing the number of redundancy packets during deep fading is necessary. References [7–11] offer solid explanations for the basics of network coding.

3.3   Concluding Remarks

We have just scratched the surface of error control coding so as to provide you with an idea of how it relates to tactical networks. This book does not specialize in error control coding. You are encouraged to check references such as [12], which is a classic book which explains the basis of error control coding (although it lacks the last few decades' discoveries in areas such as turbo codes, trellis coded modulation, network coding, etc.). You are also encouraged to check more recent references in the areas of network coding, transport layer coding, and tactical edge proxy techniques to learn as much as you can about information flow in tactical networks. Appendix 3.A offers a case study regarding the use of RS codes in the transport layer of tactical networks with High Assurance Internet Protocol Encryption (HAIPE) encrypted core.

3.3.1   *The Role of Information Theory and Coding in Tactical Wireless Communications and Networking*

You may be wondering why we are discussing information theory and coding in this book. In fact, we have hardly delved into these fields. It is essential to understand the essence of information theory in order to understand the depth of our tactical wireless networking problems. You have seen how transmission over a single channel can be optimized. However, keep in mind that many models and assumption are used for this optimization. We assumed the presence of AWGN, and the independence of source symbols. We used spectrum models that follow a specific pattern. We assumed the statistical independence in signal estimation. We used models that are stationary so as to simplify analysis. We treated events as if they were discrete with no correlations in order to express and analyze the communications channel; we even assumed symmetry of the binary channel. With all of these assumptions in mind, and given the fact that a channel model's complexity pales in comparison to a tactical wireless networks model, hopefully you can see how difficult it is to bring valid mathematical models to the tactical wireless communications and networking field. Keep in mind that as you go up the networking protocol stack, some of the assumption made at the channel model may no longer be applicable. Also, the stationary models will not hold in a dynamic network. Typically, source coding can happen at the application (compression), while channel coding happens at the DLL with protocol stack layers in between that may have their form of error control coding. The protocol stack layers introduce so much overhead, and the layers can attempt to remedy the same problems independently, often making decisions that can contradict.[18]

There are many research approaches in tactical wireless networks optimization. Some promote the use of network coding to make sure that each hop adapts dynamically to the channel conditions, ensuring that all the network hops are reliable and reduce the burden on the transport layer. Some advocate cross layer signaling as the means to carry the information theory concepts of optimization to the upper layers of the protocol stack. Some present the proxy of all types of applications and control signaling over the tactical wireless core network as the means to make IP work in tactical wireless networks. These proxy protocols adapt to the changes of the wireless network while COTS application are used seamlessly by the tactical user. Some promote developing a transport layer specific for tactical networks. Others go as far as doing away with IP and developing out of the box, new tactical wireless network protocols. Some campaign the use of a super-layer where all stack layers merge into a new generation of intelligent MANET. Some promote the use of centralized intelligent process across all the stack layers to make sure that all layers work harmoniously toward optimizing information transfer, and so on. We are far away from generating mathematical models for tactical wireless networks that rival the Shannon model of a single channel. As a matter of fact, we are still debating the use of IP in tactical wireless networks. I hope that conveying an overview of the information theory concepts help you, the reader, understand the complexity of the optimization of information transmission over tactical wireless networks.

3.3.2   Historical Perspective

Ever since Claude Shannon published his work in 1948 and pioneered the science of information theory, digital communications and networking has gone through a very rapid evolution and we are now living in the information age. Some credit the Cold War for such a boost in science and technology. Many of the discoveries in this field were possible due to US government funding. Yet, it was pure individual ingenuity that made use of pre-existing theories in binary field arithmetic and Galois field arithmetic. After the introduction of linear block codes and the publications of Hamming, the 1950s and the 1960s brought a surge of energy to the field of error control coding. Elias introduced the concept of convolutional codes in 1955. Wozencraft introduced sequential decoding as a practical means to use convolutional codes in the early 1960s. In 1967 Viterbi presented maximum likelihood decoding which made the use of convolutional codes easier to implement and more powerful. In the 1970s convolutional codes were used in the space program for deep-space communications and in satellite communication. In 1977, when Voyager was launched in a space mission through the solar system, it had a Viterbi encoder built in. Turbo codes appeared soon after and the advances of hardware (memory and computational speed) made it far more usable. In the late 1970s and early 1980s, Ungerboeck published his work about trellis coded modulation,

---

[18] As an example, for a tactical MANET, during fading, the DLL can dynamically increase error correction, while the network layer can implement network coding, and the transport layer may back off after flooding the already stressed network with request for retransmission, while a call admission control preempts the session that all these layers are trying to correct independently.

which made it possible to approach the Shannon bound over band-limited channels (phone lines). The information theory field is now looking to optimize MANET networks through techniques such as network coding and by studying its relationship to transport layer packet erasure coding. Given the dynamics of MANET networks, a Shannon-like model will not be applicable. How can we optimize the performance of tactical wireless networks? The jury is still out.

Appendix 3.A: Using RS Code in Tactical Networks Transport Layer

This appendix presents a transport layer hybrid ARQ (automatic repeat request) technique that uses adaptive RS codes with packet erasure.[19] The technique considers the existence of HAIPE encryption, where decryption erases the entire IP packet if it contains any error.[20] Please refer to Section 1.4 and Figure 1.5 to see where the transport layer and the ComSec (HAIPE) encryption layers relate to each of the other layers in the tactical wireless networks protocol stack layers. This technique uses a multifaceted optimization of RS codes at the transport layer for delivery assurance, speed of service (SoS), and network throughput. The goal is for the transport layer in tactical wireless networks to fulfill the stringent requirements of QoS, imposed by the tactical network user, even under adverse conditions. These requirements define a high level of reliability (delivery assurance), a specific SoS, and optimum use of the limited bandwidth of the wireless network, where the probability of packet erasure can be very high. On one hand, focusing on network throughput alone will result in violating SoS and delivery assurance requirements. On the other hand, focusing on SoS and delivery assurance requirements can result in poor network throughput. The analysis for the optimization technique uses a homogeneous Markov chain approach to reach the desired balance between reliability, SoS, and throughput efficiency needs.

3.A.1 The Utilized RS Code

The code uses systematic RS encoding, where any sent message of length n packets consists of k data packets and $n - k$ redundancy packets. Note, the $RS(n, k)$ notation is adopted here as well. The generator matrix, used for encoding and decoding messages, is derived from a Vandermonde matrix built over a finite Galois field $GF(2^8)$. This field has 256 distinct elements. Each element can easily be mapped to a symbol of the ASCII table or to any eight bits in the data packet. Each row of a Vandermonde matrix is defined by a different element of the Galois field in power $0, 1, 2, \ldots, k - 1$. While those rows are distinct, any submatrix of the Vandermonde matrix is invertible and hence the generator matrix of size $k \times k$ is also invertible. The generation matrix also sets the range for length n of the encoded message:

$$k \leq n \leq 256. \tag{3.A.1}$$

---

[19] ARQ techniques rely on request for retransmission to recover lost packets. Hybrid ARQ techniques use FEC and when FEC fails, a request for retransmission is generated.

[20] With HAIPE, the decryption process only passes packets that are error-free. Packet erasure at the tactical MANET transport layer could be due to congestion (packets dropped at the IP queues in the network code), degradation of wireless links, and/or encryption techniques that drop tampered packets. Out-of-order and excessively delayed packets can further complicate the transport layer error recovery techniques.

The top k rows of the $n \times k$ generator matrix form an identity matrix. This ensures that the code is systematic; in other words, the first k encoded packets are identical to the data packets. The following $n - k$ rows are used for computation of the $n - k$ redundant packets that compensate for the loss of information. Such a generator matrix has rank k meaning that any submatrix of size $k \times k$ is invertible. It also means that the message encoded with such an $n \times k$ generator matrix can be correctly decoded if any k packets are received out of the n sent packets. With this technique, n can not exceed the number of elements in the $GF(2^8)$ (e.g., 256). If $n > 256$, some of the rows in the Vandermonde matrix would be identical and some of the $k \times k$ submatrices of the generator matrix would have no inverse matrix. In such a case, there is no guarantee that the message can be completely restored if, in the n transmitted packets, more than $n - k$ packets are lost in transmission. All computations used to build the generator matrix coding/decoding of the message are indeed performed according to $GF(2^8)$ arithmetic operations.

All the elements of the generator matrix are derived from an $n \times k$ Vandermonde matrix and depend only on the parameter k. That is, generator matrices can be easily precalculated for the given Galois field $GF(2^8)$. The necessary range of message size, k, can be calculated and stored at the source and the destination locations. The message size can then be used for encoding and decoding for any n encoded packets that satisfy the inequality in Equation 3.A.1. This technique provides a means to determine the encoding rate k/n (or the minimum number of redundancy packets) required to allow the delivery of a message in ever-changing, unreliable network conditions, while meeting reliability and SoS requirements.

3.A.2   Packet Erasure Analysis

With RS encoding, the sent message consists of k data packets and $n - k$ redundancy packets. Depending on the network conditions, all n packets, a subset of n packets, or no packets could reach the destination. There are $n + 1$ possible outcomes when a message of length n packets is transmitted over the network, meaning that after the first transmission, the destination can get $0, 1, 2, \ldots, k, \ldots,$ or n packets. In other words, the destination can be in one out of $n + 1$ possible states, as defined by the number of packets that reach the destination.

The probability of each possible outcome after the first transmission depends on the network conditions. This can be given by a binomial distribution, assuming that the packet drop events are independent (interleaving the source packet stream can help making packet drop random, increasing its independencies). As such, the probability for the destination getting any i packets (to be in state i) out of n sent packets is:

$$P_i = C_n^i \cdot p^i \cdot (1 - p)^{n-i}, 0 \leq i \leq n, \tag{3.A.2}$$

where,

p = probability that a single packet will be delivered
C_n^i = binomial coefficients
P_i = probability that the destination is in state i after the first transmission.

If any k out of n encoded packets are received, the RS decoder allows the complete restoration of the message. If fewer than k packets are received, a selective acknowledgment is sent, which indicates the lost packets' sequence numbers. The probability of successful feedback delivery also impacts the overall reliability of communication. The analysis here focuses on optimizing the transport layer RS coding, assuming that selective acknowledgments are received successfully. In the actual implementation of such techniques, a lost acknowledgment will trigger a timeout at the sender, which prompts sending extra redundancy packet(s). These redundancy packets will help recover any lost packets and trigger a new acknowledgment. Thus, the packets dropped in the first transmission define the packets to be retransmitted in the second transmission. Following the initial transmission, if no packets are received then the destination would be in state 0. In this case, all n packets should be retransmitted, and this time around, the destination can transit into any of $0, 1, 2, \ldots, k, \ldots, n$ states. If only one packet is received after the initial transmission, the destination would then be in state 1. In this case $n - 1$ packets are retransmitted. As a result, the destination could move to any of the states $1, 2, \ldots, k, \ldots, n$. Notice that state 0 is not considered because one packet has already been successfully received. If the destination receives $k, k + 1, \ldots, n$ packets, any further retransmissions are not needed and the state of the destination would remain constant.

The destination's state transitions can be fully described by the following transition probability matrix:

$$T(p, n, k) = \begin{pmatrix} P_{0,0} & P_{0,1} & P_{0,2} & \cdots & P_{0,k} & \cdots & P_{0,n} \\ 0 & P_{1,1} & P_{1,2} & \cdots & P_{1,k} & \cdots & P_{1,n} \\ 0 & 0 & P_{2,2} & \cdots & P_{2,k} & \cdots & P_{2,n} \\ \cdots & \cdots & \cdots & \cdots & \cdots & \cdots & \cdots \\ 0 & 0 & 0 & \cdots & 1 & \cdots & 0 \\ \cdots & \cdots & \cdots & \cdots & \cdots & \cdots & \cdots \\ 0 & 0 & 0 & \cdots & 0 & \cdots & 1 \end{pmatrix}. \tag{3.A.3}$$

The $(n + 1) \times (n + 1)$ matrix in Equation 3.A.3 represents the probabilities $P_{i,j}$ for the system (destination) to move from state i (i packets were received in all previous transmissions) to state j (an additional $j - i$ packets are received in the current transmission). Matrix $T(p, n, k)$ depends on the probability p of a single packet to be delivered, and is defined by the number of data packets k and the overall number of sent packets n. Matrix $T(p, n, k)$ has all zeros below the main diagonal. This reflects the fact that the number of received packets cannot be reduced, and a transition to a state with fewer received packets is impossible. The elements $P_{i,j}$ are also defined by the following binomial distribution:

$$P_{i,j} = C_{n-i}^{j-i} \cdot p^{j-i} \cdot (1 - p)^{n-j}, 0 \leq i < k, i \leq j \leq n. \tag{3.A.4}$$

All diagonal elements $P_{i,i}$ with $k \leq i \leq n$, are equal to 1, and all other elements in the same row are 0. That is, in these states all required packets are received, and the message is fully decoded. As a result, no retransmission is required, and the destination's states are unchanged.

All elements of matrix $T(p, n, k)$ are non-negative. The sum of all elements in each row of this matrix (Equation 3.A.3) is 1. Hence matrix $T(p, n, k)$ represents the stochastic communication process, and together with initial distribution (Equation 3.A.2), it defines a homogeneous Markov chain.[21] The parameters of this process are the probability of delivering a packet p, the number of data packets in a message k, and the overall message size n.

If the network conditions are unchanged (over a retransmission time of the same message), the probability of delivering a packet p is constant, as is the transition matrix $T(p, n, k)$. The evolution of the technique following m transmissions is described by the following $n + 1$-dimensional vector probability density function:

$$\overline{P}(m) = \overline{P}_{init} \times T^{m-1}(p, n, k).\qquad\text{(3.A.5)}$$

In Equation 3.A.5:

$\overline{P}(m) = $ vector $\{P_0(m), P_1(m), \ldots, P_n(m)\}$

$P_i(m) = $ probability for the destination to get exactly i packets after m transmissions and $0 \le i \le n$

$\overline{P}_{init} = $ vector of initial probabilities $\{P_0, P_1, \ldots, P_n\}$ (see Equation 3.A.2).

The second factor on the right side of Equation 3.A.5 is the transition matrix $T(p, n, k)$ to the power $m - 1$. Note that at $m = 1$, $T^0(p, n, k)$ is an identity matrix, and the distribution in Equation 3.A.5 is identical to the initial distribution in Equation 3.A.2. Figure 3.A.1 demonstrates how the probability density function $\overline{P}(m)$ for an eight-packet message ($k = 8$) evolves with retransmission at $p = 0.5$.[22] $RS(8,8)$ indicates that no redundancy packets are added to the message. $RS(10,8)$ means that two redundancy packets are added to the message (hybrid ARQ with packet erasure). These results show how the positive skew of the $\overline{P}(m)$ curves increase with retransmission and become especially prominent when the number of redundancy packets is greater than zero (Figure 3.A.1), dashed lines). As a result, adding two redundant packets causes the probability of receiving eight or more packets ($k = 8$) on the second transmission to increase from 0.1 to 0.31 (consider the area under the curve from the number of delivered packets equals 8, to the number of delivered packets equals 10). This also means that adding redundancy packets increases the likelihood of decoding the message in a shorter time (meeting SoS requirements at such high packet loss of 50% ratio). Thus, the probability that an $RS(n, k)$ encoded message of overall length n containing k data packets and $n - k$ redundancy packets would be successfully delivered and decoded after m transmissions (or less than m transmissions) is defined as:

$$R(n, k, m, p) = \sum_{i=k}^{n} P_i(m),\qquad\text{(3.A.6)}$$

where, $R(n, k, m, p)$ is the message delivery assurance or reliability (probability of successfully delivering and decoding a message), and $P_i(m)$ are the probabilities defined by Equation 3.A.5.

In Equation 3.A.6, k is the given message size, and m is the number of transmissions, which is usually limited due to SoS requirements. In order to satisfy a desired SoS requirement, variations of network condition p could be compensated by only changing the number of redundancy packets $n - k$ (and hence changing n).

---

[21] The only steady state probability distribution $\bar{q}$, which satisfies the equation $\bar{q} \times T = \bar{q}$, describes the status of communications when there are no packets in transmission, and the destination gets and decodes the entire sent message.

[22] At such high ratio of packet loss, TCP is likely to fail to deliver the message and either back off for a long period or abort the session.

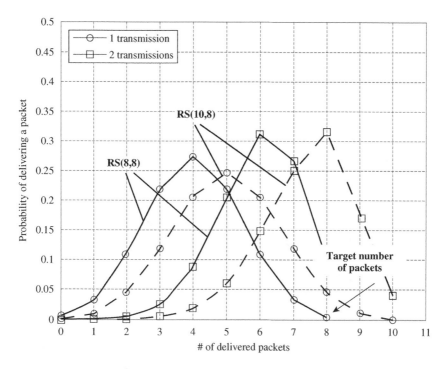

Figure 3.A.1 Evolution of $\bar{P}(m)$ with transmissions for RS (8,8) and RS (10,8) coding at $p = 0.5$.

The summation from k to n in Equation 3.A.6 represents the fact that any $RS(n, k)$ encoded message could be decoded if at least k out of n packets are received. Figure 3.A.2 depicts the probability of delivering an eight-packet message ($k = 8$) in one transmission ($m = 1$) when the number of redundancy packets, ($n - k$), varies. The presented data illustrates that increasing the number of redundancy packets significantly improves message delivery assurance, even for the case where the probability of delivering a packet p is as low as 0.5.

Figure 3.A.3 depicts the effect of retransmission on the probability of delivering an eight-packet message with no redundancy (solid lines) and with two redundancy packets (dashed lines). Thus, the solid lines represent $RS(8,8)$, and the dashed lines represent $RS(10,8)$ encoding. Note that the $RS(8,8)$ case is an ARQ protocol.

Let us consider the efficiency of $RS(n, k)$ coding with compensation for packet erasure. We will define the throughput efficiency η as the ratio between the number of data packets k to the expected number of all packets (data packets + redundancy packets) N sent over the network to achieve message delivery. The expected number of packets N is greater than or equal to n. Note that in the very first transmission ($m = 1$), the number of transmitted packets is always equal to the encoded messages size. Thus,

$$E\left[N_1\right] = n. \tag{3.A.7}$$

In the following transmissions, only dropped packets are retransmitted. This means that if the system is in state i and $i < k$ after $m - 1$ transmissions, then only $n - i$ packets

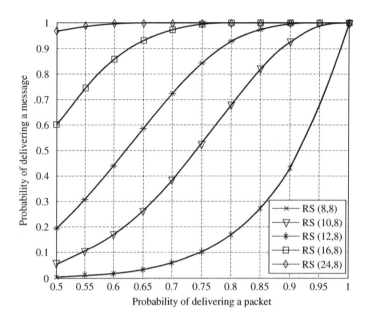

Figure 3.A.2   Probability of delivering an eight-packet message in one transmission.

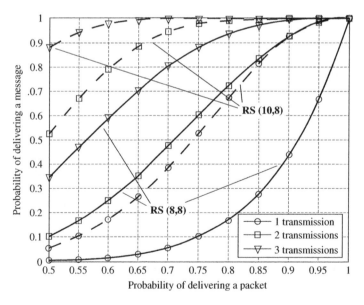

Figure 3.A.3   Effect of retransmission on RS encoded messages as a function of the probability of delivering a packet.

are retransmitted. The number of retransmitted packets is a random number defined by the probability $P_i (m - 1)$. So, the expected number of packets to be sent on the m-th transmission is defined by the expression:

$$E\left[N_m\right] = \sum_{i=0}^{k-1} (n - i) \cdot P_i (m - 1), m > 1. \tag{3.A.8}$$

Hence, the overall expected number of packets sent in m transmissions is defined by the expression:

$$E\left[N (m)\right] = N_1 + \sum_{j=2}^{m} E\left(N_j\right) = n + \sum_{j=2}^{m} \sum_{i=0}^{k-1} (n - i) \cdot P_i (j - 1), m > 1. \tag{3.A.9}$$

In Equation 3.A.9 above, the first term corresponds to the initial transmission where $j = 1$, which has the explicit value of N_1 or n packets. Figure 3.A.4 depicts the effect of redundancy packets and the number of transmissions on the expected throughput efficiency defined as:

$$\eta = k/E\left[N (m)\right]. \tag{3.A.10}$$

Figure 3.A.4 demonstrates the variation of the throughput efficiency over one, two, and three transmissions while trying to deliver an eight-packet message in various network conditions. The data on the upper plot show that $RS(8,8)$ with ARQ is sending approximately $8/0.67 \approx 12$ packets in two transmissions, and $8/0.57 \approx 14$ packets in three transmissions when $p = 0.5$. On the other hand, the bottom plot shows that $RS(10,8)$ sends $8/0.54 \approx 15$ packets in two transmissions, and $8/0.48 \approx 17$ packets in three transmissions. Let us compare this with the results presented in the Figures 3.A.2 and 3.A.3. Consider the case when $p = 0.5$. One can see how the three extra packets sent with $RS(10,8)$ encoding provide more than a 0.4 gain in terms of probability of successful delivery of a message. For some high-importance messages (such as a call-for-fire), the addition of redundancy packets is a small price to assure the delivery within the SoS requirements.

For additional comparisons between ARQ and RS-based hybrid ARQ protocols, let us compute the expected number of transmissions required to successfully deliver a message. The message could be delivered in one transmission, if on the very first transmission, the destination gets $k, k + 1, \ldots, n$ packets. This is quite an unlikely event, even if the probability of delivering a packet is around 0.9 (see Figure 3.A.2). In general, the number of transmissions needed to deliver a message in m or fewer transmissions is a random number with the cumulative probability function $R(n, k, m, p)$, as described by Equation 3.A.6. As a result, the expected number of transmissions $E[m]$ needed to completely deliver a message is defined by the following formula:

$$E[m] = 1. \sum_{i=k}^{n} P_i + \sum m. \left[R(k, n, m, p) - R(k, n, m - 1, p)\right]. \tag{3.A.11}$$

Note that the expression in the square brackets on the right side of Equation 3.A.11 represents the probability that the message delivery was completed exactly on the m-th transmission.

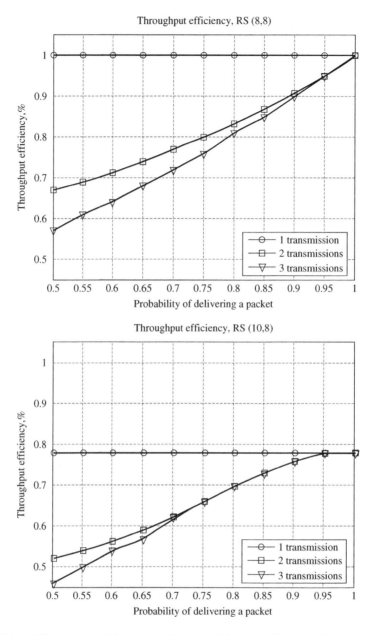

Figure 3.A.4   Throughput efficiency as a function of the probability of delivering a packet.

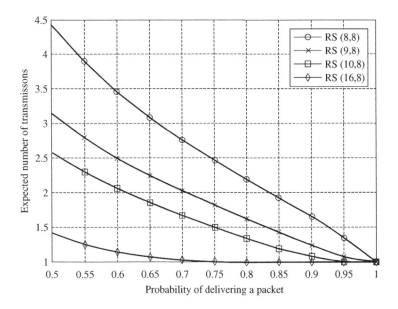

Figure 3.A.5 Expected number of transmissions to deliver a message.

We can assign the expected number of transmissions $E[m]$ as one of the major parameters that defines message delivery time. Therefore, compliance with the SoS requirements for high-importance messages in tactical networks is satisfied. The results in Figure 3.A.5 illustrate that at $p = 0.5$, even one redundancy packet reduces the expected (average) number of required transmissions by almost 1.5 times. Two extra packets decrease the expected number of transmissions from 4.5 to 2.5. Note that on the one hand, the added redundancy means a decrease in throughput, but on the other hand, reducing retransmission eliminates the need for the negative acknowledgment (NACK) packets and the retransmission of lost packets. Most importantly, the added redundancy allows for the specified SoS requirements to be satisfied.

One can see that simple ARQ requires an average of two transmissions to deliver an eight-packet message, assuming that the network condition, p, is about 0.85. On the other hand, hybrid ARQ with $RS(10,8)$ encoding needs an average of two transmissions if p is only around 0.6. However, the reliability of the delivery in either case is only around 80% (see Figure 3.A.3).

The results here indicate that in order to stabilize QoS and SoS the encoding rate must be adjusted to the network conditions.

3.A.3 Imposed Tactical Requirements

The above analysis of packet erasure studies the effect of retransmissions and RS encoding on the assurance of message delivery and throughput. These results can be used to optimize message delivery in environments with high erasure rates. For example, in a mobile, tactical environment, the survivability requirement to deliver high-importance, short messages (e.g., call for fire, missile launching, etc.) within the SoS necessitates trading off some throughput with delivery time and assurance. We will clarify this with the following example. It takes one second to deliver a non-real-time single data packet from a source to a destination over a large-scale network, where the path between the source and destination could have multiple satellite hops. Thus, if the source sends an eight-packet message in a burst, one can expect to receive the acknowledgment within 2 seconds of the initial transmission. A packet loss extends the message completion time from 2 seconds to 4, 6, or more seconds.

Let us hypothesis that the design of tactical networks requires the delivery of 99.5% of these high-precedence messages within four seconds, which bounds us to two transmissions ($m = 2$). Given the analysis above (see Figure 3.A.3), one can see that an ARQ protocol will fail to meet these requirements as the path reliability degrades. The presented hybrid ARQ with RS encoding, which takes the path-packet-loss measurements into consideration, allows the optimization of delivery assurance, SoS requirements, and throughput efficiency.

The results verify that an increase in the number of redundancy packets and retransmission both positively affect the reliability of delivery. Increasing the number of packets negatively impacts the throughput, and additional retransmissions increase message completion time.

To meet the transmission delay requirement m_{req}, and reduce the impact of throughput (while still providing the required level of reliability R_{req}) let us try to find the smallest possible message size $n \geq k$, where k is the fixed number of data packets which satisfies the following inequality for the given network condition p:

$$R\left(n, k, m_{req}, p\right) \geq R_{req}. \qquad (3.A.12)$$

The left side of the inequality in Equation 3.A.12 is the reliability of message delivery as defined by Equation 3.A.6. The solution of this inequality allows us to adapt an encoding rate for the variable network conditions.

Note from Figures 3.A.2 and 3.A.3 that $R(n, k, m, p)$ is a monotonically increasing function of n and p. This allows the use of the following algorithm to satisfy the criterion in Equation 3.A.12 for any given message length k. The algorithm starts with the probability $p = 1$. There is no need for redundancy in this situation, and we set $n = k$. Obviously, $R(n, k, m, 1) = 1$ and the condition in Equation 3.A.12 is satisfied. Then, we reduce p by 0.01 and calculate the probability of delivering the message in m_{req} transmissions $R\left(n, k, m_{req}, p\right)$ according to Equation 3.A.12. If $R\left(n, k, m_{req}, p\right)$ is greater than or equal to the required message completion rate R_{req}, then the table that keeps values $n - k$ for every probability p, and every given data size k, is created/updated. Then, the probability p is reduced by another percentage point. If $R\left(n, k, m_{req}, p\right) < R_{req}$, the previous p that satisfied the inequality in Equation 3.A.12 is selected and the message size n is incremented by one. This means that one redundancy packet should be added to the encoded message. This process continues until p is greater than or equal to P_{min}, which is 0.5. Some of the calculation results for $4 \leq k \leq 120$, $m = 2$, $R_{req} = 99.5\%$, and $0.5 \leq p \leq 1$ are presented in Table 3.A.1 which shows that an eight-packet message can be sent as is, for almost perfect network conditions characterized by the probability $p \geq 0.98$. Only one redundancy packet is required to guarantee successful message delivery if $0.97 \geq p > 0.89$, and so on. To preserve the same message completion rate in poor network conditions, $p \approx 0.5$, the number of redundancy packets has to be increased to nine. In general, to preserve message delivery assurance with worsening network conditions, more and more redundancy packets are required. Implementing the described RS code would result in the table being stored at the

Table 3.A.1 Number of redundancy packets needed for different message size and different probability of delivering a packet

Message size, packets	Probability of delivering a packet								
	1	0.99	0.98	...	0.75	0.74	...	0.51	0.50
4	0	0	0	...	3	3	...	5	5
8	0	0	0	...	3	3	...	9	9
40	0	0	1	...	5	6	...	25	26
120	0	1	1	...	16	18	...	59	60

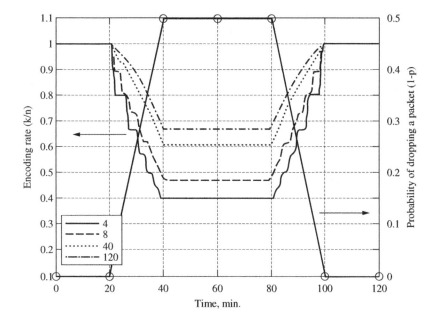

Figure 3.A.6 Encoding rates for 99.5% assurance at variable network conditions.

source. The source would then use it to adapt the number of redundancy packets according to network conditions in order to achieve the desired reliability of delivery.

The results in Figure 3.A.6 illustrate how the encoding rate would change over time in the simulation, where the network condition (probability to drop a packet) follows the marked line. One possible example would be a mobile antenna as it moves toward, in, and out of a forest, There is no need for any encoding ($k/n = 1$) for the first 20 minutes. As the network condition worsens, the probability of delivering a packet decreases, and to compensate for losses, the number of redundancy packets ($n - k$) has to be increased, which causes a decrease in the encoding rate (k/n). The pattern lines depict how the encoding rate for messages of various sizes varies over time. The encoding rate is constant for stable network condition between 40 and 80 minutes. As network conditions improve, the number of redundancy packets decreases, to adapt to better communications. This preserves the reliability and improves the throughput efficiency of the code. As a result of using this adaptive RS code for the entire 120 minutes (Figure 3.A.6), the delivery assurance was never worse than 99.5%, and the SoS requirement $m \leq 2$ was met as the network condition shifted from ideal ($p = 1$) to poor ($p = 0.5$). Note that longer messages require relatively fewer redundancy packets for the same packet loss. To explain this, let us point out that the skew of binomial distribution (Equation 3.A.2) is:

$$\gamma_1 = \frac{1 - 2 \cdot p}{\sqrt{n \cdot p \cdot (1 - p)}}. \tag{3.A.13}$$

This is a non-positive function for the discussed range of probabilities $0.5 \leq p \leq 1$. This indicates that most messages will be delivered in the first transmission with a probability less than the required R_{req} and will need retransmission. When the number of packets n in the message increases, the skew γ_1 decreases, which means that a greater number of messages will be delivered with a higher probability; hence, fewer packets will require retransmission.

Bibliography

1. Brillouin, L. (2004) *Science and Information Theory*, 2nd edn, Dover, Phoenix.
2. Ziv, J. and Lempel, A. (1978) Compression of individual sequences by variable rate coding. *IEEE Transactions on Information Theory*, **IT-24**, 530–536.
3. Landgon, G. (1979) Arithmetic coding. *IBM Journal on Research and Development*, **23**, 149–162.
4. Grushevsky, Y.L. and Elmasry, G.F. (2009) Adaptive RS codes for message delivery over encrypted mobile networks. *IET Proceedings-Communications*, **3** (6), 1041–1049.
5. Byers, J., Luby, M., and Mitzenmacher, M. (2002) A digital fountain approach to asynchronous reliable multicast. *IEEE Journal on Selected Areas in Communications*, **20** (8), 1528–1540.
6. Hagenauer, J. (1988) Rate-compatible punctured convolutional codes (RCPC codes) and their applications. *IEEE Transactions on Communications*, **36** (4), 389–400.
7. Ahlswede, R., Cai, N., Li, S.-Y.R., and Yeung, R.W. (2000) Network information flow. *IEEE Transactions on Information Theory*, **46**, 1204–1216.
8. Koetter, R. and Médard, M. (2003) An algebraic approach to network coding. *IEEE/ACM Transactions on Networking*, **11**, 782–795.
9. Li, S., Yeung, R., and Cai, N. (2003) Linear network coding. *IEEE Transactions on Information Theory*, **49**, 371–381.
10. Ho, T., Médard, M., Koetter, R. *et al.* (2006) A random linear coding approach to multicast. *IEEE Transactions on Information Theory*, **52**, 4413–4430.
11. Wu, Y., Chou, P.A., and Kung, S. (2005) Minimum-energy multicast in mobile ad hoc networks using network coding. *IEEE Transactions on Communications*, **53** (11), 1906–1918.
12. Lin, S. and Costello, D.J. Jr. (1985) *Error Control Coding: Fundamentals and Applications*, Prentice Hall.
13. Shannon, C. (1948) A mathematical theory of communications. *Bell System Technical Journal*, **27**, 379–423.
14. Gallager, R. (1968) *Information Theory and Reliable Communications*, John Wiley & Sons, Inc.
15. Blahut, R. (1972) Computation of channel capacity and rate-distortion functions. *IEEE Transactions on Information Theory*, **18**, 460–473.
16. Fano, R. (1961) *Transmission of Information*, MIT Press.
17. Larson, H. and Shubert, B. (1979) *Probabilistic Models in Engineering Sciences*, vols. I and II, John Wiley & Sons, Inc.
18. Carlson, A. (1986) *Communications Systems: An Introduction to Signals and Noise in Electrical Communications*, 3rd edn, McGraw-Hill.
19. Wicker, S. (1995) *Error Control Systems for Digital Communication and Storage*, Prentice Hall.
20. Wicker, S. and Bhargava, V. (1994) *Reed Solomon Codes and their Applications*, Edited, IEEE Press, New York.
21. Blahut, R. (1987) *Principles and Practice of Information Theory*, Addison-Wesley.

22. Wozencraft, J.M. and Jacobs, I.M. (1965) *Principles of Communications Engineering*, John Wiley & Sons, Inc., New York.
23. Simon, M., Hinedi, S., and Lindsey, W. (1995) *Digital Communications Techniques: Signal Design and Detection*, Prentice Hall.
24. Elmasry, G. (1999) Joint lossless-source and channel coding using automatic repeat request. *IEEE Transactions on Communications*, **47** (7), 953–955.
25. Elmasry, G. (1997) Arithmetic coding algorithm with embedded channel coding. *Electronics Letters*, **33** (20), 1687–1688.
26. Elmasry, G. (1999) Embedding channel coding in arithmetic coding. *IEE Proceedings-Communications*, **146** (2), 73–78.
27. Wilson, S.G. (1996) *Digital Modulation and Coding*, Prentice Hall.
28. Proakis, J. and Salehi, M. (1994) *Communications Systems Engineering*, Prentice Hall.
29. Haykin, S. (1994) *Communications Systems*, John Wiley & Sons, Ltd.
30. Ross, S.M. (2009) *Introduction to Probability and Statistics for Engineers and Scientists*, 4th edn, Academic Press.
31. Leon-Garcia, A. (1994) *Probability and Random Processes for Electrical Engineering*, 2nd edn, Addison-Wesley.
32. Popoulis, A. (1991) *Probability, Random Variables, and Stochastic Processes*, 3rd edn, McGraw-Hill.

4

MAC and Network Layers in Tactical Networks

So far we have been focusing on the lower layers of the protocol stack, which include the physical layer and the data link layer (DLL). These were presented within the context of tactical wireless communications and the tactical networking information theory construct. At the end of Chapter 3 we touched upon network coding and presented a case study for transport layer coding. In this chapter, we will review some of the important concepts of the network layer. Essentially we are now addressing packets that flow up and down the protocol stack layers and between different nodes. We are now facing a range of different issues, in addition to the physical limitations of the communications channel, which we addressed in the previous chapters. In this chapter we address issues such as how the transmission media resources are shared among different nodes in a tactical wireless subnetwork; what design concepts should be followed to manage these shared resources; what is the best way to model packets arrival and departure in queues; how queue overflow can cause packet loss; and how to overcome catastrophic events which can occur in tactical wireless networks.

4.1 MAC Layer and Multiple Access Techniques

As you will see in Chapter 5 onwards, legacy tactical radios were (non-IP) implemented with static resource allocation techniques in order to manage access to the physical layer resources. Each node was designated a specific configuration, delineating when and how to use the physical layer resources, such that no interference occurred. The problem with static configuration is the poor utilization of the physical layer resources. This is especially problematic when traffic demand between nodes in the same subnet is dynamic and can vary drastically. With the move to IP-based tactical radios, dynamic resource allocation techniques for sharing the physical layers' resources had to be implemented at the MAC layer. While this chapter presents some of the challenges facing the design of MAC layer protocols, Chapter 5 onwards further explains the fundamentals of managing physical layer resources for tactical radios.

The protocol stack layering presented in Chapter 1, Figure 1.1, showed the MAC layer as a stand-alone layer (although some communications and networking books show the MAC

Tactical Wireless Communications and Networks: Design Concepts and Challenges, First Edition. George F. Elmasry.
© 2012 John Wiley & Sons, Ltd. Published 2012 by John Wiley & Sons, Ltd.

layer as a DLL sub-layer). In tactical wireless networks, the MAC layer plays an important role in satellite communications and multiple access radio, where the physical layer resources are shared between multiple nodes. The separation of functions between the MAC, data link, and network layers is not as clear cut when it comes to multiple access tactical radios as it is with point-to-point links. Figure 1.2 presented the IP model of the protocol stack layers where the MAC and DLL are presented as one layer (layer 2). The inseparable aspect of MAC and DLL can be understood in the context of request for retransmission. The error control coding at the DLL can initiate this request, and the MAC layer could then see this request as being central to the resource allocation of the physical medium (could be a result of collision) and use it for flow control. At the IP layer (layer 3), each node has a queue of packets to be transmitted over the multiple access media. Ideally, resource sharing of the multiple access media should have knowledge of what type of packets and how much queue buildup is happening at each node.

Tactical radios have some unique challenges when it comes to multiple access solutions. One needs to address layer 2 formations and the layer 3 subnet as two different (yet inseparable) hierarchies. When a MANET is formed, two layers of hierarchical subnets can be formed. The upper layer is the IP-based subnet (layer 3) and the lower layer is composed of one or more layer 2 subnets that belong to the same IP subnet. This can occur because in a given IP-based subnet each node could be in the reception range of a group of nodes (a subset of all the nodes in the IP subnet). The same node can transmit to a different subset of nodes than it can receive from.

Intuitively, one can think of two extreme cases with multiple access media. One extreme case is a "free-for-all" where nodes send packets as they arrive and hope that no other nodes are transmitting at the time (no collision[1]). The other extreme case is to use a perfect schedule where nodes have access to the physical medium in order and relinquish access after an interval of time.

With the first case, a technique called carrier sense multiple access (CSMA) is used, where packet transmission is not allowed to start if the channel is sensed to be busy. CSMA can prevent some collisions but not all collisions. The ratio of propagation delay to packet transmission time (referred to as β) plays a major role in deciding the probability of collision. Consider the case where node i starts transmitting, and node j starts transmitting before sensing that node i has already started (due to propagation delay). If the propagation delay is small, the chance of collision is low. CSMA has other challenges: if node i is transmitting to node j and node k also wants to transmit to node j, then there is no guarantee that node k can hear node i (distance between nodes i and k, or terrain can make this scenario possible). CSMA can utilize a busy tone, from a tiny portion of the spectrum which has been dedicated to busy tones. In this example, node j, once it has started to receive from node i (and all the nodes that can hear i), will send the busy tone so that node k will then refrain from transmitting. Busy tone usage can reduce throughput of the radio. When node i starts to send signals and all the nodes that can hear node i start sending the busy tone, then all the nodes within the range of the nodes which are in the range of node i, will then refrain from transmitting. In most of these cases, there may be no chance of collision, which reduces throughput. CSMA is known to result in bandwidth inefficiency under high contention environments.

---

[1] In terms of the physical layer, when two nodes transmit at the same time, we use the term interference. In terms of packets or MAC layer, we use the term collision.

In the commercial wireless arena, where we have a single focal point, the IEEE 802.11 standards use the CSMA approach, while the 802.11e standards added QoS support through a protocol called enhanced distributed channel access (EDCA) which provides differentiated service for different priorities of data. EDCA can starve lower priority data of physical media resources in order to give higher priority data better QoS. You are encouraged to reference commercial technology MAC layer implementations. In this book we focus on the tactical radio implementations where all nodes are mobile and there is no focal point (wireless router or base station).

With the second case, one would need to design a scheduling approach that determines:

- what the scheduling order is – it could be dynamic based on each node's demand.
- how long the reserved interval is.
- how to inform nodes of their turns.

With tactical radios, time division multiple access (TDMA) is commonly used to schedule node access to the transmission media. Recall from Chapter 2 that spread spectrum modulation (which requires a wide frequency band) is essential for jamming resistance in tactical radios and hence frequency division multiple access (FDMA) is less desirable.

To understand the depth of TDMA scheduling in tactical radios, one would need to first understand the concept of reachability. In Chapter 2, we discussed how tactical radios at times need their spectrum emission to be as low as possible (to extend battery life and to resist eavesdropping by the enemy). The distance between many pairs of nodes in a deployed tactical radio area of coverage (AoC) can be much larger than the spectrum reach.

The graph in Figure 4.1 shows a simple tactical radio subnet with different node reachability. This graph can be expressed as $G = (N, L)$, where N is the number of nodes and L is the number of possible reachability (links[2]). Each link in L is associated with the reachability of a pair of nodes (i, j), to indicate that transmission from i can be heard at j. Note that in some cases, $(i, j) \in L$ but $(j, i) \notin L$. In other words, node i can reach node j, but node j cannot reach node i. This is common with tactical radios where some nodes (vehicle mounted) can have more battery power than other nodes (soldier carried) and hence can have a wider spectrum coverage area. For simplicity, let us assume that in the graph in Figure 4.1, there is symmetry, and each line denotes two links.

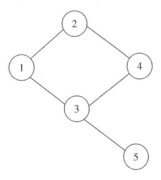

Figure 4.1 Node reachability graph for a five-node example of a tactical radio.

---

[2] The term link is used here loosely to express reachability and should not be confused with point-to-point links.

In Figure 4.1, if node i transmits a packet, this packet will be received by node j correctly if:

- $(i, j) \in L$. That is, if there is a link from i to j,
- while i is transmitting, j is not transmitting,
- while i is transmitting, no other node k for which $(k, j) \in L$ is transmitting.

Let us consider some scenarios based on the graph in Figure 4.1 as follows:

- Nodes 1 and 3 are transmitting simultaneously. Node 2 will receive the packet sent by node 1 correctly, and nodes 4 and 5 will receive the packet sent by node 3 correctly. If the packet is intended for node 4, node 5 will discard it.
- Nodes 2 and 3 are transmitting simultaneously. Collision will occur for nodes 1 and 4, while node 5 will receive the packet sent by node 3.

One can see that as the number of nodes in a tactical radio subnet increases, reachability gets complex and the topology graph equivalent to the one in Figure 4.1 becomes more complex. The positive side of increasing the number of nodes in the subnets is that the number of possible links increases (the number of pairs of nodes that can communicate increases). The downside is that the likelihood of collision increases. The design of tactical radios considers the number of nodes in a subnet, the spectrum emitted by each node (to decide the reachability distance of individual nodes), and the total area of coverage of the deployed subnet.

The amount of traffic that can be carried by a tactical subnet is not fixed. As nodes move around, the graph expressing the reachability could change. At a given moment, there exists a set (or multiple sets) of links that can carry traffic simultaneously with no packet collision. The two scenarios above can generate the following sets: {(1,2), (3,4)} and {(3,5)}. If all nodes are idle (no transmission) at a given time, the subnet throughput depends on which node starts to transmit first, since the set of links that can carry traffic simultaneously can be defined.

TDMA based approaches are especially preferred in multihop tactical radios since TDMA can provide better network capacity and can help with QoS prioritization. Non-real-time packets can be allowed to wait for the next round access to the physical media, while real-time packets are given instant access, once the node scheduled turn starts. The challenges we face with TDMA based MAC layers are that we require tight clock synchronization and constant exchange of control information. Distributed time synchronization techniques are used for access control.

A tactical MANET subnet is expected to be self-organized in the sense that it can operate in dynamic environments with peer-to-peer, multihop, multicast, and broadcast capabilities. One of the known protocols in tactical MANET MAC layer implementation is called unifying slot assignment protocol (USAP), which manages the TDMA slot and channel assignments[3] in dynamic environments where subnet formation (reachability) and traffic demand are constantly changing. The application of USAP to the Joint Tactical Radio Systems (JTRS) wideband networking waveform (WNW) is covered in details in Chapter 6.

---

[3] Some tactical radios can use multichannel, increasing the complexity of the resource management protocols.

Now that you have a general idea about MAC layer challenges in tactical radios, let us review some of the basic networking concepts and see how they apply to tactical wireless networks.

4.2 Queuing Theory

Queuing theory is an essential concept to review in light of tactical wireless communications and networks. Understanding queuing theory will help you address some of the challenges facing the optimization of tactical networks performance. Understanding queuing theory will help you address critical areas of networking, such as packet delay, packet loss due to queue overflow, and the advantages of statistical multiplexing. As we have done with the physical layer and DLL analysis, we will use simplified assumptions regarding queuing theory to conduct meaningful analysis,[4] since realistic assumptions inhibit clear theoretical analysis.

When a packet traverses a network hop, it can experience the following delays:

- **Processing delay** – the time from the packet's arrival at the network layer of the node until it is assigned to an outgoing queue (if routed at the IP layer). If the packet is transmitted over the air, we must add processing delay at the MAC, DLL, and physical layers.
- **Queuing delay** – the time from when the packet arrives to the queue until it is transmitted. Other packets waiting in the queue ahead of this packet can cause it to wait until they are transmitted.
- **Transmission delay** – the time between the transmission of the first and last bits of the packet (the capacity of the physical media is a decisive factor in this delay).
- **Propagation delay** – the time from when the last bit is transmitted until the last bit is received at the destination node. In satellite communications, this is a substantial amount of delay.

Note that other delays can occur if a packet requires retransmission (think of network coding or transport layer coding discussed in Chapter 3). Queuing theory focuses on queuing delay, which is a major contributor to packet delay and when queue depth is limited, queuing theory provides a good insight into packet loss due to queue overflow.

4.2.1 Statistical Multiplexing of Packets

We have seen in Chapters 2 and 3 how the channel has a limit to the number of bits per second it can transmit. From a channel point of view, channel capacity decides the limit of information that can be transmitted. From the networking point of view, there is a *transmission capacity*, that is decided by the channel's physical limitations and by the way that a queue, with waiting packets, accesses the channel. If we have a multiple-access medium, each point of access transmits over the channel, sharing media resources. The ways by which the resources of the physical media are shared decide the *transmission capacity*.

---

[4] Although accurate quantities of delay and packet loss analysis can be better obtained through discrete event simulation, queuing theory provides adequate delay and queue depth analysis and can be conducted in a much shorter time than discrete event simulations. Most importantly, queuing theory can be used as an aid in protocol design.

With packet switched networks, packets of multiple streams are merged into a single queuing mechanism before accessing the transmission media. The simplest form of queue access is to have a single queue for all packets, where packets are served on a first-in-first-out (FIFO) basis. If the entire transmission capacity C (bits per second) is allocated to a packet until that packet is completely transmitted, and if the packet length is L bits, then the packet transmission time is L/C seconds.

With queuing theory models, packets randomly arrive at a queue and wait to be serviced. To understand the effect of a hop over the packet stream, and to design parameters such as queue depth and to assign transmission capacity to queues, the following metrics are needed:

- the average number of packets waiting in the queue or being processed.
- the average delay per packet – the expected value of time a packet spends in the queue plus it service time.

We are interested in estimating these metrics in terms of known information such as packet arrival rate and packet service rate. The way packets arrive to a queue can differ. Packets can arrive in a burst, in a random fashion, or clocked (even periods of inter-arrival time). Burst arrival causes the average queuing delay to increase.[5] Thus, packet arrival should be defined not only in terms of the rate of packets per unit of time, but also in terms of its *statistical* information. The same applies to service rate. In systems like TDMA, a queue can be serviced during the slot(s) of time assigned to it, and the service rate is zero while other queues are being serviced. Even the way slot allocation is delineated can affect the packet service time. The behavior differs when the time slot allocated to the queue is long enough to service the entire packet, as opposed to the slot time being short and a single packet having to wait through the idle time before it is completely serviced. Thus, we need to know *statistical* information about packet arrival and service rate. Let us start first simply try to understand the basics of queuing theory.

Little's Theorem

Let us observe a sample queue starting from time $t = 0$ to the indefinite future and observe the following parameters:

$N(t) = $ number of packets in the queue at time t

$\alpha(t) = $ number of packets arrived during the time interval $[0, t]$

$T_i = $ time that the ith arriving packet spent in the queue

We will use two terms here. One is the *average* number of packets in the queue, and the other is the *typical* number of packets in the queue. The *typical* number of packets observed in the queue up to time t is:

$$N_t = \frac{1}{t} \int_0^t N(T)dT, \tag{4.1}$$

which is basically the *time average* of $N(t)$ up to time t.

---

[5] Sometimes it is easier to conceptualize queuing theory by thinking of customers arriving at a queue in places like supermarkets.

Notice that N_t changes with time as the general case. In a stable queuing system, N_t tends to reach a *steady state* N as t increases. That is,

$$N = \lim_{t \to \infty} N_t. \tag{4.2}$$

We can refer to N as the *steady-state time average* or as the *time average* of $N(t)$.

Now let us think of the packet arrival statistics. One can express the *time-average arrival rate* over the time interval $[0, t]$ as:

$$\lambda_t = \frac{\alpha(t)}{t}. \tag{4.3}$$

The *steady-state arrival rate* of packets in a stable queue is:

$$\lambda = \lim_{t \to \infty} \lambda_t. \tag{4.4}$$

Now let us define the *time average of packet delay up to time t* as:

$$T_t = \frac{\sum_{i=0}^{\alpha(t)} T_i}{\alpha(t)}. \tag{4.5}$$

Think of T_t as the average time a packet spends in the queue up to time t. We should have a *steady-state time-average packet delay* defined as:

$$T = \lim_{t \to \infty} T_t. \tag{4.6}$$

Notice that we have made some substantial assumptions here. We are assuming that the limits defined above exist. In the case of tactical wireless networks, where everything is dynamic, this may be a stretch. However, understanding queuing theory and getting a feel for how packets traverse a system should help you make better assumptions and follow more solid design concepts even with dynamic networks.

Little's theorem relates N, λ, and T as follows:

$$N = \lambda T. \tag{4.7}$$

Notice that λ is in packets per second and T is in seconds resulting in N units being packets. Intuitively, in any queuing system, when the queue is long (N is large), waiting time T is high.[6]

The proof of Little's theorem can be done through the graph in Figure 4.2. In the graph, the vertical axis represents the number of packet arrivals $\alpha(\tau)$ as well as the number of packet departures $\beta(\tau)$. Packet i arrives at time t_i and is delayed T_i. At any time τ, the number of packets in the queue is $N(\tau)$, which is the difference between packets arrived and packets serviced at time τ, that is, $(\alpha(\tau) - \beta(\tau))$. At time zero, the queue is assumed to be empty ($N(0) = 0$). This is a FIFO queue where packets are serviced in the order they

---

[6] Queuing theory applies to many areas such as highway design, when traffic moves slowly (meaning larger T), the highway is more congested (larger N). Interestingly, road engineers know that increasing a highway from one lane per direction to two lanes per direction; more than doubles the highway capacity since faster cars can use one lane and slow cars use the other. In computer networking this is parallel to creating a queue for "expedited forwarding" for real-time short packets.

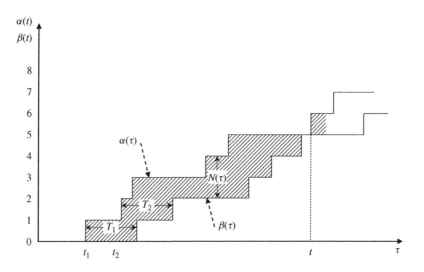

Figure 4.2 Proof of Little's theorem.

arrive and packet arrival is random (inter-arrival time is not equal). Packets' service time can also differ. Notice how the queue in this example can be empty at times and can have up to three packets at a time.

The shaded area between $\alpha(\tau)$ and $\beta(\tau)$ can be expressed as:

$$\int_0^t N(\tau)d\tau. \tag{4.8}$$

If we start our analysis from $\tau = 0$, with the queue being empty, and if we select t such that the queue is also empty at time t, the shaded are can also be given by:

$$\sum_{i=1}^{\alpha(t)} T_i. \tag{4.9}$$

Equating Equations 4.8 and 4.9 and dividing both sides by t. We get:

$$\frac{1}{t}\int_0^t N(\tau)d\tau = \frac{1}{t}\sum_{i=1}^{\alpha(t)} T_i = \frac{\alpha(t)}{t}\frac{\sum_{i=1}^{\alpha(t)} T_i}{\alpha(t)}.$$

Using Equations 4.1, 4.3, and 4.5, the above equality gives:

$$N_t = \lambda_t T_t. \tag{4.10}$$

Taking the limits $N_t \to N$, $\lambda_t \to \lambda$, $T_t \to T$, we reach Little's theorem in Equation 4.7 Note that this queue is assumed to eventually become empty since the service rate is greater than the arrival rate. Asymptotically, the queue will become empty at some given time.

If the service rate is much higher than the arrival rate, then the queue will become empty very often. If the arrival rate is closer to the service rate, then the queue will rarely become empty. To understand this assumption in a little more depth, one can see in Figure 4.2 that the shaded area is between $\sum_{i=1}^{\alpha(t)} T_i$ and $\sum_{i=1}^{\beta(t)} T_i$. One can also see that we have the following bound:

$$\frac{\sum_{i=1}^{\beta(t)} T_i}{\beta(t)} \leq \frac{\sum_{i=1}^{\alpha(t)} T_i}{\alpha(t)} \quad \text{or} \quad \frac{\sum_{i=1}^{\beta(t)} T_i}{\beta(t)} \leq N_t.$$

When this queue (that becomes empty at arbitrary times) reaches its steady state, we can say that $\lim_{t \to \infty} \alpha(t)$ approaches $\lim_{t \to \beta} \beta(t)$. In this queue, the order of packets serviced may not match the order of packets arrived and we do not have to necessarily start with an empty queue. Simply, one can say that the arrival rate (in packets per second) multiplied by the average time a packet spends in the queue (in seconds) equals the steady state average of the number of packets in the queue.[7] Little's theorem can be applied further to say that:

1. The arrival rate (in packets per second) multiplied by the average time a packet spends waiting in the queue (before it gets transmitted) equals the average number of packets waiting in the queue (not being transmitted). This is expressed as:

$$N_Q = \lambda W. \tag{4.11a}$$

2. The arrival rate (in packets per second) multiplied by the average transmission time of a packet (in second per packet) $\bar{X}$ gives us the utilization factor ρ or the proportion of time packets are transmitted (non idle time). This is expressed as:

$$\rho = \lambda \bar{X}. \tag{4.11b}$$

3. In a network of nodes that reaches a steady state, we can say that the arrival rate (in packets per second) at node i multiplied by the average time a packet spends in node i (in second) equals the steady state average of the number of packets in node i ($N_i = \lambda_i T_i$). Furthermore, one can say that for this network in a steady state, we have,

$$N = \sum_{i=1}^{n} \lambda_i T, \tag{4.11c}$$

where N is the average total number of packets in the network and T is the average delay per packet in the network.

This generalization applies to the network regardless of the packet length distribution or how packets are routed in the network.

---

[7] Intuitively, if the service rate in this queue is much higher than the arrival rate, the average time that a packet spends in the queue is small and the steady state average of the number of packets in the queue becomes small (the queue does not or rarely builds up). On the other hand if the service rate in this queue is closer to the arrival rate (still has to be higher), the average time that a packet spends in the queue is larger, and the steady state average of the number of packets in the queue becomes larger (queue build up).

4.2.2   Queuing Models

There are different queuing models known in queuing theory literature. Most have a naming convention of "letter/letter/number," where:

1. The first letter indicates the type of the arrival process, which could be:
   (a) M, which indicates a memoryless arrival process with Poisson distribution (which means an exponential distribution for inter-arrival time)
   (b) G, which means general distribution of inter-arrival times
   (c) D, which stands for deterministic inter-arrival time.
2. The second letter indicates the service rate probability distribution, which can be M, G, or D as in the arrival process.
3. The last number indicates the number of servers.

In this chapter we will cover the M/M/1 and the M/M/m queue models. You are encouraged to study other queuing models in references such as [1]. Our interest in this chapter is to correlate these queue models to good design concepts in tactical networks.

4.2.2.1   M/M/1 Queue

The M/M/1 queue model is the simplest model and is used here to clarify some essential concepts in tactical networks. In defining this model, let us use the following notations:

- N = the average number of packets waiting in the queuing system
- T = the average time a packet spends in the queuing system
- N_Q = the average number of packets waiting in the queue
- W = the average time a packet waits in the queue.

Let us use Little's theorem to differentiate between the queuing delay (queue wait W) and the queuing system delay T, which encompasses the different delay factors. That is,

$$N = \lambda T \quad \text{and} \quad N_Q = \lambda W. \tag{4.12}$$

If we define p_n as the probability that n packets are in the queue, $n = 0, 1, \ldots$, which can lead to defining the average number of packets waiting in the queuing system as:

$$N = \sum_{n=0}^{\infty} n p_n.$$

Notice that the statistics of the queuing system (how packet arrives, queue depth, how packets are served) can further specify N, T, N_Q, and W, which in turn, can define the steady-state probabilities.

M/M/1 Queue Arrival Statistics

The M/M/1 queue model assumes that packets arrive to the queue following a Poisson process, which can be defined using a counting process $A(t)$ that represents the total number of packet arrivals from time 0 to time t. At time 0, $A(0) = 0$, and for $s < t$, and $A(t) - A(0)$ corresponds to the number of arrivals in the interval $(s, t]$.

Notice that the Poisson process assumes that the numbers of arrivals occurring at disjoint time intervals are independent. Although we know that in actual network queues, there is a form of time correlation in packet arrivals at a given queue, the Poisson arrival approximation still provides us with good insight into the dynamics of the queues.

The number of packet arrivals in any time interval of length τ follows a Poisson distribution with parameter $\lambda\tau$, where λ is the packet arrival rate. That is,

$$P\{A(t + \tau) - A(t) = n\} = e^{-\lambda t}\frac{(\lambda t)^n}{n!}, n = 0, 1, \ldots \text{ for all } t, \tau > 0. \tag{4.13}$$

With M/M/1 queue, the Poisson process suggests that λ is the average number of packet arrivals per unit time and $\lambda\tau$ is the average number of packet arrivals in an interval of length τ. The Poisson process has some of the following properties:

1. Packet inter-arrival times are independent and have exponential distribution defined by λ. In other words, if the nth packet arrives at time t_n (and the packet after arrives at time t_{n+1}), then the time interval $\tau_n = t_{n+1} - t_n$ will have the following probability distribution:

$$P\{\tau_n \leq s\} = 1 - e^{-\lambda s}, s \geq 0. \tag{4.14}$$

The inter-arrival time intervals $(\ldots, \tau_{n-2}, \tau_{n-1}, \tau_n, \tau_{n+1}, \tau_{n+2}, \ldots)$ are mutually independent and have the following pdf:

$$p(\tau_n) = \lambda e^{-\lambda \tau_n}, \tag{4.15}$$

where the mean of this pdf is $\frac{1}{\lambda}$ and the variance is $\frac{1}{\lambda^2}$.

2. Studying the Poisson process to understand the number of packet arrivals in the queue during an interval of length δ yields the following, given that $t \geq 0$ and $\delta \geq 0$:

$$P\{A(t + \delta) - A(t) = 0\} = 1 - \lambda\delta + o(\delta), \tag{4.16}$$

$$P\{A(t + \delta) - A(t) = 1\} = \lambda\delta + o(\delta), \tag{4.17}$$

$$P\{A(t + \delta) - A(t) \geq 2\} = o(\delta). \tag{4.18}$$

The equations above allow us to see the number of packets arriving in the queue during the interval δ, in terms of the interval length, the average arrival rate λ, and the function $o(\delta)$ which is denoted as follows:

$$\lim_{\delta \to 0} \frac{o(\delta)}{\delta} = 0. \tag{4.19}$$

3. This property relates to the case when we have a queue that is fed by multiple traffic sources (this can happen in the network core where a link is shared by multiple sources of traffic). The M/M/1 queue models this case as having multiple Poisson processes $(A_1, A_2, \ldots, A_k)$ that are merged into a single process $A = A_1 + A_2 + \ldots + A_k$. Also, A is a Poisson process with the arrival rate equaling the sum of the arrival rates of its components. This property should give you a sense of how queues in the network core can be conceptualized.

4. This property relates to the case when we have a queue that routes traffic over multiple links (opposite of property 3). This is essentially a split of a Poisson process into multiple processes. If we have a split into two processes, and these two processes are independent of each other, and the probability of the first split to be assigned a packet is p (the probability that the second split to be assigned a packet is $1 - p$), the split processes are Poisson as well.[8]

Although the M/M/1 queue arrival model is a reliable approximation, which provides insight onto queue behavior, it does have many assumptions in place. The Poisson process allows us to conceptualize traffic aggregation of large numbers of users but we have to assume that these sources of traffic are independent and have some similarities. With the Poisson process, we can merge n independent packet arrival processes if they have identical pdfs. We can assume that each of these processes has a packet arrival rate of λ/n and the aggregate process has a packet arrival rate of λ (as $n \to \infty$). Keep in mind that with this model, the inter-arrival times τ, between packets from the same process, follow an exponential pdf as indicated in Equation 4.14. This exponential pdf can be expressed as:

$$f(s) = P\{\tau \le s\}, \text{ where } f(0) = 0, \text{ and } \frac{df(0)}{ds} > 0. \tag{4.20}$$

Please refer to probability references for more details regarding the properties of the exponential pdf and how it relates to the Poisson process.

M/M/1 Queue Service Statistics

The M/M/1 model assumes that the packet service process follows an exponential distribution with parameter μ (corresponding to the packet service rate) where if the nth packet is serviced at time s_n, we have,

$$P\{s_n \le s\} = 1 - e^{-\mu s}, s \ge 0. \tag{4.21}$$

Here, the pdf of s_n is:

$$p(s_n) = \mu e^{-\mu s_n}, \tag{4.22}$$

with a mean of $\frac{1}{\mu}$ and variance of $\frac{1}{\mu^2}$.

Notice that the M/M/1 queue model assumes that:

- packet service times s_n are independent of packet arrival times t_n (naturally a packet cannot be serviced before it arrives at the queue);
- packet service times s_n are mutually independent;
- packet service rate μ expresses packets served per unit time. Intuitively, the server will be busy serving the nth packet for a time interval before it serves the next packet.

---

[8] It is essential in this model that the assignment of each packet arrival to a split process is independent of the assignment of other packet arrivals. If, for example, we do assignment by alternation where even numbered packets go to one process and odd number packets go to the other, the split processes are *not* Poisson. This is significant in modeling network queues.

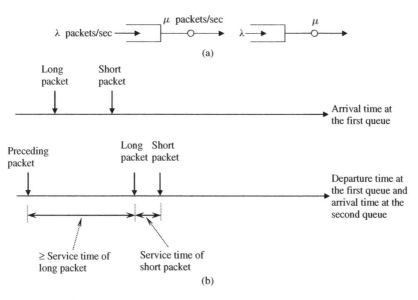

(a)

(b)

Figure 4.3   (a) Above are two identical queues, in tandem, both of which have a service rate of μ packets per second, and the packets arriving at the first queue all have equal length. Provided that $\mu > \lambda$ and that the first packet arriving at the second queue finds the queue empty, there should not be any queuing delay at the second queue. (b) Here we have two identical queues in tandem, both with a service rate of μ packets per second and the packets arriving at the first queue *do not* have equal length. We assume that a long packet has arrived, followed by a short packet, after a preceding packet is already in service at the second queue. The inter-arrival time between the preceding packet and the long packet is greater than or equal to the service time of the long packet. Whereas, the inter-arrival time between the long and short packets is equal to the service time of the short packet. Hence, the packet inter-arrival time at the second queue is correlated to the packet length.

The assumptions made by the M/M/1 model are not perfectly maintainable in actual network queues. For example, the M/M/1 model assumes that packet inter-arrival times and packet service times are independent. This implies that the length of an arriving packet should not affect the arrival time of the next packet. One can realize that this is not true specifically at the network core where queues can be in tandem. Consider Figure 4.3a where we have two queues in tandem. Let us assume that both queues have equal capacity (have the same service rate μ). If all packets have the same length, there should be no queuing delay at the second queue (provided that the first packet finds the second queue empty). If packets have different length, the timing diagram in Figure 4.3b shows that the packet arrival time at the second queue is affected by packet length.

Notice that the assumed exponential distribution for both packet inter-arrival time τ_n and packet service time s_n has a memoryless property which can be expressed as follows:

$$P\{\tau_n > r + t | \tau_n > t\} = P\{\tau_n > r\} \quad r,\ t \geq 0, \tag{4.23}$$

$$P\{s_n > r + t | s_n > t\} = P\{s_n > r\}, \quad r,\ t \geq 0. \tag{4.24}$$

This implies that the time up to the next packet arrival is independent of when the preceding arrival occurred and the time needed to finish a packet being serviced is independent of when the service started.[9]

The memoryless property of the exponential distribution can be calculated as follows:

$$P\{\tau_n > r + t | \tau_n > t\} = \frac{P\{\tau_n > r + t\}}{P\{\tau_n > t\}} = \frac{e^{-\lambda(r+t)}}{e^{-\lambda t}} = e^{-\lambda r} = P\{\tau_n > r\}, \tag{4.25}$$

$$P\{s_n > r + t | s_n > t\} = \frac{P\{s_n > r + t\}}{P\{s_n > t\}} = \frac{e^{-\mu(r+t)}}{e^{-\mu t}} = e^{-\mu r} = P\{s_n > r\}. \tag{4.26}$$

Markov Chain Representation of the M/M/1 Queue

As we have shown, the M/M/1 queue model has a memoryless property for the exponential distribution of inter-arrival and service times. Since inter-arrival and service times are independent, then at any given time t, if given the number of packets in the queue $N(t)$, then we can know that a correlation does not exist between how current packets in the queue are being serviced and the time at which future packets arrive and are serviced. Only $N(t)$ (the current number of packets in the queue) decides the future number of packets in the queue. In other words, we can express the number of packets in the queue at any given time $\{N(t)|t \geq 0\}$ as a Markov chain. To simplify this model further, we will assume that it is a discrete-time Markov chain with the following discrete times:

$\{0, \delta, 2\delta, \ldots, k\delta, \ldots\}$, where δ is a small positive number.

The number of packets in the queue at time $k\delta$ can be expressed as $N_k = N(k\delta)$.[10] Let us also denote the transition probability from state i to state j as P_{ij}, which can be expressed as follows:

$$P_{ij} = P\{N_{k+1} = j | N_k = i\}. \tag{4.27}$$

It is important to understand how P_{ij} depends on the transmission capacity. This dependency was omitted from the equation above for simplicity. Now, let us utilize Equations 4.16–4.18 to create the following transition probabilities:

$$P_{00} = 1 - \lambda\delta + o(\delta), \tag{4.28}$$

$$P_{ii} = 1 - \lambda\delta - \mu\delta + o(\delta), \quad i \geq 1, \tag{4.29}$$

$$P_{i,i+1} = \lambda\delta + o(\delta), \quad i \geq 0, \tag{4.30}$$

$$P_{i,i-1} = \mu\delta + o(\delta), \quad i \geq 1, \tag{4.31}$$

$$P_{ij} = o(\delta), \quad j \neq i, \; j \neq i - 1, \; j \neq i + 1. \tag{4.32}$$

---

[9] Refer to Figure 4.3b to see how the transmission capacity in bits per second and the packet length in bits can be used to estimate the time needed to service a packet, which is independent of when service starts.

[10] The notation $N(t)$ is used to express the continuous-time Markov chain while the notation $N_k = N(k\delta)$ is used to express the discrete Markov chain. In the M/M/1 queue model, the steady-state probabilities (of the queue occupancy) of the assumed discrete Markov chain are the same as that of the continuous Markov chain, which justifies the use of a discrete-time Markov chain.

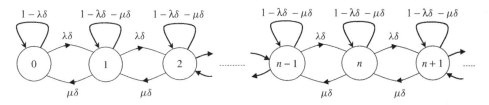

Figure 4.4 Discrete-time Markov chain representation of the M/M/1 queue model, where n corresponds to n packets in the queue.

Equations 4.18 and 4.32 quantify $o(\delta)$ as the probability that two or more packets are added to the queue during the small time interval δ, which is very small considering the exponential distribution. If we substitute $o(\delta)$ as zero in Equations 4.28–4.32, we can draw the discrete-time Markov chain of the M/M/1 queue model where the state n corresponds to having n packets in the queue as shown in Figure 4.4.

Let us consider some cases of this Markov chain in order to show how Equations 4.28–4.32 are reached.

Case I

At state $i \geq 1$, the probability of zero packet arrival and zero packet departure in the interval δ can be expressed as follows:

$$P\{0 \text{ packet arrival and packet departure in } \delta - \text{interval } \} = \left(e^{-\lambda\delta}\right)\left(e^{-\mu\delta}\right). \tag{4.33}$$

Equation 4.33 is based on the fact that the number of packet arrivals and the number of packet departures follow a Poisson distribution and are independent of each other. If we expand Equation 4.33 in a power series as a function of δ, we get:

$$P\{0 \text{ packet arrivals and } 1 \text{ packet departures in } \delta - \text{interval}\} = 1 - \lambda\delta - \mu\delta + o(\delta), \tag{4.34}$$

which corresponds to Equation 4.29.

Case II

At state $i \geq 1$, the probability of zero packet arrivals and one packet departure in the interval δ can be expressed as follows:

$$P\{0 \text{ packet arrivals and } 1 \text{ packet departure in } \delta\} = \left(e^{-\lambda\delta}\right)\left(1 - e^{-\mu\delta}\right), \text{ if } i = 1, \tag{4.35}$$

$$P\{0 \text{ packet arrivals and } 1 \text{ packet departure in } \delta\} = \left(e^{-\lambda\delta}\right)\left(\mu\delta e^{-\mu\delta}\right), \text{ if } i > 1. \tag{4.36}$$

In Equation 4.35, $\left(1 - e^{-\mu\delta}\right)$ is the probability that the only packet in the queue will be serviced in the interval δ. In Equation 4.36, $\left(\mu\delta e^{-\mu\delta}\right)$ is the probability that the packet in service will be completed in the interval δ, while the subsequent packets will not be serviced in that time. The expansion of both equations in the power series, as a function of δ gives:

$$P\{0 \text{ packet arrivals and } 1 \text{ packet departure in } \delta - \text{interval}\} = \mu\delta + o(\delta). \tag{4.37}$$

Case III

The probability of one packet arrival and zero packet departures in the interval δ can be expressed as follows:

$$P\{1 \text{ packet arrival and } 0 \text{ packet departures in } \delta - \text{interval}\} = \left(\lambda\delta e^{-\lambda\delta}\right)\left(e^{-\mu\delta}\right). \quad (4.38)$$

The above equation can be expressed in the power series as a function of δ as:

$$P\{1 \text{ packet arrival and } 0 \text{ packet departures in } \delta - \text{interval}\} = \lambda\delta + o(\delta). \quad (4.39)$$

Notice that the sum of the probabilities in the three cases above equals one plus a small value, that is in the order of $o(\delta)$. Considering that $o(\delta)$ is very small, the discrete-time Markov chain in Figure 4.4 shows these three cases substituting zero for $o(\delta)$.

We could also study a fourth case at state $i \geq 1$, where the probability of equal packet arrival and packet departure in the interval δ is within the same value as the first case (within $o(\delta)$) difference from Equation 4.34, which correlates to Equation 4.29.

Referring back to the discrete-time Markov chain in Figure 4.4, one can say that during a given time interval, the total number of transitions from state n to state $n + 1$ should differ from the total number of transitions from state $n + 1$ to state n, by zero or one (we only can stay in the same state or move to a neighboring state). This means that, asymptotically, the frequency of transition from state n to state $n + 1$ is the same as the frequency of transition from state $n + 1$ to state n (recall that we know that the queue will be empty after a long period of time since $\lambda < \mu$). This can lead us to state that in a given transition interval, the probability that the queue is in state n and transitions to state $n + 1$ is the same as the probability that the queue in state $n + 1$ and transitions to state n. Thus, we can reach the following equality:

$$p_n\lambda\delta + \delta(0) = p_{n+1}\mu\delta + \delta(0). \quad (4.40)$$

As $\delta \to 0$, we get:

$$p_n\lambda = p_{n+1}\mu. \quad (4.41)$$

Equation 4.41 is sometimes referred to as the *global balance equation* of the states of the discrete-time Markov chain. Now let define a new parameter ρ, the ratio of packet arrival rate to packet service rate. That is,

$$\rho = \frac{\lambda}{\mu}. \quad (4.42)$$

Thus we can obtain:

$$p_{n+1} = \rho p_n, \ n = 0, 1, 2, \ldots. \quad (4.43)$$

As a result of applying Equation 4.43 to all values of n, we obtain,

$$p_{n+1} = \rho^{n+1} p_0, \ n = 0, 1, 2, \ldots. \quad (4.44)$$

The sum of all the probabilities p_n for all values of n adds up to one, that is,

$$1 = \sum_{n=0}^{\infty} p_n = \sum_{n=0}^{\infty} \rho^n p_0 = p_0 \sum_{n=0}^{\infty} \rho^n. \quad (4.45)$$

Recall our assumption that the arrival rate is less than the service rate ($\rho < 1$), Thus Equation 4.45 can be expressed as:

$$1 = \frac{p_0}{1 - \rho} \text{ or } p_0 = 1 - \rho.^{11} \tag{4.46}$$

Then Equation 4.44 can be expressed as:

$$p_n = \rho^n(1 - \rho). \tag{4.47}$$

Now, let us calculate the average number of packets in the queue of a steady state as follows:

$$N = \sum_{n=0}^{\infty} np_n = \sum_{n=0}^{\infty} n\rho^n(1 - p) = \rho(1 - \rho)\sum_{n=0}^{\infty} n\rho^{n-1}$$

$$= \rho(1 - \rho)\sum_{n=0}^{\infty} \frac{\partial}{\partial\rho}\rho^n = \rho(1 - \rho)\frac{\partial}{\partial\rho}\sum_{n=0}^{\infty} \rho^n. \tag{4.48}$$

Since ρ is a fraction, the term $\sum_{n=0}^{\infty} \rho^n$ is the series expansion of $\frac{1}{1-\rho}$. Thus,

$$N = \rho(1 - \rho)\frac{\partial}{\partial\rho}\left(\frac{1}{1 - \rho}\right) = \rho(1 - \rho)\frac{1}{(1 - \rho)^2} = \frac{\rho}{1 - \rho}. \tag{4.49}$$

Since $\rho = \frac{\lambda}{\mu}$,

$$N = \frac{\lambda}{\mu - \lambda}. \tag{4.50}$$

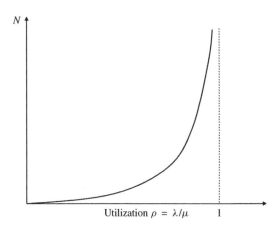

Figure 4.5 The M/M/1 queue model estimation of the average number of packets in the queue N as a function of the utilization factor ρ.

---

[11] One can think of ρ as the utilization factor, percent utilization, or the long-term proportion of time the server is actually serving packets. If we refer back to the Markov chain in Figure 4.4 and recall that p_0 is the probability of having no packets in the queue, we can say that $\rho = 1 - p_0$.

The curve in Figure 4.5 shows N as a function of ρ. This illustrates how N increases as ρ increases, and that $N \to \infty$ as $\rho \to 1$. It is important to emphasize that we must have $\rho < 1$ in all the analysis for the M/M/1 queue model. In the case that $\rho > 1$, the service rate of the packets will not be sustainable with the arrival rate of the packets, and the queue length cannot be bounded. If you recall our study of channel capacity in Chapter 3, you will notice that if we transmit at a rate higher than the channel capacity, we lose information, so we can say here that if $\rho > 1$, and C is our transmission capacity in bits per second, then $\lambda L > C$, where L is the average packet length in bits (recall that λ is defined in packets per second).

Now let us find the average delay per packet (waiting time in the queue plus service time) using Little's theorem.

$$T = \frac{N}{\lambda} = \frac{\rho}{\lambda(1 - \rho)}. \tag{4.51}$$

Substituting $\rho = \frac{\lambda}{\mu}$, we express the average packet delay as:

$$T = \frac{1}{\mu - \lambda}. \tag{4.52}$$

It is very important to understand that while the average number of packets in the queue, N, depends on the utilization factor ρ (regardless of the actual values of λ or μ), the average delay per packet T, as shown in Equation 4.52 above, is a function of $\mu - \lambda$ and can be expressed as $T = \frac{1}{\mu} \cdot \frac{1}{1-\rho}$. Figure 4.6 demonstrates how the relationship between T and ρ differs for each μ. As seen in Figure 4.6, if we have a specific constraint for the average delay per packet, then this constraint can be met with higher utilization as μ increases.

The average packet waiting time W, defined as the average delay T minus the packet service time $1/\mu$, can be calculated as:

$$W = \frac{1}{\mu - \lambda} - \frac{1}{\mu} = \frac{\rho}{\mu - \lambda}. \tag{4.53}$$

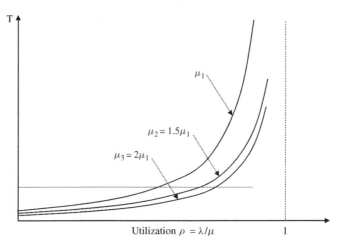

Figure 4.6 Plotting the average delay per packet as a function of the utilization for the M/M/1 queue model. As the service rate μ increases, the average delay per packet decreases for the same utilization factor. Higher μ means that we can tolerate higher utilization to meet a specific average delay per packet.

Also, the average number of packets in the queue (excluding the packet in service) can be given by:

$$N_Q = \lambda W = \frac{\rho^2}{1-\rho}. \tag{4.54}$$

Example: Scaling up both packet arrival rate and packet service rate by the same factor Let us suppose that we have a queue where we increase the packet arrival rate from λ to $k\lambda$, where k is a scaling factor greater than 1. The packet length distribution is assumed to stay the same. The service rate is increased by the same factor k (this makes the average transmission time reduce from $\frac{1}{\mu}$ to $\frac{1}{k\mu}$). In this case, the utilization factor ρ and the average number of packets in the queue stay the same since $N = \frac{\rho}{1-\rho}$. The average delay per packet, on the other hand, does not remain unchanged, but rather it will become $T = \frac{N}{k\lambda}$ and will therefore decrease by a factor of k. Figure 4.7 illustrates this phenomenon, showing that the statistical characteristics of the queue stay the same, except for the timescale. In other words, the processing is sped up by the factor k. When a packet arrives at the queue, it will find the same number of packets ahead of it in the queue; however, packets are moving out of the queue k times faster. Recall the highway example in footnote 6: freeways, where cars move faster, have more capacity per lane than roads where cars move slower.

Tactical MANET and Statistical Multiplexing in Light of the M/M/1 Model

The design of tactical MANET protocols, that addresses how to share the physical layer resources, faces a critical challenge in how each node gets access to transmit packets waiting in its queue. Let us discuss two possible scenarios in modeling packet transmission in light of the M/M/1 queue model.

- First scenario: We model the physical layer resources for the entire subnet as a single M/M/1 queue where the number of nodes in this subnet is m and each node feeds that single M/M/1 queue with a packet stream at the rate of λ/m packets per second. Packet streams are assumed to have identical and independent Poisson distribution. The average transmission time over the MANET subnet is $1/\mu$ seconds and the aggregate arrival rate of packets is also a Poisson stream with rate λ packet per second. In this case the average delay per packet should be:

$$T = \frac{1}{\mu - \lambda}. \tag{4.55}$$

- Second scenario: We model the physical layer resources to be divided equally between the different nodes in the subnet, creating m separate M/M/1 queues each with λ/m packets per second arrival rate and μ/m packets per second service rate. In this case, the average delay per packet should be:

$$T = \frac{m}{\mu - \lambda}, \tag{4.56}$$

which is m times longer than the first scenario.

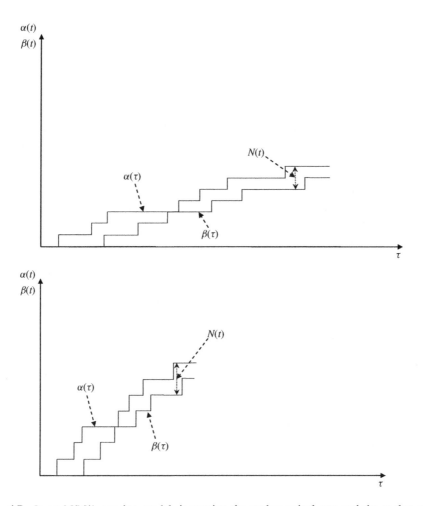

Figure 4.7 In an M/M/1 queuing model, increasing the packet arrival rate and the packet service rate by the same factor of 2 means that the average wait time is reduced by a factor of 2, even though the sample path of packet arrival $\alpha(t)$ and packet departure $\beta(t)$ remain constant. When a packet arrives in the sped up queue, it will find the same number of packets ahead of it, but because packets are moving twice as fast, the average wait time is reduced by a factor of 2.

The above scenarios emphasize the importance of statistical multiplexing in tactical MANET physical layer resource management. It is essential that we build resource management techniques that utilize statistical multiplexing. Recall the discussion mentioned in Section 4.1. If the TDMA technique used in tactical MANET simply divides the time slots among the nodes sharing the physical layer resources equally and used a simple technique such as a round robin, the tactical MANET subnet will perform very poorly in terms of transmission delay and physical layer resources utilization. The situation worsens when the packet arrival rates at the different nodes accessing the physical layer resources are not proportional to TDMA slot allocation. However, if we build a protocol that (i) allocates the TDMA slots proportion to the arrival rate at each node; (ii) creates the

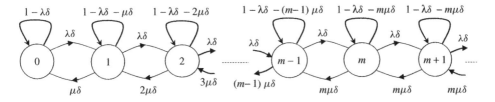

Figure 4.8 Discrete-time Markov chain representation of the M/M/m queue model.

effect of statistical multiplexing such that the physical layer never idles when a node has a packet to be transmitted (aggregate queue effect); and (iii) is able to prioritize packets based on their real-time and survivability needs (QoS capabilities) – that protocol will give us close to optimum performance in terms of physical layer resources utilization, average delay per packet, and QoS capabilities.

There are different queuing models that are similar to the M/M/1 queue model, where the arrival process is assumed to follow a Poisson distribution and service time is independent from inter-arrival time and follows an exponential distribution. These models are expressed as a discrete-time Markov chains as the M/M/1 model. The state transition diagram can show us the queue occupancy probabilities and we can find the average delay per packet in the queue by applying Little's theorem. In this chapter, we will cover the M/M/m queue model.

4.2.2.2 The M/M/m Queue Model

This model is similar to the M/M/1 queue model except that it assumes m servers.[12] For simplicity, this model assumes that a packet at the head of the queue (it is about to be serviced) is routed to any link (where service is available). The Markov chain expressing the state transition for this model is shown in Figure 4.8.

As with the M/M/1 queue model, we can reach the steady-state probabilities considering that $o(\delta)$ is very small. In the M/M/m model we have the packet service rate as:

$$\rho = \frac{\lambda}{m\mu} < 1. \tag{4.57}$$

As with Equation 4.41, we can find the global balance equation of the M/M/m queue as:

$$p_n\lambda = p_{n+1}\mu n \qquad n \leq m, \tag{4.58}$$

$$p_n\lambda = p_{n+1}\mu m \qquad n > m. \tag{4.59}$$

Equations 4.58 and 4.59 can be expressed as:

$$p_n = p_{n-1}\frac{\rho}{n} \qquad n \leq m, \tag{4.60}$$

$$p_n = p_{n-1}\frac{\rho}{m} \qquad n > m. \tag{4.61}$$

---

[12] The M/M/m queue model was originally developed by Erlang at the turn of the 20th century for the telephone systems when human operators were needed to make phone connections. Erlang intended to study how many telephone operators (servers) were needed, such that when a call arrives it will find an operator available (not busy serving another connection) within a defined probability.

Equations 4.60 and 4.61 can yield:

$$p_n = p_0 \frac{(m\rho)^n}{n!} \quad n \leq m, \tag{4.62}$$

$$p_n = p_0 \frac{m^m \rho^n}{m!} \quad n > m. \tag{4.63}$$

Considering that $\sum_{n=0}^{\infty} p_n = 1$, as with the M/M/1 queue model we can reach the probability that the queue is empty, p_0, as:

$$p_0 = \frac{1}{1 + \sum_{n=1}^{m-1} \frac{(m\rho)^n}{n!} + \sum_{n=m}^{\infty} \frac{(m\rho)^n}{m!} \frac{1}{m^{n-m}}} = \frac{1}{\sum_{n=0}^{m-1} \frac{(m\rho)^n}{n!} + \frac{(m\rho)^m}{m!(1-\rho)}}. \tag{4.64}$$

A relevant measure of the performance is the probability that a packet arrives and finds all m servers busy. The arriving packet in this case will have to wait in a queue. This is referred to as the probability of queuing and can be calculated as:

$$P_Q = \sum_{n=m}^{\infty} p_n = \sum_{n=m}^{\infty} \frac{p_0 m^m \rho^n}{m!} = \frac{p_0 m^m \rho^m}{m!} \sum_{n=m}^{\infty} \rho^{n-m} \cong \frac{p_0 m^m \rho^m}{m!(1-\rho)}. \tag{4.65}$$

In Equation 4.65, p_0 is given by Equation 4.64.[13]
The number of packets in the queue can be calculated as:

$$N_Q = \sum_{n=0}^{\infty} n p_{(m+n)}. \tag{4.66}$$

Applying Equation 4.66 for the value of $p_{(m+n)}$, we get,

$$N_Q = \sum_{n=0}^{\infty} n p_0 \frac{m^m \rho^{m+n}}{m!} = \frac{p_0 m^m \rho^m}{m!} \sum_{n=0}^{\infty} n \rho^n. \tag{4.67}$$

From Equation 4.65, we can express N_Q as:

$$N_Q = P_Q \frac{\rho}{1-\rho}. \tag{4.68}$$

Equation 4.68 suggests that $\frac{N_Q}{P_Q} = \frac{\rho}{1-\rho}$. This means that we have a conditional probability of the expected number of packets found in the queue by an arriving packet. The condition here is that the arriving packet will be forced to wait in the queue. Refer to the M/M/1 queue model in Equation 4.49. The average number of packets in the queue upon arrival is $\frac{\rho}{1-\rho}$. One can see that in the case of the M/M/m queue model, when a packet arrives and is forced to wait in the queue, we should expect identical behavior as with the M/M/1 queue model (remember that $\rho = \frac{\lambda}{m\mu}$ in the M/M/m model).

---

[13] Equation 4.65 is known as the Erlang C formula; it applies to circuit switched networks for call arrival finding the line busy. It is used in call admission control to estimate the number of circuits needed for an expected call load. It applies to packet switched networks, considering that an arrived packet, finding all servers busy, will wait in a queue.

Now let us find the average time a packet has to wait in the queue, W, using Little's theorem:

$$W = \frac{N_Q}{\lambda} = \frac{\rho P_Q}{\lambda(1 - \rho)}. \tag{4.69}$$

The average delay per packet, including service time, is:

$$T = \frac{1}{\mu} + W = \frac{1}{\mu} + \frac{\rho P_Q}{\lambda(1 - \rho)} = \frac{1}{\mu} + \frac{P_Q}{m\mu - \lambda}. \tag{4.70}$$

Using Little's theorem $N = \lambda T$, we can calculate the average number of packets in the queue as:

$$N = \frac{\lambda}{\mu} + \frac{\lambda P_Q}{m\mu - \lambda} = m\rho + \frac{\rho P_Q}{1 - \rho}. \tag{4.71}$$

Statistical Multiplexing with the M/M/m Model

Consider a MAC layer protocol in a MANET subnet designed to serve m different nodes over the same physical layer resources. Each node has a Poisson arrival λ_m and the overall arrival from all nodes is $\lambda = \sum_{i=0}^{m} \lambda_m$, which is also a Poisson arrival. Let us assume that the subnet's physical capacity can be divided into m separate channels and that one channel can be assigned to each node. If a specific node has no packets waiting in its queue, then the corresponding capacity is used to transmit packets from another node. The service time of all packets from all nodes has an exponential distribution with a mean equaling $1/\mu$. There are two ways to implement this protocol. One way is to have the service rate assigned to each channel as the fraction μ (think of FDMA). The second possibility is to have statistical multiplexing where the service rate is aggregated for the packet in service as $m\mu$ (as if all the channels are combined to serve one packet at a time).

The first case gives us the following average delay:

$$T_1 = \frac{1}{\mu} + \frac{P_{Q1}}{m\mu - \lambda}. \tag{4.72}$$

The second case gives us the following average delay:

$$T_2 = \frac{1}{m\mu} + \frac{P_{Q2}}{m\mu - \lambda}. \tag{4.73}$$

When the subnet is lightly loaded ($\rho \ll 1$), both P_{Q1} and P_{Q2} are zero and $\frac{T_1}{T_2} \cong m$. That is, statistical multiplexing reduces delay by a factor of m. However, if the queuing system is loaded (ρ is slightly less than one), both P_{Q1} and P_{Q2} approach 1. As the dependence of $\frac{T_1}{T_2}$ on the value of m decreases, the advantage of statistical multiplexing reduces.

You are encouraged to study queuing models in more depth, especially if you are specializing in designing MAC and network layers protocols, or in the modeling and simulation field, where queue models can serve as an eye-opener for understanding simulation results. The material covered by this chapter is intended to help you practice good design concepts based on the conceptualization of how packets are dynamically queued and serviced.

4.3   Concluding Remarks

This chapter emphasized queue modeling in order to give you an understanding of how tactical networks get congested at the IP layer or at the MAC layer. Other books, covering the network layer, may focus on protocols such as IP or OSPF (open shortest path first). However, in tactical wireless networks, we not only have connectivity at layers 2 and 3 (which cause us to look at routing differently from wired and point-to-point networks), but we also face specific challenges with fluctuation of bandwidth that can cause queue build-up. This can lead to packet drops, excessive delays, and delay variation (jitter). For the purpose of addressing tactical wireless network challenges, we would rather you acquire a firm grip on the causes of degradation of service, and the recommended design concepts, instead of explaining basic protocols that can be found in other books.

4.3.1   How Congestion Happens in Tactical Wireless Networks

The M/M/1 queue model, which we discussed in detail, albeit the simplest queue model, can provide us with a good sense of some of the challenges that we are facing with tactical wireless networks. Let us look at a common dangerous engineering practice that correlates percentage utilization to congestion in a linear fashion. Going through the M/M/1 queue model you can see that if we correlate congestion to: (i) the expected number of packets waiting in the queue (which can cause packet drop, since we have a limited queue size and cannot tolerate excessive delay as a result of a queue buildup); and (ii) the expected wait time in the queue; you will come up with a different understanding of what causes congestion. You will find that $\mu - \lambda$ plays a major role (this value represents the difference between the service rate and the arrival rate – the headroom). Figures 4.5–4.7 demonstrate how we should regard congestion as a factor of percentage utilization and the head room $\mu - \lambda$ and know it is non-linear. The concavity of the curves in Figures 4.5 and 4.6 suggests that there is a breakpoint where the curves start to increase exponentially (congestion is not tolerable). The higher the value of the available bandwidth (the higher the service rate), the later we reach this breakpoint (the higher percentage utilization we have before delay becomes intolerable). Very often in a tactical MANET subnet, an event happens (such as poor connectivity due to mobility or increased activities of control traffic when a node joins a subnet) that causes the bandwidth available for user traffic to decrease. As a result, queues, at their corresponding nodes, are served at a much lower service rate. This drop in service rate can cause catastrophic effects. Even though the transport layer can react by reducing flow, it is still hard to recover from these events. Recovery is difficult when the service rate becomes so low that it is necessary to reduce arrival rate drastically (to have a very low ratio of λ/μ) in order help the queue to reach a steady state again.

Another lesson we can learn from the M/M/1 queue model is how we can recover from congestion. If you think of limited queue size (which is used for packets from real-time applications), congestion can cause a high packet loss ratio. Transport layers have no way of knowing if packet loss happens due to physical layer issues (jamming, loss of connectivity) or due to queue overflow. Real-time applications will not retransmit the lost packets and will have a period of blackout (in VoIP, you can detect when the voice quality is so bad due to the loss of packets). If we think of an infinite queue size (or a large queue size for non-real-time applications), transport layers can retransmit delayed packets (after waiting for an acknowledgment for a certain period of time), which makes a bad situation even worse and increases the load of an already loaded MANET subnet. Recovery from these catastrophic events, with non-real-time queues, can take a very long time. Think of the Markov chain in Figure 4.4: if we reach a large value of n, we will need a longer period of time to come back to a low value of n such that the transport layer can work under acceptable path quality. The larger μ and $\mu - \lambda$ become, the faster we can recover from a catastrophic congestion event.

You need to also develop a feeling for burst packet arrivals (which deviate from the M/M/1 model). Burst packet arrivals are common, especially at the network edge (at the network core, the tandem queues can regulate burst arrival, getting us closer to the Poisson model we assumed). Essentially, burst packet arrival can cause a sudden jump in the Markov chain, making it reach the larger n states. A reliable metric to keep an eye on is the delay variation (jitter), which can be increased by burst arrivals. While some applications can tolerate a certain level of jitter,[14] excessive jitter can be problematic. The receiving end of a packet stream, suffering from high delay variations, will have no use for packets that are received out of their expected time frame.

This chapter should have provided you with reasons as to why we need cross layer signaling. Research in tactical MANET has shown that queue overflow is one of the major reasons of degraded QoS. Cross layer signaling can prevent queue overflow. Think of how the MAC layer queue is in tandem to the IP layer queue and how the IP layer queue server feeds the MAC layer queue. If the MAC layer decreases its service rate to its queue without informing the IP layer, the IP layer can flood the MAC layer queue, causing a high ratio of packet loss. Signaling between the MAC layer and the IP layer is known to increase QoS drastically in tactical MANET. If you can imagine that each layer of the protocol stack has a state machine similar to the same Markov chain explained above, which needs to be in a steady state, then you can think of how cross layer signaling can help all these state machines stay in their steady states or recover from deviations quickly. Researchers who advocate increased cross layer signaling in tactical MANET are looking to create harmony between the protocol stack layers, allowing them to be in a steady state or ready to react to any event that causes deviation at any layer, thus avoiding catastrophic events.

4.3.2 Historical Perspective

Many of the techniques used in computer networking today are based on mathematical concepts which were established long before the foundation of computer networking as a science. The roots of traffic engineering and queuing theory were pioneered by the Danish statistician Agner Krarup Erlang at the turn of the 20th century. Erlang's work focused on finding how many telephone circuits were needed to provide acceptable telephone service. His mathematical principles are the foundation for studying queuing models in packet switched networks. The notation to describe the characteristics of queuing models was first suggested by David G. Kendall in 1953, with the purpose of enhancing circuit switched networks. Kendall's ideas introduced an A/B/C queuing notation that is now used in all standard works on queuing theory. Most of the theoretical basics that applied to telephone systems also applied to circuit switched networks and later to packet switched networks.

---

[14] VoIP applications have a jitter buffer at the decoder that can trade jitter (delay variation) for delay. There is a limit on how much jitter can be absorbed before the VoIP call quality degrades drastically.

The explosion in computer networking owes it progress to advances in solid-state technology. The leaps in very large scale integration (VLSI) led to faster, larger, less expensive memory and storage devices. This, in conjunction with leaps in microprocessor design, led to the progressive shift from centralized, time-shared computing to the personal computer and workstations. The growing need by organizations to have their computers share information led to the development of networks. This resulted in an explosive demand for packet switched networks, where resources are shared, instead of circuit switched networks where dedicated resources for a specific connectivity (that can be idle at a given time) can lead to poor utilization of the most precious resource, bandwidth (given that processing power is becoming a non-issue).

While many networking approaches were budding in the 1970s, (e.g., ARPANET and TYMNET as wide-area networks, and local area Ethernet), the defense industry made great leaps, building wireless multiple access technologies such as Link 16, enhanced position location reporting system (EPLRS), and single channel ground and airborne radio system (SINCGARS), where airborne vehicles, ground vehicles, and soldiers on the ground were able to form a wireless subnet and exchange information (voice and data) within their subnet. These wireless subnets were islands of subnets and a human-in-the-loop was needed in joint missions. The defense industry's use of tactical core networks led to building gateways between the core network and these peripheral wireless subnets in order to create better communications capabilities.

The move to an all-IP infrastructure and the creation of the global information grid (GIG) vision, at the turn of the millennium, led to developing a new generation of IP-based multiple access radios that could be connected to each other, and to an IP core network, seamlessly. The future of warfare is changing drastically. In the old days, those who had the big guns won the war. Today, one can say those who have the communications dominance win the war. With the ever-increasing use of technologies, such as armed unmanned aerial and ground vehicles (even armed satellites), sensors, and cruise missiles, communications is becoming the cornerstone to winning a war. The warfighter is no longer needed to be deployed physically to a war theater. A warfighter can fly his drone from an office in the USA to conduct a mission halfway across the world, where he/she can collect intelligence through aerial sensors (images and video), relay information from ground sensors, define a target, and fire on the target. Without seamless reliable communications between the warfighter devices in the USA and the unmanned deployed equipment, such missions cannot be executed.

4.3.3 Remarks Regarding the First Part of the Book

Having come to the conclusion of the first part of this book, you should now have a good grasp on some of the challenges facing the design of tactical wireless networking protocols at the different layers of the protocol stack and how they may relate to each other. You have now reviewed some of the basic wireless communication and networking concepts and you should feel ready to move on to the details of the legacy and IP-based tactical radios, as well as design concepts at the upper protocol stack layers that are necessary for tactical networks. These concepts will be introduced in the next two parts respectively.

It is important to note that the utilization of simple modeling at the protocol stack layers was essential to an understanding of which engineering philosophy should be adopted in our design. However, please keep in mind that models do deviate from reality. A tactical MANET subnet is very dynamic and it is hard to develop a realistic model that reflects its states accurately and simply. The challenge here was to start from an appropriate model, which provided us with the right engineering sense of the problem, but developed protocols that are gracefully adaptive to the state of the network as it changes dynamically. In Chapter 2, we learned the value of spectrum modeling, although, we know that signal spectrum can deviate from the models used. The use of the additive white Gaussian noise (AWGN) model allowed us to develop good signal detection techniques, although we know that sometimes the noise is not exactly AWGN. In Chapter 3, we emphasized the value of Shannon's theory; this helped us to find the capacity bound of the communications channel, even though we made many assumptions. In this chapter, we emphasized the advantages of exploiting statistical multiplexing in MANET physical layer resource management. We started from an M/M/1 queue model that showed us the advantages of statistical multiplexing. Even though, in reality, the M/M/1 is an approximation, the advantages of statistical multiplexing still apply. The development of a protocol such as USAP, is based on this understanding. However, it is not rigidly based on an exact M/M/1 model. Another example is the use of a Markov chain, which gives us a good sense of how the protocol stack layers need to reach a steady state. The development of cross layer signaling will help us to ensure that protocol stack layers reach a level of harmony and heal from deviations from their steady states gracefully. The cross layer signaling technique should not be based on an exact discrete-event Markov chain at each layer, but it still has the steady state concept behind it. Researchers who bring concepts of the theory of governing dynamics to computer networks attempt to bring equilibrium to the protocol stack layers of the different nodes sharing a common resource (e.g., bandwidth).

Another lesson we learned from this part of the book (Chapter 3) is not to burden one layer of the protocol stack just to overcome some odd moments. In Chapter 3, we showed that although network coding can enhance the performance of the tactical MANET under adverse conditions of high packet loss, we cannot design network coding in a vacuum. Network coding has to be designed in lieu of the DLL capabilities and the transport layer coding. The same good engineering practice should be carried forward in creating tactical network design concepts all across the protocol stack layers. Putting the burden on a single layer of the protocol stack, to overcome a specific challenge, can lead to poor performance and we would rather have the whole protocol stack working in harmony to overcome challenges. Yet another lesson comes from the fading example in Section 2.4, which points to how relaxing constraints is sometimes better engineering practice than using brute force.

Whatever your area of specialty (RF, detection and estimation, error control coding, DLL and MAC layers protocols, IP layer and routing, QoS and resource management, network management, transport layer protocols, application development, etc.), good engineering practices will come from understanding the overall picture, from an application generating data, to a bit-by-bit modulation over the physical layer. There is a great deal of dependency in tactical wireless networks. In wired networks, we may be able to have engineers specialize in one protocol stack layer where layers can be independent. In tactical wireless networks, we *need* to maintain a good understanding of the overall picture before we can specialize effectively in one area. I hope that the first part of this book serves that purpose.

Bibliography

1. Bertsekas, D. and Gallager, R. (1994) *Data Networks*, 2nd edn, Prentice Hall.
2. Schwartz, M. (1987) *Telecommunication Networks Protocols, Modeling and Analysis*, Addison-Wesley.
3. Tanenbaum, A. (1996) *Computer Networks*, 3rd edn, Prentice Hall.
4. Park, S. and Sy, D. (2008) Dynamic control slot scheduling algorithms for TDMA based mobile ad hoc networks. Proceedings of Milcom 2008, WSN-1.
5. IEEE Standard 802.11-2007. *Wireless LAN Medium Access Control (MAC) and Physical Layer (PHY) Specifications*, IEEE LAN MAN Standards Committee.
6. Zhu, C. and Corson, M.S. (2001) A five-phase reservation protocol (FPRP) for mobile ad hoc networks. *Wireless Networks*, **7** (4), 371–384.
7. Zhu, C. and Corson, M.S. (2000) An evolutionary-TDMA scheduling protocol (E-TDMA) for mobile ad hoc networks. Proceedings of Advanced Telecommunications and Information Distribution Research Program (ATIRP).
8. Keshav, S. (1997) *An Engineering Approach to Computer Networking*, Addison-Wesley, Reading, MA, pp. 150–153.
9. David Young, C. (1999) USAP multiple access: dynamic resource allocation for mobile multihop multichannel wireless networking. Proceedings IEEE MILCOM '99, October 1999.
10. Marshall, P. (2006) Adaptation and integration across the layers of self organizing wireless networks. MILCOM 2006.
11. Brehmer, J. and Utschick, W. (2005) Modular cross layer optimization based on layer descriptions. WPMC 2005, Aalborg, Denmark, September 2005.
12. Christopher Ramming, J. (2005) Control-Based Mobile Ad-hoc Networking (CBMANET) program motivation and overview. DARPA, 30 August, 2005.

Part Two

The Evolution of Tactical Radios

Part Two

The Evolution of
Bacterial Plasmids

5

Non-IP Tactical Radios and the Move toward IP

In the first part of this book, we reviewed some basic theoretical concepts of tactical wireless communications and networking. The second part of this book will focus on legacy non-IP radios; introducing the seamless IP concept; some of the important aspects of IP-based tactical radios; software defined radios (SDRs); and cognitive radios.

This chapter reviews some legacy non-IP tactical radios and addresses the issues and concerns arising from the integration of legacy, non-IP, radios with IP-based network core services.

5.1 Multistep Evolution to the Global Information Grid

The transition from push-to-talk radios to the global information grid (GIG) vision is a multistep, gradual process. This evolution is ongoing because it is economically impractical to replace all the deployed technological equipment every decade or so. Instead, new technologies are deployed alongside old technologies and gateway capabilities are constructed to allow both technologies to function simultaneously (often in a less than optimal way). Some of the steps leading to the GIG vision include the introduction of IP over circuit switched networks (known as digitizing) to form an IP core network, and the introduction of IP gateways between the IP core and legacy non-IP nets. These steps make it possible for near-real-time applications, such as friendly and enemy forces position tracking, to work effectively. This tracking technology shows the positions of all friendly forces on a digital map on a scope while making updates in near real time. This allows convoys to stay together and reduces friendly fire casualties. The explanation of the Tactical Internet (TI) in this chapter should provide you with a sense of how the gradual migration to the GIG is taking place.

The drive for technological advancement during the Cold War furthered the development of many critical defense technologies that are considered modern marvels. One such example would be the defense industry's introduction of code division multiple access (CDMA), which now serves as the commercial wireless world's basis for 3G/4G air interfacing technologies. The defense industry also constructed outstanding waveforms such

Tactical Wireless Communications and Networks: Design Concepts and Challenges, First Edition. George F. Elmasry.
© 2012 John Wiley & Sons, Ltd. Published 2012 by John Wiley & Sons, Ltd.

as Link-16. This waveform was ahead of its time with anti-jamming capabilities, mobility at high speed, and long ranges between nodes. This waveform is currently in use by many US, joint, and coalition forces worldwide and is expected to stay with us for a while. This chapter will cover some of the important aspects of Link-16.

You will see how the previous chapters apply to this legacy waveform (Link-16) where we cover how modulation, error control coding, frequency hopping (FH), sharing of the physical layer resources, and so on, apply to the Link-16 waveform. This chapter will also briefly introduce you to the enhanced position location reporting system (EPLRS) waveform and the single channel ground and airborne radio system (SINCGARS) waveform. You will be introduced to the TI and see how these legacy waveforms fit in an end-to-end IP-based architecture. We will also discuss the challenges of building gateways between an IP backbone (core) network and a non-IP legacy net, as well as creating IP touch points between different legacy non-IP nets.

Although this is not the main focus of the book, this chapter introduces you to non-IP waveforms to demonstrate how the defense industry has been ahead of the commercial industry in all aspects of wireless communications and networking leading up to the Internet and wireless boom. An essential objective of this text is for you to be able to identify the cyclic nature of science. Although the defense industry's work has facilitated the success of the commercial sector, it can still benefit from the recent success of commercial wireless. The defense industry would likewise gain in borrowing mature commercial technologies and fitting them within tactical requirements and giving the warfighter much-needed rich applications.

5.2 Link-16 Waveform

Link-16 is one of the most sophisticated wireless tactical data systems in use by the US Joint and Coalition Services. It supports time-critical, robust radio communications and is integrated into many platforms, including fast movers. Link-16 supports voice, free text, and variable format messages. It is a complex waveform that deploys TDMA (time division multiple access), CDMA, and FDMA (frequency division multiple access) schemes.

As a legacy waveform, Link-16 was conceived before the standardization of the protocol stack layers discussed in Chapter 1. The core functionalities of this waveform are in the physical and data link layers. Link-16's equivalent of the MAC layer (where the apportionment of physical layer resources is handled) is done in the planning phase. In this phase, each Link-16 terminal is assigned a set of time blocks. Also, instead of MAC frames and IP packets, a block of 75 bits (known as a J-word) is used for transmitting and receiving data over the Link-16 net.

Let us delve further into the unique physical and data link features of the Link-16 waveform. Figure 5.1 details the slot structure of Link-16. Figure 5.1a illustrates the 127 different possible nets in Link-16 (horizontal rings numbered 0–126, with net 127 representing a stacked net); with the vertical axis representing the FDMA dimension of the waveform (each of these nets has its own FH pattern). Figure 5.1b demonstrates how a single ring (net) employs TDMA, creating 98 304 timeslots during a period of 768 seconds. Figure 5.1c shows how separate nets can work simultaneously, while a unique hopping pattern is assigned to each terminal in the net. These figures are taken from reference [1] and demonstrate some

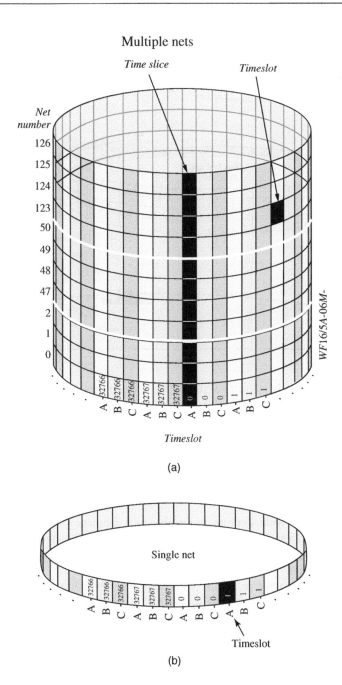

Figure 5.1   (a) Link-16 stacked nets – vertical axes represent FDMA. (b) TDMA in a Link-16 net.

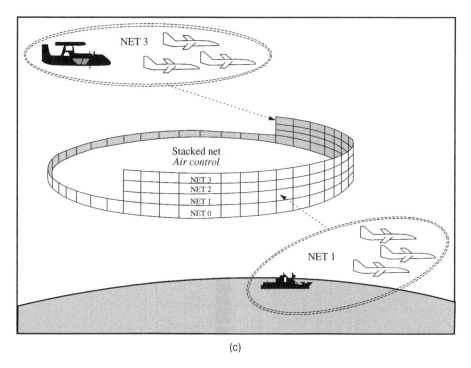

(c)

Figure 5.1 (c) Simultaneous use of Link-16 nets. Reproduced from Northrop Grumman Corporation, Understanding Link-16: A Guidebook for New User, San Diego, CA, September 2001. Figures originally from US government archive.

of the important design concepts of Link-16 that are detailed in this chapter. These design concepts contribute to the great success of this waveform.

Link-16 also has the following criteria:

- It operates within a wide spectrum in the ultra high frequency (UHF) Lx band between 960 MHz and 1215 MHz. There are 51 frequencies assigned to the joint tactical information distribution system (JTIDS) explained below. These 51 frequencies are between 969 and 1206 MHz.
- The slot structure shown in Figure 5.1 creates a combination of FDMA and TDMA timeslot structure. The structure is split into 12.8 minute (768 seconds) epochs. Each epoch contains three sets of timeslots (i.e., set A, B, and C) numbered from 0 to 32767. A, B, and C are used as a numbering scheme such that timeslots can be numbered A0, B0, C0, A1, B1, C1, ... A32767, B32767, and C32767. Each timeslot is 7.8125 ms and consists of a train of pulses each of 6.4 μs duration separated by an interval of 6.6 μs.
- The secure data unit (SDU) explained below holds up to eight crypto-variables stored as today/tomorrow pairs. Each of these pairs could be used with 127 possible net numbers to allow a potential of 508 hopping patterns for any given timeslot block.
- The use of FH creates resilience against jamming. In addition, the use of CDMA makes it possible to detect a signal at low signal-to-noise ratio (SNR). Jammers that work on a specific frequency can be mitigated by FH, and jammers that spread their

spectrum will be combated by CDMA and error correction. In Link-16, hopping occurs within the same timeslot, every 13 µs during every pulse, approximately 600 times per timeslot. Refer to Section 2.6.2 to review the advantages of fast hopping.

- The use of two types of encryption security, transmission security (TSEC) and message security (MSEC) provide excellent multilayered defense. An enemy attempting to listen to a received Link-16 signal has to zero in on the hopping pattern and decipher the MSEC code, and so on, in order to get useful information.

The predecessor to Link-16 (Link 11) relied on a network control station (NCS) to poll each terminal in the net. It did so in a round-robin sequence to transmit at a timeslot. Link-16 eliminated the need for an NCS by giving each terminal a pre-assigned set of timeslots. Link-16 also uses a higher data rate than its predecessors (Link-4 and Link-11) depending on the transmission mode. These data rates are 28.8, 57.6, 115.2, and 238 kbps. Voice communications over Link-16 uses two-channel digitized voice at 16 kbps with push-to-talk protocol. Although the voice communication mode is encrypted, it has no error correction coding. Depending on the transmission mode, data-transmission can use comprehensive error protection measures that include:

- error detection using parity bits;
- error correction and detection using multiple modes of Reed–Solomon (RS) encoding expressed as (n, k) – here k is the number of information symbols and n is the number of information and redundancy symbols added, and the symbol size is 5 bits;
- symbol interleaving for block error reduction, where a burst error can be made to have the effect of a random error;
- symbol repetition (double pulse) where a 5-bit symbol is transmitted twice.

With error detection, 12 parity bits are used within each 3-word block of 225 bits, with 201 of these bits being data. The parity bits are generated using (237, 225) polynomial code. Error detection can be applied to selected bits of the messages.

The error correction and detection uses RS code with 5 bits per symbol and can use rate (16, 7) or rate (31, 15). That is, each transmitted codeword can be either 80 bits, with 35 information bits and 45 redundancy bits, or 155 bits with 75 information bits and 80 redundancy bits. The various uses of RS code in Link-16 and its error and erasure capabilities are explained below.

Symbol interleaving is used to combat burst errors by turning them into random errors. It also contributes to the MSEC and the anti-jamming margin of the JTIDS signal. Symbol redundancy, where a pulse carrying one symbol is transmitted twice, increases error resilience at low SNR (from jamming or long distance links).

JTIDS is the communications architecture over Link-16. JTIDS makes it possible to form networks over the Link-16 physical and data link layers. Figure 5.2 (taken from Reference [1]) further explains how Link-16 data is packaged to fit more information onto the JTIDS TDMA slot architecture. There are four packing options to transmit JTIDS messages over Link-16. Jamming resistance decreases as the density of the data packing increases. Data packed in a Link-16 timeslot is transmitted such that interference is minimized and synchronization is assured between the transmitting and receiving terminals. For Link-16, the most reliable data packing structure uses a single timeslot consisting of the following intervals: jitter, synchronization, time refinement, message (header and data), and propagation.

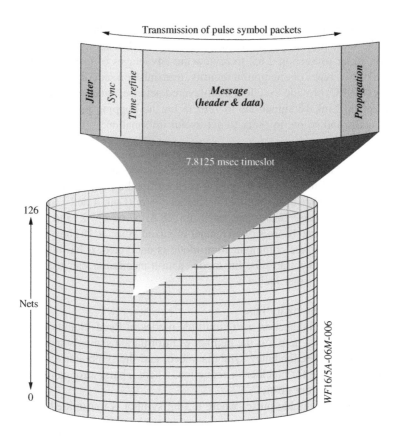

Figure 5.2 JTIDS utilization of Link-16 timeslots. Reproduced from Northrop Grumman Corporation, *Understanding Link-16: A Guidebook for New User*, San Diego, CA, September 2001. Figures originally from US government archive.

Jitter, synchronization, and time refinement ensure that the receiving terminal is synchronized with the beginning of the message without ambiguity. The propagation period ensures that there will not be any interference from the subsequent transmission.[1]

The amount of jitter varies from timeslot to timeslot in a pseudorandom pattern determined by the TSEC crypto-variable. Jitter contributes to the anti-jam nature of the JTIDS signal because an effective jammer needs to know when to turn the jamming signal on/off. After the jitter interval, the synchronization and the time refinement intervals are used by the receiver to recognize and synchronize with the transmitted signal. The message interval is then divided into a header and data symbols. Lastly, the propagation interval, which is a guard time, is needed to ensure that signal propagation occurs without overlapping of neighboring pulse symbol packets. There are two types of guard times: one is for normal range ($\cong 300$ nautical miles), and the other is for extended range ($\cong 500$ nautical miles). Note that the jitter and propagation times are dead times in which no signal is transmitted.

---

[1] Considering the speed at which the RF signal propagates and the waveform's long range coverage, the propagation period with Link-16 could be as long as 5 ms.

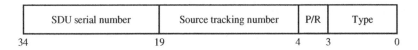

SDU serial number	Source tracking number	P/R	Type
34 19	4 3		0

Figure 5.3 Link-16 message header.

5.2.1 Link-16 Messages

The Link-16 message shown in Figure 5.2 consists of a header and a data part. The header, containing 35 bits, follows the format in Figure 5.3. This format is as follows: bits 0–2 specify the type of data; bit 3 indicates the relayed or packed nature of the message; bits 4–18 identify the tracking number of the transmitting terminal and has embedded parity bits; bits 19–34 carry the serial number of the SDU associated with the transmitting terminal (only a decrypted message can reveal this serial number).

The data part of the Link-16 message can be any of four types: fixed-format, variable-format, free text, or round-trip timing (RTT). A fixed-format type is used to exchange J-Series messages, while a variable-format type is used to exchange any type of user-defined message. However, the variable-format types are not used by the USA. Free-text types handle digitized voice, and the RTT type messages are for terminal synchronization. A terminal must be synchronized with the net in order to receive and transmit on a JTIDS network.

In a J-Series message, there are three types of J-Series words: initial, extension, and continuation, as shown in Figure 5.4. Each word consists of 75 bits. Of these bits, 70 bits are J-Series message data, four bits are parity checks, and one bit is reserved as a spare. J-Series messages may contain an initial word, one or more extension words, and one or more continuation words. Up to eight words may be used to form a single J-Series message. If the standard-double-pulse (STD-DP) packing structure is used, three J-Series words (225 bits) are formed and transmitted in each timeslot. If the Pack-4 format is used, twelve J-words are transmitted in one timeslot. Messages are taken in groups of three, six or twelve J-words for transmission.

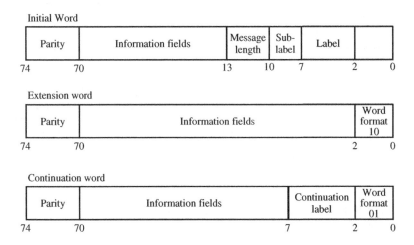

Figure 5.4 Link-16 fixed-format message words.

5.2.2   Link Layer Operations of Link-16

At the transmitting terminal for Link-16, the J-Series words are assembled at the link layer, before they are forwarded to the physical layer for transmission. First, the data packing structure – STD-DP, Pack-2 single pulse (P2SP), Pack-2 double pulse (P2DP), or Pack-4 single pulse (P4SP) – is determined and recorded in the header. Second, the twelve parity bits are generated using a binary (237,225) cyclic redundancy check (CRC) error detection code. Here the 210 data bits of the 30-word blocks and the 15-bit source track number from the header are used to calculate a 12-bit parity value. The twelve parity bits are then distributed at bit positions 71–74 of each 75-bit word as shown in Figure 5.4. Third, the SDU serial number is encrypted and stored in the header (bits 19–34). Lastly, the header and message data are forwarded to the physical layer for transmission. At the receiving terminal, the link layer performs the preceding operations in reverse, and only the J-Series messages that pass the parity check continue to the protocol stack. Note that the header is interpreted at the link layer, but the J-Series message data are processed at the physical layer; that is, the entire J-Series message (header and data) cannot be passed to the protocol stack in one step. The data symbols must be buffered and demodulated after the packing structure in the header is determined at the link layer.

5.2.3   JTIDS/LINK-16 Modulation and Coding

JTIDS uses: RS coding, symbol interleaving, cyclic code shift keying (CCSK) for M-ary baseband symbol modulation, minimum shift keying (MSK) chip modulation for transmission, double-pulse diversity, and combined FH/DS (direct sequence) spread spectrum. The physical layer of a JTIDS-type system is illustrated in Figure 5.5 for both the transmitters a receiver.

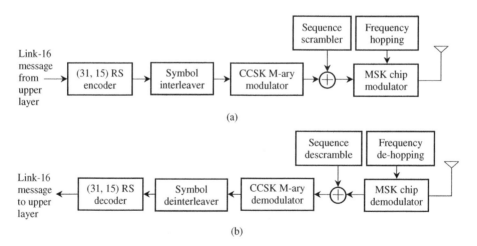

Figure 5.5   (a) JTIDS/Link-16 message function model at the transmitter. (b) JTIDS/Link-16 message function model at the receiver.

5.2.3.1 Reed–Solomon Codes

Here, RS codes use m bits at a time, which are combined to form a symbol. All the possible combinations of m bits, M symbols, can be represented as $M = 2^m$. Please refer to Sections 3.2.1 and 3.2.2 where error control coding is introduced. The RS(31,15) code used in the JTIDS takes 15 information symbols and generates 31 coded symbols. This code can correct up to eight-symbol-error in a codeword of 31 symbols. In the JTIDS, each symbol is five bits. Thus, this RS code can correct a burst error of up to 40 bits.

When a Link-16 message arrives at the physical layer, it is mapped onto 5-bit symbols. The 35-bit message header becomes seven symbols, and each 75-bit J-Series word (message data) becomes 15 symbols. The seven symbols of the message header are encoded with a (16,7) block code (this code is related to the RS(31,15) code by shortening and/or puncturing).[2] The 15 symbols of message data are encoded with the RS(31,15) code as indicated in Figure 5.5. After encoding, the Link-16 message header consists of 16 coded symbols, while each J-Series word consists of 31 coded symbols.

Note that an RS(16,7) code cannot exist since the number of RS coded symbols n must be of length $n = 2^m - 1$. Such a code could be obtained from puncturing an RS(31,7) code. Punctured RS codes are maximum distance separable (MDS) codes and have the error correction capabilities explained in Section 3.2.1, where $d_{min} \geq 2t + 1$. This means that the (16,7) code used for the header has an error correction capability of $t = \frac{d_{min}-1}{2} = 4$. That is, it can correct four symbol errors.[3]

Another possible method to obtain the equivalent of a (16,7) code is by shortening and puncturing another RS code. Shortened RS codes are also MDS codes. An (n, k) RS code is shortened by setting i information symbols equal to zero at the encoder and decoder, resulting in a $(n - i, k - i)$ codeword. The shortened code corrects at least as many errors as the original, but the codewords are shorter. If an RS(31,15) code is first shortened by eight information symbols, we get a (23,7) shortened RS code. If the shortened code is then punctured by seven parity symbols, the result is a (16,7) shortened, punctured RS code.

Whether the first or the second method is used to obtain a (16,7) code, the header coding can still be considered less robust than the message data, considering the fading characteristics of the channel. With JTIDS, the message header is always transmitted in a double-pulse format to combat fading channels and/or narrowband interference. Note that most packet-based communications systems require that headers are more resilient than packet payloads. A header error causes packet drop, while a packet payload error can be tolerated by the application. With JTIDS, the header contains the SDU serial number and the source tracking number. This makes delivering the packet to its intended destination possible; hence, the double pulse is used with the header.

---

[2] Please refer to the error control coding references in Chapter 3 for details about shortening and puncturing of RS codes.

[3] In the JTIDS header, we can correct four out of 16 symbols; in the JTIDS message, we can correct eight out of 31 symbols. For a random error pattern, the correction capability in both cases is comparable. Considering the fading characteristics of the channel, the ability to correct a burst of eight symbols in the JTIDS message is more robust than correcting a burst of four symbols in the JTIDS header.

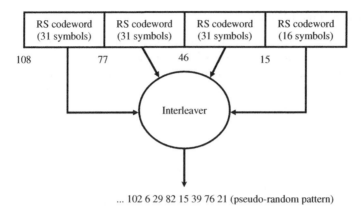

Figure 5.6   Symbol interleaving in JTIDS.

5.2.3.2   Symbol Interleaver

Symbol interleaving shuffles the symbols from several different codewords. As a result, symbols from a specific codeword are not transmitted sequentially. De-interleaving at the receiving end reverses the process, putting the received symbols back into the proper order before passing them on to the decoder. For JTIDS, the interleaver is used to interleave both the header symbols and the data symbols. Since the header specifies the type of data and identifies the source track number of the transmitting terminal, the communications link could be significantly degraded if the header symbols are jammed.

The size of the symbol interleaver depends on the number of codewords in the packing structure. For example, an STD-DP packing structure has an interleaver which contains 109 symbols. The symbol interleaving process is shown in Figure 5.6. The interleaver read-in sequence is always fixed, but the starting point of the read-out is pseudorandom; this provides the first layer of TSEC.

5.2.3.3   CCSK Baseband Symbol Modulation

After the symbols are interleaved, 32 synchronization symbols and eight time refinement symbols are appended to the beginning of the interleaved, coded symbols. In JTIDS, modulation is done at the baseband level using CCSK and at the chip level using MSK. CCSK provides M-ary baseband modulation and spreading, since each 5-bit symbol is represented by a 32-chip sequence. Table 5.1 provides an example of how 32-chip CCSK sequences can be derived by cyclically shifting a sequence, S0, one place to the left n times (where n is between 1 and 31) to obtain a unique sequence for all the possible combinations of five bits. As explained in Section 2.6, the demodulator determines which 5-bit symbol was received by computing the cross-correlation between the received 32-chip sequence and all possible 32 sequences. The 5-bit symbol corresponding to the largest cross-correlation is decoded. It can be shown from Table 5.1 that the original 5-bit symbol can be recovered, with up to six chip errors occurring, for the 32-chip initial sequence chosen for JTIDS.

Note that after the CCSK symbol-to-chips spreading, each 32-chip CCSK sequence is scrambled with a 32-chip pseudo-noise (PN) sequence. This process provides a uniform

Table 5.1 32-chip CCSK sequences used for JTIDS – notice the left-hand side direction of shifting of the CCSK codewords

5-Bit symbol	CCSK codeword
00000	S0 = 01111100111010010000101011101100
00001	S1 = 11111001110100100001010101011011000
00010	S2 = 11110011101001000010101110110001
00011	S3 = 11100111010010000101011101100011
...	
...	
11111	S31 = 00111110011101001000010101110110

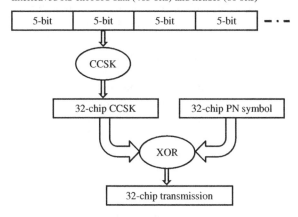

Figure 5.7 Generating 32-chip transmission symbols in JTIDS.

spreading of the baseband waveform and works as a second layer of transmission security. The resulting 32-chip sequence is called a 32-chip transmission symbol. Figure 5.7 demonstrates this process.

5.2.3.4 Minimum-Shift Keying (MSK) Chip Modulation

After scrambling, each chip is modulated for transmission with a special case of continuous phase frequency-shift keying (CPFSK), also known as MSK. Please refer to Section 2.2 for shift keying modulation techniques. Some advantages of MSK include constant envelope (to combat amplitude change with fading), efficient spectrum utilization, error rate performance (similar to that of BPSK), and simple implementation. MSK can also be considered a form of offset quadrature phase shift keying (OQPSK) with sinusoidal pulse weighting. When viewed as an OQPSK signal with sinusoidal pulse weighting, the MSK waveform can be expressed as:

$$S(t) = a_1(t) \cos\left(\frac{\pi t}{2T}\right) \cos(2\pi f_c t) \sin\left(\frac{\pi t}{2T}\right) \sin(2\pi f_c t). \tag{5.1}$$

In Equation 5.1, $a_1(t)\cos\left(\frac{\pi t}{2T}\right)$ is the in-phase chip stream waveform with sinusoidal pulse weighting. $a_Q(t)\sin\left(\frac{\pi t}{2T}\right)$ is the quadrature chip stream waveform with sinusoidal pulse weighting and f_c is the carrier frequency. a_1 represents the even chips of the chip stream and a_Q represents the odd chips of the chip stream. a_1 and a_Q can be either $+1$ or -1. Equation 5.1 can also be expressed as:

$$S(t) = \cos\left[2\pi\left(f_c + \frac{b_k}{4T}\right)t + \phi_k\right],\tag{5.2}$$

where $b_k = -a_1 \times a_Q = \pm1$ and the phase ϕ_k, is zero or π, corresponding to $a_1 = 1$ or $a_1 = -1$.

One can see from Equation 5.2 that the MSK waveform has a constant envelope with two signaling frequencies. The higher signaling frequency is $f_c + \frac{1}{4T}$, and the lower signaling frequency is $f_c - \frac{1}{4T}$. The frequency deviation between these two signaling frequencies is $\Delta f = \frac{1}{2T}$, which is the same as that of coherent binary frequency shift keying (BFSK). This signaling technique is referred to as *minimum shift* keying.

In MSK chip modulation, two signaling frequencies are used to represent the change in chip value between successive chips rather than modulating the absolute chip value. For example, if the chip binary stream to be modulated is 100 …, the higher signaling frequency is used to represent the transition from the first to the second chip, whereas, from the second to the third chip, the lower signaling frequency is used since the chip value stays at 0. For MSK, this results in phase continuity in the RF carrier at the chip transition instants. This phase continuity can mitigate interference. MSK modulation schemes can be used when efficient amplitude-saturating transmitters are required.

5.2.3.5 JTIDS Pulse Structures

The Link-16 message data can be sent as either a single pulse or a double pulse, depending on the packing structure. The single-pulse structure consists of 6.4 μs pulse of modulated carrier and 6.6 μs dead time, making a total duration of 13 μs. The double-pulse structure consists of two single pulses which carry the same data but use different carrier frequencies. This repetition provides robustness against fading and/or narrowband interference. The double-pulse structure is illustrated in Figure 5.8. Note that an MSK chip modulator is used for both the single-pulse and the double-pulse structure.

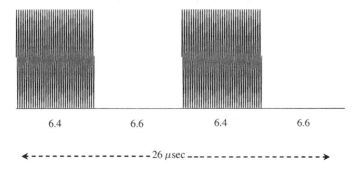

| 6.4 | 6.6 | 6.4 | 6.6 |

←- - - - - - - - - - - - - - - -26 μsec - - - - - - - - - - - - - - - - →

Figure 5.8   JTIDS standard double-pulse structures.

While, the data rate for the double-pulse structure is half that of the single-pulse structure, the average energy per bit is doubled with the double-pulse structure. Keep in mind that JTIDS is not a constant average energy per bit system. In order to create and maintain a more robust transmission for the message header (where the tracking number of the transmitting terminal and the associated SDU information are transmitted) versus the message data, the JTIDS cannot maintain a constant average energy per bit system. Refer to Section 2.2 to see how the probability of bit error detection is related to the energy per bit.

If we start from the chip level, we can express the probability of a chip error in the presence of AWGN as follows (refer to Equation 2.11):

$$pc = Q\left[\left(\frac{2E_c}{N_0}\right)^{1/2}\right],\tag{5.3}$$

where E_c is the average energy per chip.

Since each 5-bit symbol is converted into 32 chips, one can conclude that $E_s = 5E_b = 32E_c$. It is not a straightforward analytic procedure to obtain symbol error probability, p_s in terms of the chip error probability, p_c. However, one can obtain an upper bound, based on a permutation approach that considers how the CCSK sequences are constructed. When dealing with AWGN, the correlation receiver integrates over multiple chips to decode one bit. This results in a probability of bit error rate equivalent to that of the binary antipodal signal, expressed in terms of bit-energy-(not chip energy)-to-noise power density ratio. The probability of bit error rate in terms of bit energy can be expressed similar to Equation 5.3 as $p_b = Q\left[\left(\frac{2E_b}{N_0}\right)^{1/2}\right]$. However, the same concept can't be carried to the symbol error probability since the cross-correlation performed by the demodulator at the symbol level can produce any of the other 31 valid CCSK sequences. Permutation can only produce an upper bound of the symbol error probability.

5.2.3.6 JTIDS Frequency Hopping (FH)

As explained in Section 2.6, in a frequency-hopping spread-spectrum system, the carrier changes pseudorandomly according to a pre-designated PN sequence. FH increased resilience to jamming can come at the cost of lower throughput. At any given moment, slots remain underutilized to allow hopping within the slots at different frequencies. JTIDS uses fast FH; as a refresher, please refer to Section 2.6.2 for the advantages of fast FH. The JTIDS waveform hops pseudorandomly over 51 frequency bins at a rate of around 77 000 hops per second. The 51 carrier frequencies are shown in Table 5.2.

5.2.3.7 JTIDS Security Layers

JTIDS messages have multiple layers of TSEC that include:

- the data and header symbols interleaved;
- 32-chip CCSK sequences scrambled with a 32-chip PN sequence;
- FH over the 51 possible carrier frequencies;
- the starting point of the pulse train pseudorandomly jittered, providing a fourth layer of TSEC – this jitter makes it difficult for a jammer to decide when to turn on the jamming signal.

Table 5.2 The 51 JTIDS hopping frequencies

Frequency number	Frequency (MHz)	Frequency number	Frequency (MHz)	Frequency number	Frequency (MHz)
0	969	17	1062	34	1158
1	972	18	1065	35	1161
2	975	19	1113	36	1164
3	978	20	1116	37	1167
4	981	21	1119	38	1170
5	984	22	1122	39	1173
6	987	23	1125	40	1176
7	990	24	1128	41	1179
8	993	25	1131	42	1182
9	996	26	1134	43	1185
10	999	27	1137	44	1188
11	1002	28	1140	45	1191
12	1005	29	1143	46	1194
13	1008	30	1146	47	1197
14	1053	31	1149	48	1200
15	1056	32	1152	49	1203
16	1059	33	1155	50	1206

5.2.4   Enhancements to Link-16

The success of Link-16 resulted in constant enhancements to its implementation over the years. These enhancements include the leveraging of RS code erasure correction capabilities, and the use of different mutations of the 32-chip CCSK shown in Table 5.1. In this section, the increase of the waveform robustness from these two enhancements is presented.

5.2.4.1   Error and Erasure Coding

The concept of erasure coding was explained in Section 3.2.4 in the context of concatenated codes (Figure 3.11). With concatenated codes, the failure of the inner code (convolutional code) introduces erasure bursts to a stream of symbols. The interleaving process randomizes the erasure bursts and the outer code (being RS code with erasures) can then correct the symbol erasure.[4] When it comes to Link-16, RS coding with erasure derives from the context of soft-decision demodulation. Please refer to Section 2.2, where we discussed how binary antipodal signals can be decoded with soft-decision demodulation (see Figure 2.11).

The introduction of soft-decision demodulation to Link-16 resulted in the CCSK symbol demodulator behaving in one of two ways. It either decides which symbol was received or the demodulator is unable to make a decision and marks the symbol as an erasure.[5] RS code is then utilized to correct errors and erasures simultaneously, resulting in more error

---

[4] We also discussed packet erasure coding in Section 3.2.5 and Appendix 3.A.
[5] Recall from Section 2.2 that marking a symbol as an erasure requires defining an erasure threshold. The relationship between soft-decision demodulation and RS erasure capability is an interesting research area of study.

control capabilities.[6] Notice that since Link-16 is commonly used with fast movers, where channel fading is a common occurrence, the introduction of a combined error and erasure RS coding and soft-decision demodulation yields considerable gain.

Recall from Section 3.2 that the error correction capability of a linear binary block code is characterized by the minimum Hamming distance d_{min}. If a codeword is received with a single erased bit, one can consider that all valid codewords are separated by a Hamming distance $d \geq (d_{min} - 1)$. If there are e erasures in a received codeword, all valid codewords are then separated by a Hamming distance of $d \geq (d_{min} - e)$. Hence, the effective minimum Hamming distance between valid codewords, when there are e erasures in the received codeword, is $d = (d_{min} - e)$ and the number of non-erasure errors in that codeword that be corrected is:

$$t_e = \left\lfloor \frac{d_{min} - e - 1}{2} \right\rfloor. \tag{5.4}$$

In other words, a linear block code can correct t_e errors and e erasures as long as $(2t_e + e) < d_{min}$. Hence, for a given value of d_{min}, twice as many erasures as errors can be corrected. The amount of coding gain that can be obtained from soft-decision demodulation and erasure coding is dependent on how symbol erasure is determined. With JTIDS, the RS(31,15) code can correct any combination of symbol errors, t_e, and erasures, e, where $(t_e + e) \leq 16$. The (16,7) code can correct any combination of errors and erasures where $(t_e + e) \leq 9$. The RS(31,15) code is generated by the function G(X) whose coefficients are elements of the Galois field of (5^2). The root of the primitive polynomial $X5 + X2 + 1 = 0$, and the generating function is given by[7]:

$$G(X) = \prod_{i=1}^{i=16} (X + a^i). \tag{5.5}$$

At the receiving side, the decision of which 5-bit symbol was originally sent is based on the cross-correlation values obtained when the de-scrambled sequence is cross-correlated with the 32 valid CCSK sequences. Figure 5.9 shows this process, where the de-scrambled received sequence is $\hat{S}$ and the 32 valid CCSK sequences are $S_0 - S_{31}$. Suppose symbol S_0 was sent and there is no chip error in the de-scrambled sequence, then the cross-correlation is given by:

$$\Re i = \begin{cases} 32, & i = 0 \\ h_i, & i = 1, 2, \ldots 31. \end{cases} \tag{5.6}$$

In Equation 5.6, i is the index of the valid CCSK sequences, h_i is the cross-correlation value between S_0 and all other valid CCSK sequences aside from the zero index sequence. $h_i \leq H$, where H is the maximum off-peak cross-correlation value (the maximum cross-correlation value of a valid CCSK sequence with another valid CCSK sequence). H is dependent on the selected 32-chip sequences. If we use the shifting approach demonstrated in Table 5.1, then the starting sequence can also influence the value of h_i. It can be shown

---

[6] The increase in error control coding capabilities with soft-decision demodulation is sometimes measured in the literature in terms of coding gain in dB, which is similar to the dB gain discussed in Section 2.2 with the use of higher-dimension signal constellations.

[7] Please refer to the references of Chapter 3 for more details about primitive polynomials and generating functions.

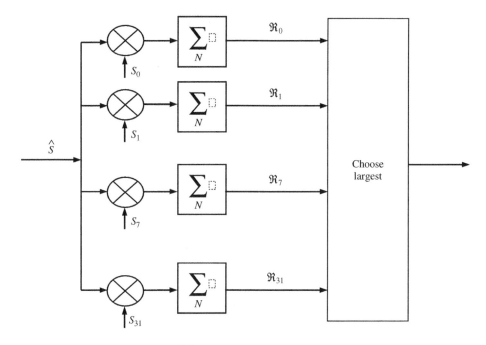

Figure 5.9   CCSK symbol demodulation.

that for the 32-chip starting sequence in Table 5.1, h_i has the values of $-4, 0, +4$. In other words, the maximum off-peak cross-correlation value $H = 4$.

As explained in Section 2.2, one needs to define an erasure threshold for the symbol demodulation. With the Link-16 CCSK case, if the erasure threshold T is too small (e.g., $T \leq 4$), the result will be similar to that of RS coding with no erasure. If T is too large (e.g., approaching 32), the loss of information (decoding a symbol as an erasure that would have been decoded correctly) will result in making RS code performance worse.[8] Therefore, an optimal erasure threshold T_{opt} is required in order to obtain the best gain. Finding this optimal erasure threshold T_{opt} depends on the noise model and the CCSK sequence used. The literature is full of material about designing soft-decision demodulation and erasure error correction codes jointly. Reference [2] shows a Link-16 specific study on finding T_{opt} based on a Nakagami fading channel model.

5.2.4.2   Orthogonal 32-Chip CCSK

In the 32-chip CCSK starting sequence presented for JTIDS in Table 5.1, S_0 and its 31 cyclically shifted counterparts, $S_1 - S_{31}$, are not orthogonal. From Equation 5.6, we can see that the off-peak cross-correlations have values other than zero (which indicate a lack

---

[8] With CCSK, even in the absence of noise, making the erasure threshold large can result in decoding a CCSK sequence incorrectly. One must approach soft-decision demodulation and erasure error correction coding as one multifaceted optimization problem.

of orthogonality). Refer to the discussion of orthogonal signals in Chapter 2, where we established that orthogonal signaling can improve dB gain. You may also reference our discussion in Section 2.6.1 about finding optimum chip codes for tactical radios. With Link-16, one can expect to find an orthogonal 32-chip CCSK set which has maximum off-peak cross-correlation values close to or less than zero. Interestingly, such orthogonal sets, allow for seven (instead of six) chip errors in the received, de-scrambled sequence without making a symbol error.

5.2.5 Concluding Remarks on Link-16 Waveform

You can find further information regarding the Link-16 in public domain literature. From this book, you should understand, and hopefully appreciate, the extent of the technological advancement made with the development of this waveform in the 1970s. Other than the US National Aeronautics and Space Administration (NASA) space program, Link-16 was the first project to use concatenated codes. The combination of FDMA, TDMA, CDMA, and different modes of interaction created such exceptional communication capabilities for fast movers (such as fighter jets) that it remains unmatched. Considering the abundant challenges that commercial wireless researchers encountered with 3G technology in the 1990s, to ensure communications between a moving end user (in a car or a train) and a cellular base station, one should appreciate the capabilities achieved by Link-16 decades earlier. The Link-16 waveform guaranteed communication between jet fighters, traveling at speeds up to Mach three and at distances up to 500 km in the presence of jamming. This was truly an astounding accomplishment.

The transition toward IP-based tactical theater, and the need to continue using non-IP waveforms such as Link-16 necessitated the use of gateway capabilities between Link-16 and IP-based core networks. Some of the challenges arising from this transition include synchronization of IP packets to the legacy waveform message structure, assuring services, packetizing, managing overhead, and handling throughput.[9] The defense industry relies on ensured IP services (e.g., using IntServ – integrated services – signaling) in order to secure delivery assurance and synchronization between non-IP peripheral nets. With the transition to HAIPE-encrypted all-IP networks, DiffServ – differentiated services – has to be used over the encrypted core instead of IntServ. This is currently generating many of its own challenges, as you will see in Part III of this book.

Link-16 is not the only successful non-IP waveform, although it is the best of its generation. The US Army maintains widespread usage of the EPLRS waveform which has excellent FH and anti-jamming capabilities. The newer generation of EPLRS is IP-capable through a gateway with a fully capable IP router that can represent each node in the non-IP net with an IP address to the IP core network. The US Army and other US and coalition forces also utilize the SINCGARS waveform. These waveforms are briefly described below and the pre-GIG architecture is presented to demonstrate how the GIG vision has been evolving over the years. The joint tactical radio system (JTRS) which is producing IP-based waveforms, is replacing all these legacy waveforms with IP-based MANET waveforms.

---

[9] The commercial sector has had similar challenges with 3GPP backhauling over IP, which created the need for synchronized IP.

5.3 EPLRS Waveform

The EPLRS waveform is a high security anti-jamming radio set that provides user-to-user data communications. A single EPLRS net can be reconfigured as a local area network (LAN), wide area network (WAN), or range extension to provide a greater area of coverage in the war theater. Configurations rely on an NCS which allocates the resources according to the commander's need. Multiple interfaces adorn the radio, allowing capabilities such as voice communications, data communications, and position location reporting.

While Link-16 met the airborne platforms' most urgent needs for communications capability between high-speed nodes, EPLRS satisfied the needs of the ground forces through building hierarchal nets and supporting situational awareness requirements.

The underlying communications capability of this radio relies on TDMA architecture as shown in Figure 5.10. In this figure, a TDMA epoch of 64 seconds (shown in the ring) is divided into 256 frames, each frame being 0.25 seconds. Each frame is then divided into 128 slots (each group of eight slots forms a logical timeslot), each of which is approximately 2 ms. Transmission takes 800–1000 μs; the remaining time is needed for processing overhead and message validation.

A major update to EPLRS radios was the development of the Internet controller (INC), an IP gateway to an EPLRS net. Through this technology, each node can be detected by any IP router. The EPLRS gateway then handles the IP address mapping to the node. This enhancement allowed for the successful integration of multiple-access on-the-move extended range hierarchal nets into an IP network while simultaneously creating the capability to flow IP traffic seamlessly to the EPLRS users.

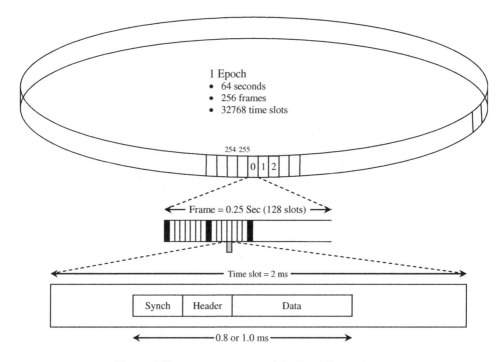

Figure 5.10 TDMA structure of the EPLRS waveform.

5.4   SINCGARS Waveform

The SINCGARS waveform uses a non-IP low bandwidth voice coding format that has been implemented in many legacy tactical radios and is known as the combat net Radio (CNR). SINCGARS also supports low rate data communications. SINCGARS radio terminals come in a variety of forms, including vehicle-mount, backpack, airborne, and handheld.

SINCGARS uses 25 kHz channels in the VHF FM band, from 30 to 87.975 MHz. It has a single frequency and can be configured to operate in a FH mode. This mode hops 111 times a second.

The precursor to SINCGARS was the Vietnam War era radios which were synthesized single frequency radios known as PRC-77 and VRC-12. The early deployments of SINC-GARS interoperated with these radios. In aircraft radio, SINCGARS phased out the older tactical air-to-ground radios (ARC-114 and ARC-131). It was the Soviet Union's ability to jam these old radios in Vietnam that pushed the USA to develop spread-spectrum FH radios such as EPLRS, SINCGARS, and Link-16.

Much like Link-16 and EPLRS, the SINCGARS waveform is a great success. There have been over half a million SINCGARS radios deployed for the US and coalition forces since their introduction in the late 1970s. Several system improvements that introduced integrated voice encryption, and additional data modes were added to the waveform implementation over the years. A lightweight handheld model was also introduced.

The radio based combat ID (RBCI) is an important addition to the legacy radios explained above. This feature is essential in reducing friendly fire incidents. Before dropping their payload, fighter pilot radios can query (send a request signal) over the area below and all friendly forces radios reply with their position locations. The pilot can then see, on a map, if the targeting area includes any friendly forces and abort the mission if needed.

5.5   Tactical Internet (TI)

By the late 1990s, the US army had reached the predecessor architecture to the GIG, which is referred to as the Tactical Internet. TI relies on the EPLRS and SINCGARS radios to create reachability of seamless IP flow to the lower echelons. This architecture is briefly described in this section.

After the success of the EPLRS radio and mobile subscriber equipment (MSE), a circuit-switch-based reach-back tactical backbone network, the US army successfully digitized the battle-space flow in the early 1990s. The motivation behind this concept is to provide seamless communications throughout the forces. This architecture required a network of horizontally and vertically integrated digital information networks that supported new systems (such as friendly and enemy forces tracking, automated weaponry systems). The TI provided reliable, seamless, and secure communications for all the US Army tactical users.

The term *Tactical Internet* was used to describe this integrated battle-space communications hierarchy of networks since it coincided with the widespread use of the commercial Internet. The term is appropriate due to functional similarities between the commercial Internet and the Tactical Internet communications infrastructure being based on moving to all-IP systems as a predecessor to the GIG. A key feature is the ability to exchange non-internet-based messages (such as that of Link-16 and the EPLRS radio) using the commercially based Internet Protocol. The use of IP was a mandate in order to cut cost and leverage the anticipated boom in commercial technologies.

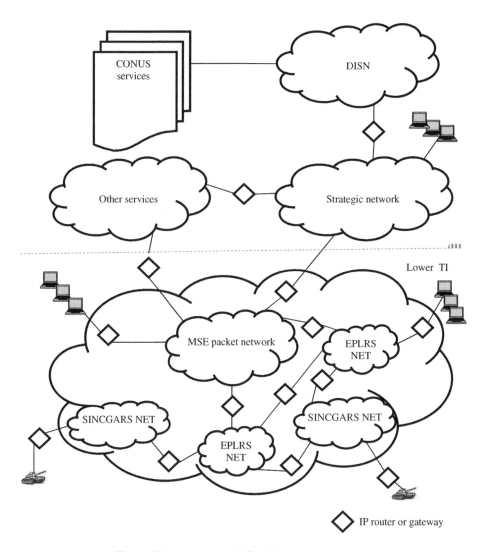

Figure 5.11   Integrated digital information network.

One needs to understand the TI in the context of the existing IP-based defense information systems network (DISN). The DISN uses commercial-off-the-shelf (COTS) IP routers, and leverages the fiber optical infrastructure in the USA and around the world (in some cases the DISN can use high capacity satellite links to create teleports anywhere in the world). The DISN has gateways to the commercial Internet and offers to the TI and other US government networks the continental United States (CONUS) network services. The DISN is the top layer in the Integrated Digitized Information Network concept and the TI is the lower layers which reach the lower echelons. As shown in Figure 5.11, the US Army has a strategic network layer that can interface to other services and can provide the lower TI with services. All the interfacing points in this architecture are COTS IP routers or IP gateways between an IP network and a non-IP network. The EPLRS radio IP gateway described above plays a major role in making this architecture possible.[10]

At the brigade level and below, the TI extends the US Army battle command systems to the soldier and weapons platform. The TI passes battle command and situational awareness data.[11] As the TI evolved, it started to provide tactical, mobile, and multimedia capabilities. These additional functionalities include voice, video, and data communications. Simultaneously, the TI began providing routing and network services between this expansive hierarchy of networks.

Notice in Figure 5.11 how the lower TI relies on an EPLRS-based backbone NET above the SINCGARS CNR layer (EPLRS can provide an area of coverage much wider than that of the SINCGARS). The MSE layer has high capacity microwave and satellite links and can be deployed in the theater to provide reach-back capabilities to the EPLRS and SINCGARS layers. All these networks are integrated using COTS routers. The integrated transport capabilities of these legacy non-IP radios is, at times, incompatible with each other. This is a key hindrance to moving information seamlessly among the different nodes and platforms. This architecture achieves seamless information transfer horizontally and vertically across the battle-space. The mandate for IP at all touch points made this integration possible. Open standards protocols such as TCP/IP can then be used seamlessly. Programs were then created to develop tactical multinet gateways (TMGs) and the EPLRS and SINCGARS INCs[12] to provide flawless connection between legacy systems and IP routers. This resulted in the ability to send messages between the different layers and nets of the tactical battle-space network and create IP-based services (e.g., client/server architecture). Today, INCs are a single board physically incorporated into an EPLRS or a SINCGARS mount radio terminal (that dedicates this radio as the INC) and hosts IP addresses for all other terminals in the net. Similarly, an INC capable EPLRS radio has an IP interface that hosts IP addresses for all nodes in the EPLRS net.

Starting with these legacy waveforms, the US Army developed and deployed enhancements which led to the creation of the TI. Enhancements to the primary army tactical communications system – EPLRS, SINCGARS, and MSE – made it possible to cleanly move the ever-increasing amount of data associated with command and control applications.

Enhancements to SINCGARS included reduced co-site interference; improved error detection and correction; reduced network access delay; and a GPS interface to obtain accurate time and position location. Collectively, these improvements greatly extended the effective data communication range and increased information throughput from 1.2 kbps to 4.8 kbps.

---

[10] The US Air Force had a parallel concept, a program named battlefield airborne communications node (BACN) which allowed real-time information exchanges between different tactical data link systems. A BACN node can interface to an IP-based network and a Link-16 net seamlessly. Currently, the US Air Force has the IP-based airborne network concept as the tactical GIG extension, which is parallel to the US Army WIN-T as the tactical side of the GIG. WIN-T is the replacement of the MSE system mentioned here.

[11] Situational awareness data can include Blue Force tracking (location of friendly forces) and Red Force tracking (locations of enemy forces). The first army-wide application of this system was in 2003 in the second Gulf War. This technology gave the US convoys the ability to move fast toward Baghdad with minimal causalities. Even during sand storms, US army vehicles had maps (similar to your car GPS map) with red dots and blue dots giving each soldier a near real-time situational awareness battlefield data.

[12] Both EPLRS and SINCGARS have their own INC version.

SINCGARS SIP (system improvement program)[13] radios are able to reliably pass data at 4.8 kbps up to a range of 35 km.[14]

EPLRS system upgrades increased the throughput of individual EPLRS users from 4 to 12 kbps.

Parallel to the SINCGARS and EPLRS upgrades described above, the MSE system has its own upgrades as well. One upgrade to the MSE system is the change of its gateway routing protocols from the exterior gateway protocol (EGP) to the border gateway protocol (BGP). This substantially reduced the bandwidth required to exchange routing information between routing devices in different networks.

The TI supports several key services such as e-mail, directory, network management, and security which are integral to its value.

Figure 5.12 (taken from reference [3]) shows a simplified view of the TI components at the brigade level. Not all the capabilities are shown in this figure. Notice the importance

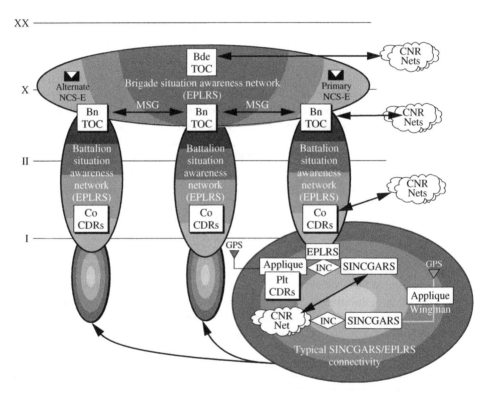

Figure 5.12 TI components at brigade level. Reproduced from http://www.globalsecurity.org/military/systems/ground/internet-t.htm. Figures originally from US government archive.

---

[13] SINCGARS enhancement programs included the SIP and advanced system improvement program (ASIP).

[14] These low data rates may seem laughable considering what we are used to in 3G and 4G commercial wireless bandwidth availabilities. In mobile tactical radio nets, spectrum is scarce and bandwidth is limited due to the lack of fixed info-structure such as base stations (which require radios to relay data to each other, thus reducing throughput efficiency). Another cause is the static configuration of how the physical layer resources are shared in these legacy radios.

of situational awareness hierarchy. At the brigade tactical operational command (TOC), there exist CNR nets (for voice communications) and a brigade situational awareness network. There is an EPLRS network manager (referred to in Figure 5.12 as the NCS-E) that controls all EPLRS networks is at the brigade level. Multiple battalions (under the same brigade) have their own battalion situational awareness networks as well as their CNR voice communications networks. The company commanders are member nodes of their battalion network. At the company level, notice the top INC (the IP gateway) being a touch point between the EPLRS and the SINCGARS nets (as mentioned above, all touch points must be IP capable). Multiple SINCGARS nets are needed at the company level for CNR voice communications as well as data communications. Notice the interface to GPS through the applique box. Each SINCGARS and EPLRS radio must report its location up to the brigade level in order for the situational awareness server to have locations of all friendly forces.

Security is critical in the TI architecture. Initially the TI was deployed at the secret level, totally isolated from the commercial Internet. The introduction of HAIPE encryptions allowed the creation of multisecurity enclaves within the TI since the HAIPE tunnels added a level of security that prevented intruders from being able to communicate to the TI users or make sense of encrypted packets if intercepted. HAIPE encryption allows unclassified users to use the TI to access unclassified computers connected to the commercial Internet. The CONUS services shown in Figure 5.11 include classified and unclassified servers. The availability of multilevel security services enabled greater network flexibility. You can see in the literature references to the secret Internet Protocol router network (SIPRNET) and the non-secure Internet Protocol router network (NIPRNET) as the CONUS parallel networks with satellite-based teleports at the tactical level for touch points with the TI.

Another key component of the TI is network operations (NetOps), which is composed of network management, planning and control of the TI. The TMGs and INCs must be technically controlled in order to successfully define the radio nets map to IP addressing. The addition of a new node to any net means the reconfiguration of the TMG and INC associated with that node. The automation of network management capabilities is introduced at TI brigade-and-below. Commercial protocols such as simple network management protocol (SNMP), with SNMP agents, are implemented by communications elements. In some cases, proxy agents (e.g., SINCGARS and EPLRS proxy agents) are hosted on the applique as shown in Figure 5.12.

The improvements to the TI paved the way for the GIG vision to be realized. The gradual replacements of TI facets (such as replacing MSE with WIN-T assets, and replacing non-IP radios with IP radios) will eventually lead to an all IP infrastructure. The use of commercial IP with the TI provides the required level of seamless connectivity between the different networks. This connectivity is so effective that the replacement of various TI components will not affect the US Army's ability to conduct war. With IP, we must increase data capacity because of increased demand and the additional header information that must be added to each message as elaborated in the following section. The Internet routers exchange routing and status information, which also increases data capacity requirements. Even with the improved data capacity of SINCGARS, EPLRS, and MSE, these systems provide relatively small communications trunks when compared to fixed-site commercial systems. In modern warfare, there is an ever-increasing demand for bandwidth. With the trend moving toward more automated unmanned military machines, machine-to-machine communication becomes vital, and bandwidth is always in short supply. The GIG vision does not solely pertain to creating seamless IP connectivity, but also to generating higher bandwidth for communications on the move at all levels of the military hierarchy.

5.6   IP Gateways

Link-16, EPLRS, and SINCGARS nets have IP gateways. These non-IP networks can be considered stub-networks[15] to an IP core. The upside is the ability to make real time applications such as voice and situational awareness function on a variety of different non-IP protocols and IP networks. The downside includes issues such as low throughput efficiency and uncertain end-to-end QoS. Let us discuss some of these issues.

5.6.1   Throughput Efficiency

As you can see from Figure 5.2, the Link-16 data or information part can occupy about one half of the timeslot, leaving room for jitter, synchronization, time refine, and propagation. The single pulse structure in Figure 5.8 illustrates how off-time can be comparable to the time of pulse transmission with the single pulse (with the double pulse, the ratio of information to total duration of the double pulse is about 0.25). The need for the overhead of forward error correction (FEC) exists. The message validation requires reducing throughput efficiency by another factor of 2. The problem of throughput efficiency within a non-IP net pertains to this range (about 6–20% at best). As gateways allow non-IP payloads to traverse an IP core network, the end-to-end throughput is drastically affected. At the IP core network, throughput efficiency can be very low since the ratio of the IP-packet payload to headers can be small. These old non-IP waveforms were designed in the 1970s (before we even imagined streaming video) and are designed for short messages and small payload voice. When these small messages are encapsulated in IP packets with large overhead, the ratio of payload to header can be small. Let us assume that a non-IP message reached the gateway with data, header, and other synchronization portions. This message is supposed to traverse the IP core over a single hop (multiple hops will further exaggerate this problem). If we assume that the message has an information portion I and a header portion $H1$, we identify throughput efficiency η for this message as $\eta = \frac{I}{I+H1}$. The non-IP message is encapsulated in an IP packet with IP, UDP, and RTP headers. If we assume the IP, UDP, and RTP headers add up to $H2$, then $\eta = \frac{I}{I+H1+H2}$. As the IP packet traverses the IP stack down, it gets divided into multiple MAC frames. If the MAC frame size is M, we will get $\left\lceil \frac{I+H1+H2}{M} \right\rceil$ frames. Each MAC frame gets its own overhead of $H3$, making $\eta = \frac{I}{I+H1+H2+MH3}$. Then the DLL can add overhead of rate $R1$ for FEC. The RF layer can add overhead of $R2$ for RF synchronization. If we assume that $R = R1 * R2$, we can have R relatively small, making,

$$\eta = \frac{I}{I + H1 + H2 + MH3} * R. \tag{5.7}$$

These are examples of overhead, and as we further delve into the topic, we can see how the end-to-end effective throughput efficiency can be further reduced. Control traffic, such as link state updates, further reduce the effective throughput of these networks. If you have

---

[15] The term "stub-network" is used to indicate the legacy network being peripheral to an IP core.

access to a testbed for wireless networks and access to a tool such as WireShark, you can run analysis that shows how inefficient wireless IP networks are. Some literature claims that for VoIP (where the actual voice information part is small in comparison to the entire packet size), traversing over multiple tactical MANET hops can result in an effective throughput efficiency that can be as low as 0.01 (only 1%). For these reasons, some researchers question the suitability of the IP stack for tactical MANET. The jury is still out on how to optimize the effective throughput in tactical MANET. In the third part of this book, you will be introduced to concepts, such as packet payload concatenation, that help to increase the effective throughput efficiency of tactical MANET.

5.6.2   End-to-End Packet Loss

Let us assume a scenario where two legacy nets are communicating over an IP core network. Here the probability of packet loss for the first legacy net is p_1, the probability of packet loss in the core network is p_2, and the probability of packet loss in the second legacy net is p_3. The probability of a packet being successfully transmitted over the first legacy net is $(1 - p_1)$. The probability of a packet to be successfully transmitted over the core network is $(1 - p_2)$ and the probability of a packet to be successfully transmitted over the second legacy net is $(1 - p_3)$. One can find that the probability, P of a packet being successfully transmitted over the two legacy nets and the core network is $P = (1 - p_1)(1 - p_2)(1 - p_3)$, which gives:

$$p = 1 - p_1 - p_2 - p_3 + p_1 p_2 + p_1 p_3 + p_2 p_3 - p_1 p_2 p_3. \tag{5.8}$$

End-to-end packet loss and large overhead-to-information ratio are some of many QoS issues that need to be considered while using IP touch points with the TI and with tactical networks in general. Other issues include end-to-end delay and delivery assurance while using techniques such as DiffServ. The introduction of HAIPE encryption adds another dimension to these challenges. The third part of this book addresses these topics in more detail.

5.7   Concluding Remarks

5.7.1   What Comes after the GIG?

For the GIG vision, we are aiming beyond the ability to achieve seamless IP flow at all levels of tactical networks. This vision includes the introduction of high bandwidth MANET networks as another stop along the continuing path toward ever-evolving battlefield communications capabilities. These evolving capabilities include the deployment of sophisticated cognitive radios which are more efficient at spectrum utilization than the JTRS generation of tactical MANET radios. A battlefield full of unmanned sensors, unmanned aerial and ground vehicles, and unmanned automated weaponry systems requires ever-evolving communications capabilities.

5.7.2   Historical Perspective

The Cold War was a major driver for developing sophisticated tactical radios. The early versions of Link-16 were attempted in the 1960s with two parallel programs for the US

Air Force and the US Navy. However, it was not until 1976 that a single US Air Force program owned a Link-16 device. Early warning aircraft were first equipped with Link-16 terminals, which were large and could not fit in a small aircraft. Smaller Link-16 terminals were finally in operational testing in the 1980s, and by the 1990s the F15s were equipped with Link-16 terminals. Different versions of the waveform were adapted by the Navy and joint forces, making Link-16 one of the most successful and widely used waveforms in networking land, sea, and air platforms.

The technological leaps of the defense communications industry have been the foundation of the successes for the commercial communications industry. For example, DARPA (Defense Advanced Research Projects Agency) net (a defense industry technology) was the basis for today's modern internet (a commercial industry technology leap). CDMA (which was a defense technology) was utilized by the 3GPP standards for the commercial wireless industry. Likewise, through the creation and development of WiFi and WiMax, the commercial industry was able to create waveforms that are accessed by mobile users. Many of today's commercial communications error resilience techniques rely on concatenated codes; these techniques originated from the US space program and the US defense industry. The list goes on and on; the developments of the defense industry are inevitably passed down and are further developed by the commercial sector.

During the 1990s and the early 2000s, the commercial sector reached many milestones in technological development. With a massive user-market[16] and heavy investment, the commercial sector advanced quickly and the defense industry appeared to lag behind in scalability, bandwidth, and rich applications. This book emphasizes how the defense industry can now utilize the advancements of the commercial sector, while adhering to tactical security constraints, to procure enhanced technology.

Bibliography

1. Northrop Grumman Corporation (2001) *Understanding Link-16: A Guidebook for New User*, San Diego, CA, Northrop Grumman, September 2001.
2. Kao, C.-H. (2008) Performance analysis of a JTIDS/Link-16 type waveform transmitted over slow, flat Nakagami fading channels in the presence of narrowband interference. PhD dissertation. Naval Post Graduate School.
3. http://www.globalsecurity.org/military/systems/ground/internet-t.htm. (accessed June 2012).
4. Gonzales, D., Norton, M., and Hura, M. (2000) Multifunctional Information Distribution System (MIDS) Program Case Study, Project AIR FORCE, RAND, DB-292-AF, April 2000.
5. Asenstorfer, J., Cox, T., and Wilksch, D. (2004) Tactical Data Link Systems and the Australian Defence Force (ADF) – Technology Developments and Interoperability Issues, Information Networks Division, Information Science Laboratory, DSTOTR 1470, revised February 2004.
6. Bell, C.R. and Conley, R.E. (1980) Navy communications overview. *IEEE Transactions on Communications*, **COM-28** (9), 1573–1579.
7. Wilson, W.J. (2001) Applying layering principles to legacy systems: link-16 as a Case Study. *Proceedings – IEEE Military Communications Conference*, **1**, 526–531.
8. Golliday, C.L. Jr. (1985) Data link communications in tactical air command and control systems. *IEEE Journal on Selected Areas in Communications*, **SAC-3** (5), 779–791.
9. Brick, D.B. and Ellersick, F.W. (1980) Future air force tactical communications. *IEEE Transactions on Communications*, **COM-28** (9), 1551–1572.

---

[16] At the time of writing this book, it is estimated that there are 4 billion cellular phones in use in our planet.

10. Pursely, M.B., Royster, T.C. IV, and Tan, M.Y. (2003) High-rate direct-sequence spread spectrum. *Proceedings IEEE Military Communication Conference*, **2**, 1101–1106.

11. Yoon, Y., Park, S., Lee, H., Kim, J., and Jee, S. (2011) Header compression method and its performance for IP over tactical data link. *International Journal of Energy, Information and Communications*, **2** (2).

12. (1997) Enhanced Position Location Reporting System (EPLRS). Scientific and Technical Reports, Raytheon Systems Company, December 1997.

13. Kopp, C. (2005) Network Centric Warfare Fundamentals – Part 3: JTIDS/MIDS. Defence Today Magazine (Sep 2005), pp. 12–19.

6

IP-Based Tactical Waveforms and the GIG[1]

We concluded Chapter 5 with a discussion of the US Army Tactical Internet (TI) architecture which consists of non-IP tactical radios, digitized core networks, and IP touch points. For the US Department of Defense (DoD) networks, the gradual transition of the TI, toward the tactical Global Information Grid (GIG), meant replacing non-IP radios with IP-based radios. The Joint Tactical Radio System (JTRS) program continues to play a major role in making IP-based tactical radios a reality within all DoD networks. Tactical core networks[2] are also gradually moving toward all IP-based wide area subnet waveforms. Figure 6.1 shows the lower echelon architecture of the GIG (we will refer to it as the tactical GIG) as envisioned by the US Army.

6.1 Tactical GIG Notional Architecture

Instead of IP touch points with the TI, the tactical GIG has a seamless IP flow with gateway nodes between multiple subnets. As you go through the detailed explanation below, please notice the parallels between the soldier radio waveform (SRW) and wideband network waveform (WNW) in the tactical GIG and the single channel ground and airborne radio system (SINCGARS) and enhanced position location reporting system (EPLRS) radios with the TI covered in Section 5.5.

The lower layers of the tactical GIG (brigade and below) have tiers of IP-based subnets (islands of MANET). Figure 6.1 depicts a notional view of such layered islands (hierarchical subnets) of MANET. There is the SRW tier. This tier can have two s, one for soldier-to-soldier communications and one for networking sensors. Above that tier is the WNW tier which also has two sub-tiers. One sub-tier forms local subnets for vehicle-to-vehicle communications and the other is for global connectivity, which generates a single subnet over the entire theater. Note that at each tier can have multiple subnets, each with

---

[1] This chapter was reviewed by C. David Young. See references [1–6].

[2] Within the US Forces, there are different networks that are considered the tactical extension of the GIG. These networks include the US Army WIN-T (Warfighter Information Network-Tactical), the US Navy ADNS (Automated Digital Network System) and the US Air Force AN (Airborne Network).

Tactical Wireless Communications and Networks: Design Concepts and Challenges, First Edition. George F. Elmasry.
© 2012 John Wiley & Sons, Ltd. Published 2012 by John Wiley & Sons, Ltd.

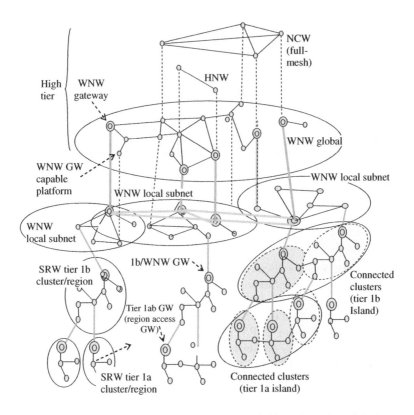

Figure 6.1 Notional view of the tactical GIG with hierarchal subnet islands.

different frequencies (except for the global subnet), which form islands of MANET. Some selected nodes can have multichannel capability to access different subnets and work as gateways between the subnets (these nodes are denoted with two circles in the Figure 6.1). Above these layered subnets, comes the division level layer. This layer has another wireless mobile core network (the US Army Warfighter Information Network-Tactical (WIN-T)) which itself can have a mix of fixed nodes and tactical command mobile nodes. This tactical core network can use waveforms, such as the high-band networking waveform (HNW)[3] and the network centric waveform (NCW), which offer satellite communications on-the-move (OTM); in addition, WIN-T utilizes microwave links for linking stationary or on-the-quick-halt nodes.

Unlike commercial wireless IP-based networks, the tactical wireless IP-based network nodes can be all mobile since there is no fixed infrastructure and the end user nodes are part of the infrastructure. Gateway nodes (with multichannel radios or multiple waveforms) are not a single point of failure since a subnet can have multiple nodes that can relay traffic between the tactical GIG hierarchal layers to create full connectivity anywhere in the theater.

---

[3] The HNW waveform offers a myriad of capabilities, including point-to-point and point-to-multipoint. An aerial vehicle, equipped with a directional antenna pointed to the ground, can be used to create a fully connected multiple access subnet.

Figure 6.2   Notional view of a tactical GIG node with the HAIPE separating the plain text and the cipher text IP layers.

The tactical GIG is much more complex than the TI. The JTRS radios, with their dynamic resource allocation, are more complex than SINCGARS, EPLRS, or Link-16, which have static resource allocation. WIN-T, with its high capacity communications OTM, is not as simple as a fairly static core network such as mobile subscriber equipment (MSE).

Another important aspect of the tactical GIG is the node architecture at all levels. As mentioned in Chapter 1, National Security Agent (NSA) has mandated the use of High Assurance Internet Protocol Encryption (HAIPE) with the GIG. A tactical node must have the internal notional architecture shown in Figure 6.2. Each security enclave (plain text subnet) must have its own HAIPE device to transfer data through a cipher text core network. This creates a plain text (red) IP layer, separated by HAIPE from a cipher text (black) IP layer. NSA requires that no information be shared between the plain text enclaves and the cipher text core, with the exception of the type of service (TOS) byte in the IP header. This is referred to as red/black separation, where passing congestion information from the black (cipher text or encrypted) core to the red (plain or unencrypted) enclaves (where admission control is needed) is prohibited. Passing topology information is also restricted. However, some topology information could be necessary for some protocols such as multicast since the cipher text core can use multiple access waveforms (a single packet over the air can reach all nodes in a multiple access waveform).[4] For multicast to work efficiently at the plain text IP layer, one may need to know which destinations in a multicast address are reachable over the same waveform.[5] Another HAIPE constraint is the encryption key (which is downloaded onto a given radio based on its intended deployment). Prior to a node joining a different subnet with a different encryption key, it needs to be shut down, a new encryption key needs to be loaded, and then the node needs to reboot. This constraint eliminated one of the best advantages of MANET radios by greatly limiting roaming.

This architecture applies to all GIG nodes. In the tactical GIG, this architecture applies only to large command and control (C2) nodes; each plain text enclave can be a LAN with a LAN router and hundreds of end users. Commercial-off-the-shelf (COTS) routers (e.g., Cisco or Juniper) are used for the cipher text IP layer where multiple waveforms, from different radio types, can be used to form multiple links to different nodes at different tiers. The same node architecture also applies to a lower echelon node. The node, using a single channel JTRS SRW with a single security enclave, becomes a single IP port. Note that within JTRS, all the layers are developed in software (SW), making JTRS a software defined radio (SDR). Also, JTRS radios (WNW and SRW) have full IP layer capability which makes them more capable than a simple radio; they are fully capable network nodes.

---

[4] The multicast over HAIPE problem is solved with the WNW waveform. However, the solution is specific to the WNW waveform and neither implemented across the GIG or in public domain.

[5] Some HAIPE implementations map plain text multicast addresses to multicast tunnels formed between HAIPE devices over the cipher text core. This, however, did not eliminate the need to tie topology formation to the multicast tunnels formed by the HAIPE devices.

6.2 Tactical GIG Waveforms

In the GIG architecture presented above, the most important and complex waveform is the WNW. The WNW and SRW waveforms of the JTRS program made IP-based tactical radios a reality. While WNW and SRW work at the brigade level, the WIN-T program at the upper echelon uses two different waveforms that are designed for the HAIPE encrypted core and form a WAN. These two waveforms are the HNW and NCW. We will briefly cover these two waveforms as well.

Since the Link-16 waveform has been in use for decades, many of the details mentioned in Chapter 5 have been made public over the years. The same does not apply to the IP-based MANET waveforms from the JTRS program. This section is constructed with the information that was made available in the public domain at the time of writing this book. We will focus on the challenging MANET problems such as cross layer signaling, physical layer resource allocations, and topology control and on how they are addressed by the JTRS program.

6.2.1 Wide-Area Network Waveform (WNW)

The WNW is an extremely complex waveform. The protocol stack of the WNW is introduced in Figure 6.3. The plain text IP layer gives the radio user access to an Ethernet port where VoIP terminals and IP COTS applications can be plugged in. One can plug an Ethernet switch into this port and form an entire plain text subnet. The HAIPE encryption layer uses an embedded processor for encryption that can adhere to multiple versions of HAIPE. A WNW node can support multisecurity enclaves (multiple plain text subnets), where the HAIPE embedded processor handles more than one parallel implementation of HAIPE in order to support these multiple security enclaves that are entirely separated (implemented

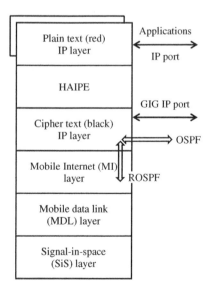

Figure 6.3 The WNW waveform protocol stack layers.

over separated hardware components).[6] Each security enclave maintains a plain text IP layer and application IP port. This is necessary for some platforms that require classified and unclassified applications which must remain separate.

Cross layer signaling is a vital aspect of the WNW design. The mobile internet (MI), mobile data link (MDL), and signal-in-space (SiS) layers have an excellent cross layer signaling design, as detailed below.

6.2.1.1 Cipher Text IP Layer

The cipher text IP layer maintains exceptional implementation of cross layer signaling with the MI layer below. With this cross layer signaling, link state updates (LSUs) are not discovered by the cipher text IP layer; rather, the open shortest path first (OSPF) protocol is altered, creating a protocol called radio open shortest path first (ROSPF). ROSPF then allows for an increase in the number of nodes in a single subnet and prevents the over-flooding of the limited physical layer resources with LSUs.[7] The cipher text IP layer has an Ethernet IP port (marked "GIG port" in the Figure 6.3) which allows the WNW subnet to communicate seamlessly with other networks of the GIG at the cipher text side (to create seamless encrypted core IP flow). The cipher text IP layer is capable of running OSPF over the GIG port and running ROSPF over the radio stack layer. It can also build route tables for internal routing in the WNW subnet and external routing over the GIG port.

Notice that the cipher text IP layer is not standardized across the GIG waveforms (neither is ROSPF) and remains specific to the WNW. In addition, the plain text side has some transport layer protocols[8] that are specific to the WNW. The plain text IP layer of WNW also implements a variation of ROSPF which uses simulated hello packets to overcome the HAIPE red/black separation and efficiently utilize the limited bandwidth with plain text LSU traffic. The application IP port in Figure 6.3 runs OSPF, while the radio stack runs ROSPF.

6.2.1.2 The MI Layer

The MI layer is responsible for the MANET networking (connectivity at the IP layer). It maintains a multilevel link state routing topology of the WNW subnet. The MI ensures that the waveform maintains a connected topology that can be dynamically mapped to the routing table for IP delivery. IP gateway selection is also managed by the MI layer. In addition, the MI layer controls data flow, prevents buffer overflow at the lower layers and implements multilevel queuing to ensure that the waveform meets QoS requirements.

The MI layer implements ROSPF, which forms a topology backbone, whereas nodes use the services of this backbone to send and receive packets. The backbone nodes use a link-state routing algorithm to maintain the integrity of their connectivity and to track the locations of cluster members. With this technique, the MI layer uses ROSPF as mentioned above. Note that ROSPF does not use the OSPF hello protocol for link discovery, LSUs, and so on. Instead, OSPF adjacencies are provided by simulated hello packets containing link state information derived from the subnet routing.

---

[6] Section 7.4 will further clarify the software architecture of the JTRS program and detail the hardware separation.
[7] With high mobility, link states change very frequently. Standard OSPF could consume all available bandwidth as the number of nodes in a subnet increases.
[8] Sometimes this transport layer is referred to as the system of systems common operation environments (SoSCOE).

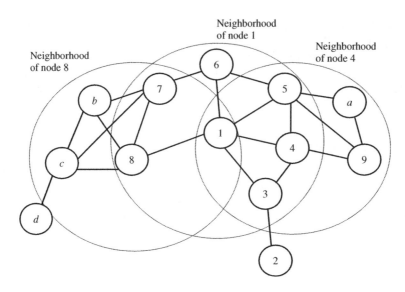

Figure 6.4 Multihop broadcast packet in MANET.

There are different cross layer signaling implementations that can be found in the literature. The use of cross layer signaling is needed to reduce the burden of mobility on the IP layer, OSPF protocol and LSU. The term "MANET extension," at layer 3, is sometimes used to indicate that layer 3 implementation does rely on lower layer information to reduce the burden of control traffic on the limited bandwidth.

6.2.1.3 MDL Layer Architecture

The WNW MDL layer implements unifying slot assignment protocol (USAP), which supports multihop broadcast of packets and controls the subnet topology. Consider Figure 6.4, where a transmitting node broadcasts to all nodes within one hop (within its area of coverage) and packets are relayed until they arrive to all intended destination nodes. Conflict resolution of TDMA slot assignment and reconfiguring slot assignment (per traffic demand of each node) becomes a challenging problem. USAP heuristically assigns the *optimum* number of TDMA slots to each neighboring node and coordinates the activation of these slots to prevent collision. USAP is tied to the heuristics of the higher layers of the protocol stack.[9] A form of cross layer signaling between the MI, MDL, and SiS layers makes traffic demand, from the upper layers, and the link condition, influence slot allocation. The manner by which USAP functions at each node affects the heuristics of the upper stack layers. This is referred to as convergence of the stack layers.

Understanding the MDL layer architecture is essential to understanding how the WNW waveform functions. The MDL layer plays a major role in flow control, topology control, creating fault tolerance topology, and minimizing the number of hops between communicating nodes. Topology control, and its relation to TDMA slot allocation, is detailed in this chapter.

---

[9] The term "unifying" was selected because the protocol is tied to the upper stack layers. It also refers to the protocol's flexibility in serving many different heuristics, with a common lower level mechanism.

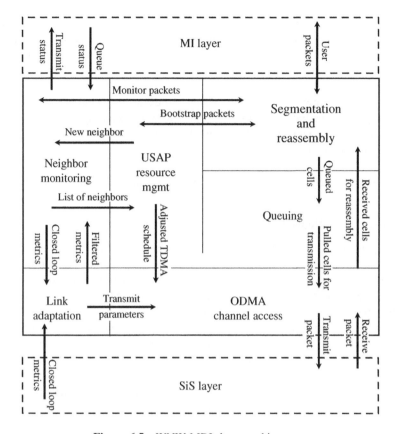

Figure 6.5   WNW MDL layer architecture.

Channel resources (TDMA slots) are reused spatially and concentrated in the nodes that are best positioned to relieve congestion. The MDL layer provides unicast and broadcast coverage through the allocation of TDMA slots.[10] The MDL layer defines operational groups of nodes on the same channel and assigns TDMA slots to optimize for channel frequency reuse. The MDL layer also allows selected nodes to bridge between channels.

Cross layer signaling allows each node to respond to the changes provided by the SiS layer; these changes refer to volatility in measurements, and reflect signal strength and symbol error rate (SER). The MDL layer uses this information to continuously optimize the use of the physical layer resources.

As shown in Figure 6.5, the MDL layer architecture consists of the following major functional modules:

- Segmentation and reassembly: This module converts (segments) user packets, internally generated monitor packets, and bootstrap packets to MDL layer transmission cells.[11] Received cells are reassembled back into their corresponding packets.

---

[10] Some MANET MAC layers can allocate some slots to be shared between network nodes in a CSMA-like protocol (contention slots). However, the WNW waveform implementation of USAP does not use contention slots.
[11] The concept of cells and bootstrap is clarified later in this chapter.

- Queuing: This module holds cells waiting to be transmitted according to priority.
- Orthogonal domain multiple access (ODMA) channel access: This module pulls cells from the queues then builds and schedules transmissions/receptions with the SiS layer.
- Link adaptation: This module adjusts transmission parameters based on the metrics gleaned from received packets.
- Neighbor monitoring: This module collects and exchanges neighbor information from monitor packets, including the metrics reported by neighboring nodes.
- USAP resource management: This module allocates slots by exchanging bootstrap packets and builds the TDMA schedule.

As we progress through this section, the role of these functions should become clearer.

6.2.1.4   Cross Layer Dependency in WNW Waveform

The cross dependency of the WNW waveform, physical layer, MDL layer, and MI layer is a core factor of the waveform's design. Figure 6.6 highlights the most important aspects of this cross layer dependency. The SiS layer throughput is determined by the slot allocation scheduler in the MDL layer. The MDL layer informs the MI layer of the service rate, ensuring that IP packet flow matches the actual packet service rate at all times. Notice that throughput is path specific. The *throughput* of a path p, at a node i, is the fraction of time that node i spends serving path p packets. The routing engines at the MI layer accurately derive (compute) the *arrival rate* for each path at each node. The MDL layer scheduler computes the *serving rate* and throughput of path p in terms of packets per second at node i. This is a function of the packet arrival rates and the link adaptation explained below. The MDL layer builds over-the-air packets from the packets queued by the MI layer (and fragments thereof when they are needed to fill a slot) and feeds these to the SiS layer.

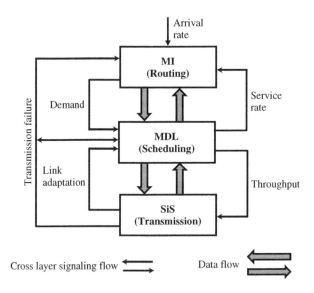

Figure 6.6   Cross layer signaling with WNW waveform stack.

Packet arrival rates to the MI layer represent traffic demand at the node (per path). This influences the USAP protocol decision regarding how many slots are allocated for each node at a given time. The SiS layer measures symbol error rate (SER); this then affects link adaptation. Link adaptation determines the transmission data rate, and influences how fast the MDL can service the packets queued by the MI layer. Thus the SER measurements can affect how fast the MI layer can feed packets to the MDL layer.

Also note how the discussion above has three sets of parameters for each node:

1. throughput of a path p at a node i.
2. arrival rate of a path p at a node i.
3. service rate of a path p at a node i.

These three sets of parameters are coupled iteratively and must converge. Here, the term "throughput" indicates the ratio between the outgoing traffic and the arriving traffic (from the upper layer) for each connection. The throughput should be relative to the convergence moment, seeing that this cross layer signaling approach influences traffic arrival.

6.2.1.5   Link Adaptation

The MDL layer adjusts the SiS layer transmission modes[12] and the power to provide robust communications under changing link conditions. These parameters are adjusted in response to the metrics provided by the SiS layer, including SNR and SER. SNR and SER are filtered and normalized using a hybrid open and closed loop blending process explained below. The filtered and blended SNR and SER drive a generic state machine, controlled by a configurable state transition table. Each entry in the table includes the transmission parameters (modulation type, FEC rate, diversity type and level, power, etc.) and thresholds for the SNR and SER. These parameters are what cause state changes. Figure 6.7 illustrates the multiple entries and threshold changes. The table entries are needed in order to make the waveform highly configurable for different deployment scenarios.[13]

The link adaptation state entries in Figure 6.7 also contain information about controlling the behavior (transitions) of the link adaptation state machine. By making the state transition table a configuration item during initialization, the link adaptation algorithm can be optimized for different operational environments and scenarios, as well as different SiS layer configurations (modulation schemes). Note that SNR and SER are measured for each link on every packet reception. For example, between nodes i and j, SNR and SER are measured for the link from node i to node j. These metrics are then filtered and become the measurements for the open loop (unidirectional from node i to node j) and are then reported from node j to node i in feedback packets called monitor packets. The feedback packets are referred to as the closed loop metrics. The MDL layer algorithms are biased toward using the closed loop metrics to compensate for links that are different in each direction. If closed loop metrics are not available, then only open loop metrics are used.[14]

---

[12] The SiS layer can use different implementation of OFDM modulation.

[13] There is a known dilemma in the defense community; soldiers do not want to be burdened with radio configurations in the middle of a mission. The wide range of deployment scenarios can force engineers to create highly configurable radios. The advances in cognitive radios can offer the solution to this predicament.

[14] Using closed-loop feedback in wireless networking is common to many protocols. The need for real-time knowledge of both the forward and reverse path can exist at the different layers of the protocol stack.

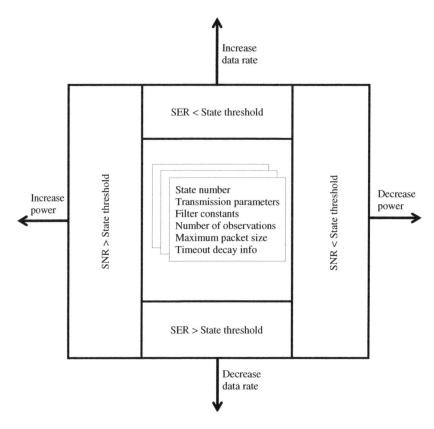

Figure 6.7   WNW waveform link adaptation.

6.2.1.6   MDL Layer Segmentation and Reassembly

The MDL layer breaks an IP packet up into fragments, as necessary, to fill the TDMA transmission slots; these fragments are queued in the TDMA slots for transmission. Fragment length is a configurable parameter and is sized to fit in a TDMA slot based on data rate. With higher data rates, more bytes can fit in a single slot. On the receiver side, fragments belonging to the same IP packet are reassembled into the original packet. Meanwhile, user data packets are delivered to the MI layer, and MDL control packets are delivered to the internal neighbor monitoring, or USAP resource management functions, accordingly (see Figure 6.5 which explains the MDL layer internal architecture).

When preparing a packet for transmission, the MDL layer pulls fragments from the broadcast queues that correspond to the current eligible receivers. As it pulls fragments, it keeps track of the intended receivers and the current lowest data rate among them. This rate will be used for the transmission. The MDL layer then has to determine how many slots correspond to an IP packet. When all the fragments are destined for one receiver, the MDL layer will use the highest possible rate that is supported by the link to that destination node.

When all fragments belonging to the same IP packet arrive over-the-air on the receiving side, they are sent to the segmentation and reassembly function (refer to Figure 6.5) to be reassembled back into the IP packet. The packet header contains link transmission parameters that are extracted and passed, along with the raw SNR and SER metrics, to the link adaptation function for processing. This process results in the open loop metrics described above.

6.2.1.7   Unifying Slot Assignment Protocol (USAP)

The carrier sense multiple access (CSMA) technique mentioned in Section 4.1 is relatively easy compared to that of the USAP protocol. CSMA requires minimal control message exchange between the nodes. As mentioned in Section 4.1, CSMA can be used when the subnet physical layer resources are high compared to traffic demand. In this case, collision is minimal. With tactical MANET, traffic demand is high relative to available physical layer resources, so USAP divides the physical layer resources into two separate portions. One is used for control traffic and the other for data communications. The control portion is used for coordination between the nodes and reservation of the data channel to avoid collisions.

USAP constructs a periodic frame structure as shown in Figure 6.8. The frame length is 1 second. Frame reservation slots (for user traffic) consist of $M \times F$ slots, where M is the number of timeslots and F is the number of frequency channels in a frame. The matrix of (timeslots $\times$ frequency channels) form the MDL cells mentioned earlier. This frame structure is sometimes referred to as the orthogonal domain multiple access frame. There are three different types of slots:

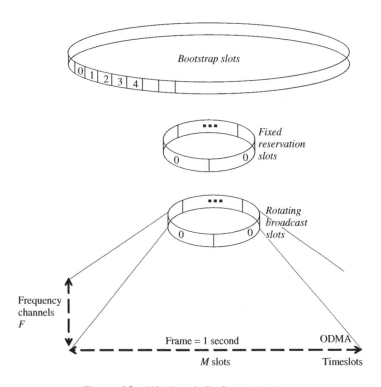

Figure 6.8   USAP periodic frame structure.

1. The *bootstrap slots* are pre-allocated to nodes for exchange of information related to network management (e.g., USAP records) and are used for reservation of slots. Each node informs neighbors about reserved slots using these slots.
2. *Fixed reservation slots* support unicast data traffic (user packets). Once a slot is assigned to a link (i, j), there remains no contention since no node in the 3-hop neighborhood transmits simultaneously. Nodes i and j cannot transmit or receive on any other frequency channel corresponding to that timeslot.
3. *Rotating broadcast slots* support unicast, multicast, and broadcast data traffic. These slots are physically the same as the fixed reservation slots but are shared in time as required by the topology and traffic load.

It is worth noting that the placement of the slots, within the frame, is not exactly as depicted in Figure 6.8, but rather spread out and intermingled as explained in Figure 6.9. For each slot type, there is a unique cycle. For a given type, the proportion of the number of slots per frame to the total number of slots in that frame determines the cycle. Typically, the cycle for a bootstrap slot is longer than that of a fixed reservation or rotating broadcast slot. The number of bootstrap slots needs to be large enough to assign each node in the network a single slot.

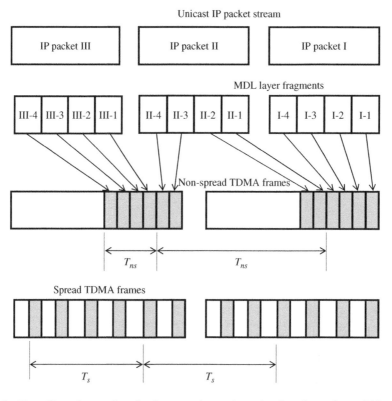

Figure 6.9 The effect of spreading fixed reservation and rotating broadcast slots within a TDMA frame.

The determination of slot type selection within a frame is made such that the maximum latency between user traffic cells is minimized. Hence, the user traffic IP packet jitter, caused by using TDMA at the MDL layer, is minimized. Let us consider the sequence of unicast packets, as shown in Figure 6.9. Each packet is assumed to be broken down into four fragments. If we do not spread the fixed reservation slots within the frame, then the time between the complete transmission of four cells belonging to one IP packet and the transmission of the four cells belonging to the next packet will be T_{ns}. However, if the fixed reservation slots are spread as shown in the lower part of Figure 6.9, then the corresponding time interval will be T_s. One can see from Figure 6.9 how T_{ns} varies drastically, while T_s does not. IP packets are assembled, from fragments, at the destination MDL layer and then delivered to the MI layer. The MI layer of the destination node will encounter less jitter (will see less delay variation) and initial latency due to spreading of the TDMA slots.

In the control channel (e.g., in the use of bootstrap slots), each node broadcasts by using the slots that are reserved for transmission and reception by itself and its neighbors. In this way, every node acquires information about the reserved slots in its 3-hop neighborhood. Let $T(l)$ and $R(l)$ denote the transmitting and the receiving node on the two ends of link l. To avoid collision, $T(l)$ reservation is based on the following rules:

1. $T(l)$ cannot reserve timeslots that are already scheduled for incoming and outgoing transmissions for $T(l)$ and $R(l)$.
2. $T(l)$ cannot reserve slots used by incoming call transmissions to its neighbors.
3. $T(l)$ cannot reserve slots used by outgoing call transmissions from the neighbors of $R(l)$.

Rule number 3 applies for 3-hop neighbors[15] as shown in Figure 6.10. Collision avoidance with simultaneous transmission has to be constrained by spatial reuse over time and frequency. Node i can have the following:

- slots where node i is transmitting ST_i;
- slots where node i is receiving SR_i;
- allocation where a node i's neighbors' neighbors are transmitting NNT_i.

USAP combines the sets of TDMA slots to create spatial and frequency reuse of physical layer resources. For example, for node i, if its neighbor j is transmitting or receiving in slot s, node i is blocked from using s on any other channel with frequency f following:

$$Bj(s) = \forall_f ST_j(s, f) \cup SR_j(s, f). \tag{6.1}$$

The 3-hop reuse is enforced by defining the set:

$$NNT_i(s) = \cup SR_j(s, f). \tag{6.2}$$

The set in Equation 6.2 applies over all neighbors j of node i on channel f only if (s, f) is not also a member of $SR_j(s, f)$, or the transmitter in (s, f) is not a neighbor of j. In other words, the rule prevents two neighbors of a given transmitter from reporting NNT_i as a result of hearing each other's SR_i. Thus, they avoid causing the transmitter to falsely

---

[15] This 3-hop constraint, instead of the 2-hop constraint, is needed to address a problem similar to the hidden terminal problem discussed with cognitive radios in the next chapter.

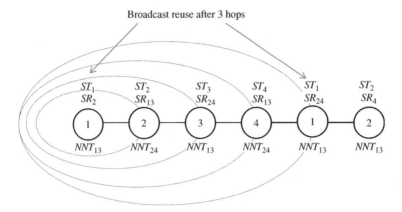

Figure 6.10 Hop reuse of TDMA slots, the subscripts of the sets ST, SR, NNT show the slots being reported.

detect a 3-hop conflict, which can happen when a node has a slot in its ST_i that is also in a neighbor's NNT_i. One can say that allocation excludes those slots that are assigned to i or any of its neighbors, or,

$$\cup\forall_{j\in\{i/s\_nbrs\}}Bj(s).\tag{6.3}$$

In other words, i's neighbor's neighbors are transmitting in,

$$\cup\forall_{j\in\{i's\_nbrs\}}NNTj(s).\tag{6.4}$$

Combining these two sets results in identifying the slot allocation where i cannot transmit as:

$$\cup\forall_{j\in\{i's_{nbrs}\}}Bj(s)\vee\cup\forall_{j\in\{i's_{nbrs}\}}NNT_j(s).\tag{6.5}$$

Fixed reservation and rotating broadcast slots can share the same pool of slots. In order for this to work, fixed reservation takes priority over rotating broadcast assignment for each slot. To mitigate the impact of this prioritization, a selected rotating broadcast slot assignment shifts one slot every frame. This successfully prevents fixed reservation from permanently interfering with any one rotating broadcast slot.

The USAP protocol can function under a connection-oriented (hard-scheduling) or connection-less (soft-scheduling) framework. Please note the following:

- In the hard-scheduling mode, slots are reserved for the duration of the call on all the links along the path which lead from the source to the destination. The performance metric, for the hard-scheduling case, is the percentage of calls blocked for each connection. A call is blocked if there is not enough available capacity (slots in frame) on all links of the path. This fixed reservation ensures that slots are served promptly, with no significant queuing delay. For this reason, delay is not an essential performance metric.
- In the soft-scheduling case, the performance metric is calculated based on both delay and throughput. Due to queuing and limited capacity on the links, we will have packet loss as a result of congestion. Throughput specifies the percentage of the delivered

traffic to the sourced traffic. Packets are queued upon arrival at the node and reservation is established, per-packet, for transmission over the link and onto the next hop. The accumulated delay over the path is a significant performance metric.

6.2.1.8 Queuing and Flow Control

Packets are queued up and serviced in FIFO order, according to priority. Prioritization means that cells belonging to an IP packet marked with a differentiated services code point (DSCP), indicating higher precedence (e.g., call for fire), are processed before new cells from a lower precedence packet. Flow control at the MDL layer utilizes cross layer signaling to and from the MI layer. The MDL layer informs the MI layer of the transmission status of each IP packet that the MDL layer has waiting in queue. The MDL layer also informs the MI layer of the total number of bytes it currently has queued for transmission. The MI layer then informs the MDL layer of its queue status (the number of bytes that are waiting in queue to be handed to the MDL layer). The MDL layer attempts to adjust its capacity in response to the offered traffic by reassigning the rotating broadcast slots and/or fixed reservation slots.

The MDL layer capacity adjustment algorithm considers factors that include:

1. the number of bytes in queue provided by the MI layer.
2. the capacity of the current rotating broadcast slots – this capacity is determined by the number of rotating reservation slots and the capacity of each slot, in bytes per second (notice how the SiS layer affects this metric as explained in Section 6.2.1.5).
3. the capacity of the MDL layer priority queues.
4. the latency that broadcast cells experience as they wait for transmission.
5. the latency that unicast cells (per destination) experience as they wait for transmission.

The factors affecting the capacity adjustment algorithm are very dynamic and the algorithm convergence changes often.

6.2.1.9 Topology Control in Tactical MANET

With tactical MANET, the underlying communication topology is constantly changing. Finding a minimum connected sub-graph (as a skeleton of the network topology) is one of the best-studied problems in the MANET MAC layer literature. There are different techniques in the literature, such as the minimum connected dominating set (MCDS). In this section, we focus on the technique used by the WNW waveform, which is known as the unifying connected dominating set (UCDS). This algorithm meets the primary objectives of a stable algorithm: simplicity of implementation, speed of execution, low time-complexity,[16] flexibility, and fault tolerance.

The implementation of the UCDS algorithm has to be distributed among the MANET subnet nodes. A distributed state machine, at each node MDL layer, defines the role of that node in the topology convergence. With UCDS, a connected dominating set (CDS) of a graph G consists of a set of dominating nodes. Each node in G is either a member of a CDS or 1-hop away from a member. UCDS defines a set of connecting nodes, which connect

---

[16] Some of the search techniques similar to the presented technique can become complex as they execute over a longer period of time. The algorithm adapted by JTRS instantiates itself (starts all over again) to maintain a high execution rate.

disjoint sets of dominating nodes. CDSs have traditionally been used in MANETs for routing applications. Despite their typical usage, the CDSs could also benefit the networks in other areas, including channel access, link adaptation, energy consumption, power control, and dynamic resource management. The minimum dominating set (MDS) problem and the MCDS problem are NP-complete. In this section, we present a simple and efficient CDS heuristic, the unifying connected dominating set (UCDS), named for its ability to simultaneously support multiple applications. UCDS works equally well for, and runs independently of, CDS applications.

CDS Algorithm

UCDS uses a rule for constructing the dominating sets (DSs) followed by rules to construct the connecting sets (CSs) such that the union of these sets forms a CDS.

DS Rule Definition

A node i is a member of the DS if

1. it has the highest dominating factor (d_i) among its neighbors, j, in which case node i designates itself as a member of the DS, but it is also possible that ...
2. a neighbor j finds that node i has the highest dominating factor among its neighbors k, in which case, neighbor j designates node i.

If the node's dominating factor equals that of a neighbor's (known as a "DS twin"), the node id is used to break the tie. In this way, a node will either designate itself or be designated by a neighboring node to be a member of the DS. Note that this algorithm has a computational complexity of (Δ) because its steps loop through the neighbors, and message complexity of $O(\Delta)$ because it shares the neighbors lists. The DS rule definition can be expressed as follows:

$$i \in DS \Rightarrow \left(\forall_j \in N_i \left(d_i > d_j\right) \vee \left(d_i = d_j \wedge i > j\right)\right) \vee$$
$$\left(\exists_j \in N_i \in \forall_j \in N_i \left(d_i > d_j\right) \vee \left(d_i = d_j \wedge i > j\right)\right), \tag{6.6}$$

with this DS rule, a node is no more than 1-hop away from a DS member.

CS Rule Definition

A node i with neighbors N_i is a *candidate* member of the CS (CS') if it is not a member of the DS, and it has neighbors j and k (at least one of which is a member of the DS) whose sets of DS neighbors (DSN_j and DSN_k, which includes j or k if they are also DS members), are disjoint. Among the common neighbors of nodes j and k that belong to CS', known as "CS twins," the one with the highest dominating factor (d) (or highest node id if d is equal), becomes the single CS node. The CS rule has computational complexity $O(\Delta 3)$ (outer neighbor loop, inner neighbor loop, and disjoint set operation) and message complexity $O(\Delta)$ (neighbor list). The CS rule can be expressed as:

$$i \in CS' \Rightarrow i \notin DS \wedge (j \in DS \vee k \in DS) \wedge (E_j, k \in N_i \in DSN_j \cap DSN_k) \equiv \emptyset)$$
$$i \in CS \Rightarrow i \in CS' \wedge \forall_n \neq i \in CS'((d_i > d_n)V(d_i = d_n \wedge i > n)). \tag{6.7}$$

In the implementation of UCDS, in order to reduce processing and speed convergence, the DS rule is applied before the CS rule.

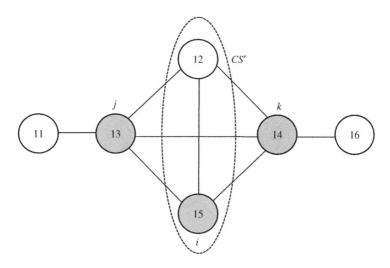

Figure 6.11   Topology showing DS and CS.

The topology shown in Figure 6.11 includes the two types of nodes discussed above: DS nodes are in gray and CS inside the dotted ring (nodes 12 and 15). Nodes 12 and 15 are CS twins. Node 15 was selected based on tie breaking.

In some cases, the CS rule can create unnecessary CS nodes, as exemplified in Figure 6.12. In the figure, node 23 has a CS twin node 21. Node 23 would have become a CS node because of its larger id. In Figure 6.12, one can see that node 21 is already a CS node because of nodes 17 and 22. Thus, there is no need for node 23 to also be a CS node. Similarly, one can see in Figure 6.12 that node 16 could be a CS node because neighbors 14 and 20 announce disjoint dominating set neighbors (DSNs), but node 16, as a CS node, is redundant due to the presence of node 19. To prevent the creation of unnecessary CS nodes, a third rule, the CS exception rule, is necessary.

CS Exception Rule Definition
The CS rule can be ignored for node i if:

1. one of the common neighbors of j or k is already a CS node, or if
2. either j or k is not a DS node, and node i shares a common DS neighbor with that node.

The CS exception rule is applied to node i unless the CS rule applies to another pair of neighbors l and m. The CE exception rule has a computational complexity of $O(\Delta 3)$ (outer neighbor loop, inner neighbor loop, and disjoint set operation) and message complexity of $O(\Delta)$ (neighbor list).

$$i \notin CS \Rightarrow (\neg \exists l, m \in N_i \Rightarrow i \in CS) \wedge \begin{pmatrix} (\exists n \in CS \ni n \in DSN_i \cap DSN_k)\vee \\ (j \notin DS \wedge DSN_i \cap DSN_j \neq \emptyset)\vee \\ (k \notin DS \wedge DSN_i \cap DSN_j \neq \emptyset) \end{pmatrix}. \quad (6.8)$$

For a given connected set of DS nodes, the CS exception rule does not disconnect them. If j and k are already connected by another CS node, i's CS status will not disconnect

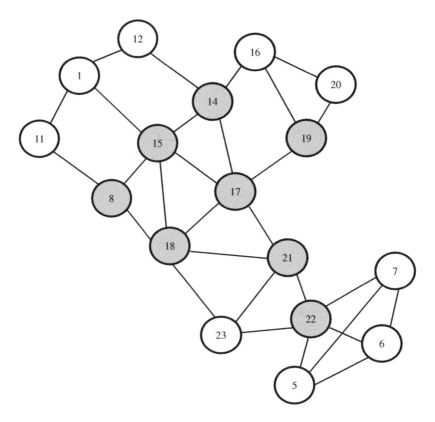

Figure 6.12 Creating unnecessary CS nodes.

j and k. In Figure 6.12, let us assume that j (node 14) is a DS node and k (node 20) is not. By the CS exception rule, i (node 16) shares a common DS neighbor (node 19) with k (node 20). Since k (node 20) is already connected via node 19, k will remain connected regardless of i's CS status.

In Figure 6.13, assume that there is a node i (node 4), that is not a DS node. The CS rule states that i will become a member of the CS if it connects two neighbors j and k which have disjoint DSNs. By the 1-hop rule, j and k must either be DS nodes or have DS neighbors. If both j and k are DS nodes, then i connects them. However, if either j or k is not a DS node (node 5 in Figure 6.13) then it must be a CS node for i to connect the disjoint DSNs. Thus, it remains to be shown that if either j or k is not a DS node, it is a CS node (or it has a CS twin).

Figure 6.14 gives another example of how the CS rule creates a connected set of CDS nodes. In this figure, node 4 is i, while node 3 is k and node 5 (which is not a DS node) is j. The 1-hop constraint suggests that there must be a DS node (node 7) on the other side of node 5. If it can be shown that nodes 4 and 7 announce disjoint DSNs then either node 5 or its CS twin (node 6) must be a CS node. Now, DS node 7 cannot be a direct neighbor of k since j and k are announcing disjoint DSNs. Thus, node 4 and DS node 7 must also be announcing disjoint DSNs, thus forcing either node 5 or node 6 to be a CS node.

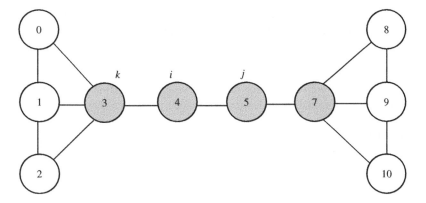

Figure 6.13 Example of the CS rule creates a connected set of CDS nodes.

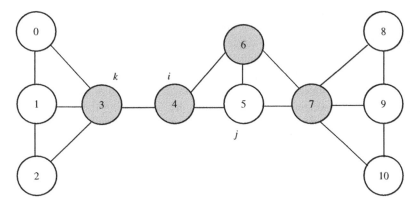

Figure 6.14 Another example of the CS rule creating a connected set of CDS nodes.

UCDS Rule Definition

The union of the CS and DS is the UCDS.

The UCDS is a connected dominating set. UCDS, as described here, has complexity $O(\Delta 3)$ due to the CS and CS exception rules. UCDS shows that it might be possible to lower the complexity if we can find a way to avoid applying all the above rules when the first applied rule (DS rule) is false (the node is not a member of the dominating set). With UCDS, the following rule can be executed after the DS rule but before the CS exception and CS rules:

Not-CS Rule Definition

A node i is not a member of the CS if all neighbors j and k are directly connected. Note that this has complexity $O(\Delta 2)$ and message complexity $O(\Delta)$. The not-CS rule can be expressed as:

$$i \notin CS \Rightarrow \forall_j, k \in N_i (j \in N_k \wedge k \in N_j). \tag{6.9}$$

Note that the not-CS rule will not prevent the CS rule from connecting the CDS since a node cannot disconnect neighbors that are directly connected.

Fault Tolerance

If redundant CDS nodes are required for fault tolerance, it is possible to modify the DS rule to allow two or more dominating nodes for every node in the network. This requires a new rule:

mDS Rule Definition

A node i is a member of the DS if:

1. it has at least the mth highest dominating factor (d_i) among its neighbors j, in which case node i designates itself, or
2. a neighbor j finds that node i has *at least the m*th *highest* dominating factor among its neighbors k, in which case neighbor j designates node i.

There is no cost increase from fault tolerance. There is no increase in computational or message complexity. The only cost is in the increased number of CDS nodes, which after all is the goal. Practically, the UCDS achieves two dominating sets at a relatively low cost. Some CS nodes, in the first dominating solution, simply become DS nodes in the second dominating solution, thereby mitigating the overall increase in CDS nodes.

The UCDS rules are more computationally efficient than similar techniques that calculate CDSs. Some of these techniques, like UCDS, are $O(\Delta 3)$, but they can take a shotgun approach of selecting (marking) every possible DS and CS node and then removing unnecessary nodes from the CDS. UCDS applies its rules in an order, DS $\rightarrow$ not-CS $\rightarrow$ CS exception $\rightarrow$ CS, which chooses the CDS nodes without toggling.

There are some cases where we need to improve performance, under mobility, with nodes switching on and off to reduce power consumption. There are different rules to handle hosts switching on and off, updating the backbone and updating the CDS. UCDS handles all of these special cases without any special rules.

UCDS Stability and Convergence Time

For a static network and a constant set of dominating factors UCDS converges to a single CDS solution. Since every node either designates itself or exactly one of its neighbors to be a DS member and every node in the network chooses exactly one DS member, the DS rule converges to a single solution.

For a given topology and constant set of dominating factors the UCDS converges to a single CDS. A CS node is determined directly or, in the case of CS exception rule 1, indirectly by the status of the DSN of nodes within 2-hops. Because the DSN must converge to a single DSN solution for each of its neighbors, the CS status of a node must also converge. Thus, the CDS must converge to a single solution.

Intuitively, the UCDS must be stable, given the dependencies between the DS and CS rules and the order in which they are applied. The DS rule depends on a constant set of factors, such as topology and the dominating factor. The CS rule depends solely on DS neighbors and is applied only if the DS rule fails; thus, it must converge. However, it is possible to define the CS rule in terms of disjoint neighbor sets, including both DS and CS neighbors. In practice, this works just as well, but the CS feedback loop makes the stability proofs more difficult.

Assuming that nodes are announcing their neighbor lists at a constant rate, it is easy to bound the number of cycles that UCDS takes for a topology change to converge to a CDS. Once the status of a neighbor link changes at a particular node, it takes one cycle (defined as the time it takes for all nodes to share their neighbor information once) for this node to announce the change. Another cycle will allow this change to propagate to the 2-hop neighbors. At that point, the DS nodes will converge. One more cycle will then allow the CS nodes (those that are not affected by CS exception rule 1), to converge. It takes one additional cycle to ensure that the latter nodes converge. Thus, after a topology change, UCDS will converge after a maximum of four cycles.

Flexible Dominating Factors
UCDS provides the ability to create redundant nodes for fault tolerance by establishing two or more dominating nodes for every node in the network. Although message processing time and computational complexity are not altered, this configuration does result in an increase of the number of nodes in the CDS. There have been many variations of the protocol that differ from the traditional goal of selecting a CDS with the richest connectivity and the minimum set of nodes. The dominating factor, explained so far in this chapter, is referred to as the "neighbor degree." One can find many proposals, in the corresponding literature, that suggest combining the neighbor degree with remaining battery power to optimize for energy consumption. This approach can be applied to sensor networks where energy consumption is an important factor. UCDS, a deviation from the CDS, generalizes the CDS approach and allows for a myriad of dominating factors. This allows the topology to be optimized. With this approach, dominating factors can be made configurable, such that all metrics are available for the network planner. The burden then lies on the network planning to select priorities and weigh factors for all the configurable metrics. This ultimately influences the selection of the UCDS and the behavior of the network. For example, a UCDS can consider the following metrics:

1. neighbor degree;
2. remaining battery life;
3. average data rate;
4. CPU – nodes with more powerful CPUs could process more traffic;
5. available queue space – nodes with more queuing capabilities will suffer from less buffer overflow during delayed transmission;
6. CDS member – this increases the stability of the selected set by giving higher weight to nodes that are current members.

Note that this algorithm could morph with time. That is, the priorities and weighting factors could change as the mission progresses to a different phase.

A key factor in this algorithm is stability. It is important that changing metrics are smoothed, in order to avoid the introduction of unnecessary thrashing of the CDS membership (where nodes join and leave a CDS quickly, thus causing instability).

Parallel CDSs

UCDS also provides the ability to create multiple CDSs in parallel. This may be necessary to support applications that have varying QoS requirements. Within the same network, one application may support sensors that generate a continuous stream of low level traffic. This traffic can tolerate high delays. This application needs to be optimized for maximum battery life. Another application aboard an automated weaponry system might generate brief bursts of traffic which is sensitive to delay. In the first case, prioritizing for remaining battery life is critical. In the second case, the key factor in generating a shorter, minimally delayed, routing path is neighbor degree. UCDS can accommodate both needs by publishing multiple flexible dominating factors along with its lists of neighbors.

One also needs to consider the price of having multiple UCDSs within a single subnet. The protocol at each node would use additional bits for each neighbor to express the flexible dominating factor and the UCDS state; this will increase the message overhead but by much less than running independent CDS algorithms.

Shortest Path Optimization

There are two parts of the CS and CS exception rules that can be relaxed to provide minimum hop routing to non-DS nodes. For example, in Figure 6.15, nodes 13 and 16 are

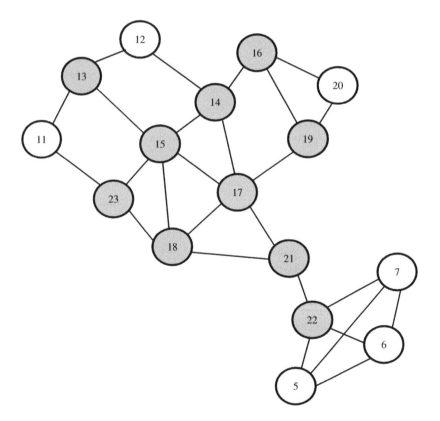

Figure 6.15 UCDS creating minimum hop routing.

both necessary for minimum hop routing to their neighboring non-DS nodes (11, 12, and 20) but would be normally removed by UCDS as CS nodes. Node 13 would not be selected due to the clause in the CS rule. In the case of node 16, it would not be selected as a result of CS exception rule 2. Without these aspects of the CS rules, a CS node could provide the shortest path to non-DS nodes.

6.2.2 Soldier Radio Waveform (SRW)

SRW is another IP-based tactical MANET waveform from the JTRS program. The SRW stack differs from that of the WNW, as shown in Figure 6.16. SRW has the legacy combat net radio (CNR) voice implementation that runs directly over both the MAC layer and the data link layer, where CNR packets fit directly into the MAC frames. CNR voice is limited to the single SRW subnet for platoon voice use. Theoretically, the standard IP data port can support VoIP; nevertheless, the radio CNR voice meets the communications needs of the dismounted soldier.

Although the early versions of the SRW waveform did not include HAIPE encryption (they had one single IP layer), the subsequent versions are HAIPE capable. This waveform is designed to provide a networked battlefield with communications capabilities for disadvantaged users engaged in land combat operations. The waveform can operate on a variety of platforms, such as vehicles, rotary wing aircraft (it cannot support high-speed movers), dismounted soldiers, automated weaponries, ground sensors, and unmanned aerial vehicles. SRW can interoperate with WNW (which has higher throughput and a larger area of coverage), by using WNW for backbone routing. WNW connectivity for SRW enables information exchange, through the GIG, to the dismounted soldier, and provides necessary capabilities for all above ground platforms. Depending on the operational mode, the waveform supports a throughput data rate of 50 kbps to 2 Mbps.

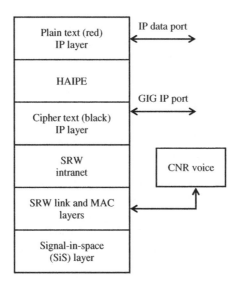

Figure 6.16 SRW data protocol stack layers.

SRW supports frequency channels with discrete bandwidths for the frequency bands 225–420 MHz and 1350–2500 MHz. The transmission ranges of the SRW waveform depend on the power amplifier it is using and the configuration of the hardware it is loaded onto. If it is configured for a line-of-site (LOS) functionality, its range can exceed 10 km.

One of the main concerns with the JTRS IP-based waveforms is the lack of standardization. There is little commonality between the different waveforms (SRW, WNW, etc.). Each waveform has different intranet layer implementations, MAC and DLL layer implementations, and SiS layer implementations. Although a deviation in design for WNW and SRW was necessary, due to different deployment requirements more attention could have been paid to standardization of the cipher text IP layer interfaces, plain text IP layer capabilities, and HAIPE versions (which differ between the two waveforms). The creation of well-defined application programming interfaces (APIs) is discussed in Part III of this book.

6.2.3   High-Band Networking Waveform (HNW)

While the JTRS waveforms described above form local area subnets, the high-tier layer in Figure 6.1 uses a different type of MANET that forms wide area subnets. The high-tier layer in Figure 6.1 relies on two waveforms for high capacity networking on the move. These two waveforms are the WNW and NCW.

HNW provides high bandwidth and long-range connectivity between mobile nodes. Notice the hierarchy in Figure 6.1. There are two types of nodes that can utilize the HNW waveform within the war theater. One type is the brigade-level nodes. These nodes have been selected as gateway nodes to the WIN-T upper hierarchy by the WNW waveform. These gateway nodes can route brigade traffic through the WIN-T for reachability to command, control, continental United States (CONUS) services, other brigades or other forces' networks for joint missions. The other type of node is the WIN-T node. These nodes could be at the tactical operational command (TOC) level with their own multisecurity level LANs. The HNW is an IP-based MANET waveform with self-forming and self-healing capabilities that use directive network technology (DNT). DNT uses directive beam antennas to extend range and improve throughput. Directive antennas also provide inherent low probability of interception (LPI) and low probability of detection (LPD).

HNW offers high throughput, long-range communications OTM and has been used with HAIPE encryption as the encrypted IP core network for long-range support (over airborne and ground nodes forming a single WAN). HNW has also been used for sensor networks. HNW supports both IPv4 and IPv6 capabilities. In the absence of satellite communications capabilities, the HNW can fulfill all tactical GIG WAN needs by utilizing airborne nodes to create full coverage of the theater.

HNW is considered to be a software programmable radio (SPR), with three layers of defined capabilities, as shown in Figure 6.17. Layer 2 is critical to the waveform design; it implements a universal media access control (UMAC) protocol.[17] Layer 3 is capable of interfacing with two open architecture routing protocols: (i) optimized link state routing protocol (OLSR)[18] and (ii) OSPF with MANET extensions. The HNW has two physical layer implementations available; one uses orthogonal frequency division multiple access

---

[17] The term "universal" is used since the waveform can universally link various nodes at layers 1 and 3 over a single WAN.

[18] While ROSPF is specific to the JTRS program, OSLR is an open architecture protocol.

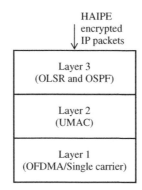

HAIPE
encrypted
IP packets

| Layer 3 (OLSR and OSPF) |
| Layer 2 (UMAC) |
| Layer 1 (OFDMA/Single carrier) |

Figure 6.17   HNW protocol stack.

(OFDMA) and the other uses a single carrier. HNW performs with ranges exceeding 40 km and 100 Mbps, for ground-to-ground links. HNW can be used as a point-to-point link or as a point-to-multipoint link, forming a WAN; when deployed on an airborne node, an HNW node can be configured as an advantage node, forming a WAN with all nodes on the ground. This can be achieved as a result of its cone-shaped directional beam; the directional antenna beam width can be adjusted and steered to form a WAN on a specific area in the war theater.

HNW has the ability to interface with a policy-based network manager in a WAN. The waveform is designed to boot from an initial startup configuration as defined by the user. Once it is booted, an external network manager can configure the waveform to meet the mission needs. For example, given the current battlefield scenario, this configuration can restrict, or permit, spectral bandwidth, modulation type, power level, data rates, number of permissible links, or other controllable parameters. Control is achieved via a defined interface to an external network manager. The formed WAN operates autonomously based on the defined configurations.

6.2.4   Network Centric Waveform (NCW)

The basic idea of the NCW waveform is to make a satellite transponder generate multiple access waveform capabilities. As shown in Figure 6.18, the transponder beam creates an area of coverage on the Earth's surface. The beam can be steered to ensure coverage of the deployment theater zone. The downlink (from satellite to Earth terminals) can reach any node on the ground that has a line of sight access to the satellite. NCW implements a full mesh IP over this satellite transponder by using TDMA architecture, designed to operate in the super high frequency (SHF) range. This range is between 3 and 30 GHz. This frequency band is also known as the centimeter band or the centimeter wave. The wavelength of this band ranges from 10 to 1 cm, which allows the use of a small-sized satellite antenna.[19] The NCW waveform implementation uses a dynamic scheduler that allocates satellite resources (bandwidth or the TDMA slot allocations) based on traffic demand and traffic priority (time sensitivity) at each node. The definition of traffic priority (which DSCP means higher priority) is configurable, making this waveform adaptable to different QoS needs.

---

[19] The OTM version of this satellite has a breakthrough antenna design that is beyond the scope of this book.

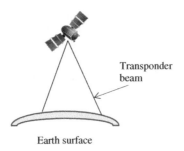

Figure 6.18   NCW transponder beam forming a multiple access subnet between ground nodes.

6.3   The Role of Commercial Satellite in the Tactical GIG

In addition to the HNW and NCW capabilities, the tactical GIG can utilize point-to-point cables, point-to-point high capacity microwave links, and commercial satellite links to create a rich topology and high capacity links, when possible. The use of HAIPE encryption makes it possible to use commercial satellites since encrypted packets can be safely sent through the commercial networks. Although the use of commercial satellites can form many types of topologies (mesh, hub/spoke, hybrid, etc.), we will focus on the hub/spoke topology, which relies on the existence of a satellite hub.

Figure 6.19 denotes a teleport hub which services both unclassified (linkage to the non-secure Internet Protocol router network (NIPRNET), DSN, and even to the public switch telephone network – PSTN) and classified (SIPRNET – secret Internet Protocol router network, defense red switch network – DRSN and DVS – DISN video services[20]) hosts. The teleport hub allows all users of the tactical GIG to access these services or any other CONUS services. With hub/spoke architecture, all nodes equipped with commercial satellite terminals have a direct link to the hub. This forms a star topology, with the hub as the focal point. The architecture of each node is discussed below.

This architecture has many advantages, including the ability of all nodes to reach the hub services in one satellite hop.[21] The main disadvantage of this topology is that node-to-node communications must use two satellite links (spoke–satellite–hub and hub–satellite–spoke), instead of one. This causes a satellite propagation delay of almost 500 ms.

6.4   Satellite Delay Analysis

Satellite delay, in the tactical GIG, is a primary challenge that applications and transport layer protocol developers face. We will analyze this satellite delay in order to better grasp its effects on networks. For geostationary orbit satellites (which are used for communications), the propagation delay from the Earth terminal to the satellite could be as long as 120 ms. For a packet to be transmitted to the satellite and back to the Earth terminal, a delay of

---

[20] The defense information systems network (DISN) has a classified video service, mainly used by classified US government video teleconferences. A teleport extends this service to the tactical GIG, where commanders on the ground can attend teleconferences with CONUS.
[21] We will use the term "satellite hop" to refer to the RF signal that propagates from one Earth terminal to a satellite and back to another Earth terminal.

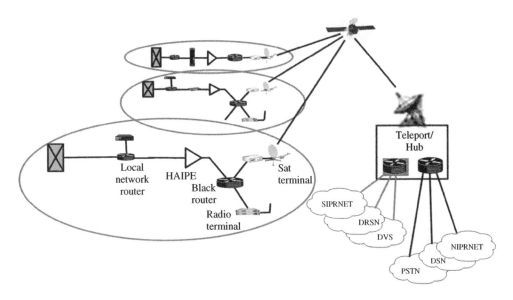

Figure 6.19   Notional view of the hub/spoke architecture with commercial satellites.

240 ms is expected. If we add the queuing delay from the NCW TDMA architecture, atop the propagation delay, we can expect a total satellite delay of nearly one second.

Example: TDMA over NCW   Let us assume that the frame period in TDMA, used with NCW, is 400 ms. Let us also assume that a large network is formed, such that the NC assigns most NCW nodes a single timeslot per frame. The nodes (assigned one timeslot per frame) will transmit for 1.5625 ms every 400 ms (we assume that 400 ms can be divided into up to 256 timeslots, since NCW can accommodate up to 255 nodes in a single network).[22] Figure 6.20a shows how a TDMA frame time of $T = 400$ ms results in the node having an active transmission period of $t = 1.5625$ ms and an idle period of $T - t = 398.4375$ ms. The active period is when the node transmits at the full transmission capacity of the NCW uplink subnet and the inactive transmission is when the node does not transmit.

If we only consider TDMA delay (the delay that a packet encounters while waiting in a queue for a TDMA slot to open), we will need to assume a light traffic load (we will discuss the more complex case of a higher traffic load later). First, imagine a packet entering an empty queue. Let us say that it *was* the node's turn to transmit (the node is in the ON phase). In this case, the lucky packet would be transmitted immediately and the TDMA delay would be zero. However, we can also imagine a packet that enters an empty queue, but at the time, the node was entering its idle phase (OFF phase). At this point, the packet would have to wait in the queue for $T - t$ ms before it could be transmitted. Such TDMA delay can be expressed as a uniform distribution probability density function, as shown in Figure 6.20b, where the density is $\frac{1}{T-t}$ over all delay values between 0 and $(T - t)$.

If we consider the propagation delay for a packet that is transmitted from an Earth station to a satellite and back to the destination node (in the Earth station), we need to

---

[22] While 255 of the 256 timeslots can be assigned to Earth terminals for TDMA access, the 256 slot is left for control signaling between terminals (network members) and the network controller.

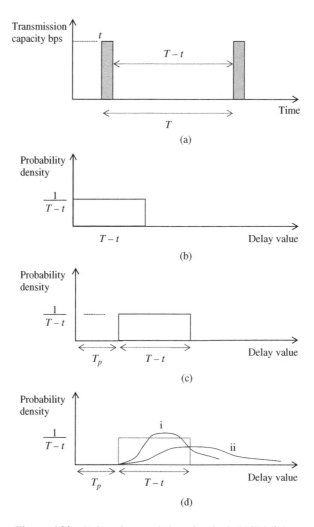

Figure 6.20   Delay characteristics of a single NCW link.

add T_p to the uniform distribution delay. This results in a shift of the pdf, as shown in Figure 6.20c.

The pdf delay in Figure 6.20c can be expected at low packet arrival rates. One should expect that at low arrival rates, the delay characteristic is dominated by the TDMA ON/OFF pattern and not by any queuing model similar to the M/M/1 queue discussed in Chapter 4.

Since the NCW network and the TDMA slot service rates are known, one can expect the pdf in Figure 6.20c to hold true at low packet arrival rates. As the traffic arrival rate increases, the pdf of delay will begin to follow Figure 6.20d, where the pdf shifts further to the right as shown in curve (i) of Figure 6.20d. This increases the average delay and creates incidents where some packets will be delayed extensively. As the ratio of arrival rate to service rate further increases, the pdf will spread more to the right as shown in curve (ii), causing even more excessive delays.[23] Notice that with TDMA models, the queue builds up during the OFF time and gets serviced during the ON time. If a packet at the end of the queue did not get serviced during the ON time, it will have to wait for the next turn. This causes it to be delayed extensively and further skews the pdf of delay (spreading the density to the right).

When traffic demand is high, the NCW scheduler may assign more than one TDMA slot to the node, thus mitigating excessive delay. You are encouraged to go through the exercise of obtaining the equivalent of Figure 6.20, where the node is assigned two TDMA slots per frame. You should assume that the two slots will be spread to minimize the delay. You can find that propagation delay remains constant, but the uniform distribution delay will change, thus decreasing the average delay. The effect of adding another TDMA slot will result in mitigating the effect of queuing delay and doubling the service rate.

A VoIP decoder at the plain text side has a buffer that trades jitter (delay variation from the TDMA and queuing delay) with fixed delay. It was noticed in TDMA based satellite networks that VoIP effective packet delay (that which the decoder encounters taking into consideration the effect of the jitter buffer), can approach one second. Users of VoIP over satellite networks need to acclimatize to this excessive delay as they communicate.

6.5   Networking at the Tactical GIG

Now that we have covered the tactical GIG main waveforms, let us explain how these waveforms may be linked together to form the hierarchy of networks shown in Figure 6.1. Figure 6.21 shows a platform[24] notional architecture. This architecture plays a major part in forming seamless IP gateways between subnets. This platform could be a communications vehicle (TOC) or a stationary platform in a command and control. At the heart of this platform is a gateway IP router (could be a COTS Cisco router). At the right-hand side are multiple HAIPE devices, each of which forms a security enclave (a local LAN in the platform). This platform can have many of these separated security enclaves. Each enclave can be at secret, top secret, or unclassified level. A platform can have one or more security enclaves that offer the network services at the corresponding classification. Notice that HAIPE allows these different encryption levels to share the core network. At the left-hand side of the COTS router, different waveform capabilities can be used, depending on the platform's role in the deployment. A WNW radio at this platform acts as the gateway node to the GIG. It does so by using the GIG port, explained in Figure 6.3, which is connected to an IP interface of the COTS router; this allows IP traffic to flow seamlessly to and from the WNW subnet. The WAN waveforms (HNW and NCW) can be present in this platform as terminals. These terminals are layer 3 capable and interface to the COTS router seamlessly, allowing WAN traffic to flow between different platforms. Keep in mind that these platforms are nodes of the corresponding WAN (e.g., HNW and NCW WANs). Notice that not all

---

[23] This excessive queuing delay can cause packet loss due to the limited size of the queue feeding the TDMA scheduler.

[24] Technically, we should differentiate between a platform and a node. A platform is a physical point where multiple waveforms are linked together, while a node serves a similar role to the subnet that it is a member of. This platform is said to have the communications capabilities offered by all waveforms on its board and is considered a node in all networks corresponding to all the radios/terminals in the platform.

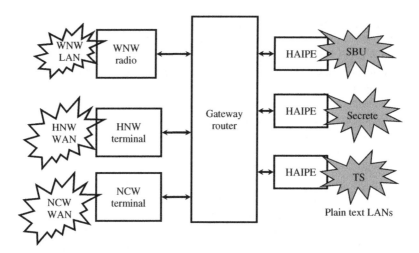

Figure 6.21 A platform notional architecture supporting both brigade-level subnet (WNW) and division-level WANs (HNW and NCW).

nodes at the upper echelon have the same capabilities. Some nodes may have just HNW terminals and are part of a single HNW WAN. Some nodes can be part of different HNW networks, while others can have NCW terminals and are part of the NCW WAN. Notice that there is a single NCW network per theater.

This platform may also have microwave links (if it was stationary) to link to other stationary platforms. It may use the commercial satellite capabilities described above to be part of a hub/spoke network. Before deployment, a planter can decide which communications capabilities are needed for each platform to ensure full connectivity and fulfill critical traffic demands.

In the notional architecture in Figure 6.21, the introduction of new waveforms should be done seamlessly, given the seamless IP connectivity. Seamless IP comes with its own challenges as discussed in the third part of this book.

6.6   Historical Perspective

We have come a long way from the days of using carrier pigeons and smoke stacks to send coded messages in the War Theater. The discovery of modulated RF signals opened the door for making military communications the ever-evolving field it is today. The transformation of radio from a scientific curiosity to a practical communications technology was the result of incremental improvements in a wide range of areas. One can say that the micrometer spark gap used by Heinrich Hertz, followed by various magnetic, electrolytic, and crystal detectors, and finally the very important improvements in three-electrode vacuum tubes made RF signal amplification practical.

One of Marconi's most important discoveries was "groundwave" radio signals, signals that resulted from the addition of a ground connection to the transmitter, which led to greatly increased transmission ranges. This was made possible by the discovery that "earthing" the transmitter antenna resulted in the ground being used as a waveguide; the signals followed the earth's plane, and spread out in only two dimensions. This was unlike a free-space transmission, such as light, which dispersed in three dimensions. This discovery, led people to see that groundwave signal strength tended to drop inversely with the distance covered. This was contrary to free-space signals, where the signal strength is proportional to the square of the distance.

Military radio communications have been around for over 100 years. There is a rich history behind this technology; military radios were even around when the US army depended on mules and horses for transportation. In 1906, the *Manual of Wireless Telegraphy* wrote about an "apparatus as compact and portable as possible so that it may be transported on the backs of mules." In the 1911 edition of *Drill Regulations for Field Companies of the Signal Corps (Provisional)*, the two main radio field units were the *pack set*, which was carried by a "section normally composed of ten mounted men and four pack mules," and the *wagon set*, whose "section is normally composed of 18 mounted men, the wagoner and engineer, ... and one wagon wireless set, drawn by four mules." The November, 1911 issue of the *Sunset Magazine*, published an article about "War as a Modern Science." This article reviewed the use of an early radiotelephone system by the Coast Artillery Corps of the California National Guard. More modern advancements were implemented with the October 1916 edition of the United States Signal Corps' Radiotelegraphy Manual which reviewed advances in the pack and wagon set designs. These advancements included the adoption of quenched spark transmitters, and a reduction of the number of mules needed to carry a field pack set. Also included in the manual was a brief section on mechanization, which included the transportation of transmitters by automobiles (instead of mules) and information on an early form of spread spectrum transmission. These advancements prepared the USA for World War I, since field units deployed on the battlefield were able to communicate with their commanders.

The 15 July 1917 issue of *Journal of Electricity* outlined research efforts by AT&T to develop a new technology that would allow two-way voice communication between ground units and airplanes. In military jargon, this meant that "squadron formations of all sorts could be maintained in the air as easily as infantry units on the ground," thus greatly empowering aerial units.

During World War I, the US military gained control of the entire US radio industry. The military then worked in conjunction with the major electrical firms to make great strides in radio engineering technology through the use of vacuum tubes. The US military had predicted that vacuum tubes, in various forms and sizes, would become widely used in every field of electrical development and applications. Following World War I, many articles were published that demonstrated a comprehensive scientific understanding of the design and operation of vacuum tubes. These articles laid out the design of vacuum tube baseband detectors and signal amplifiers. During the 1920s, vacuum tube research greatly accelerated, resulting in increasingly powerful vacuum tubes which were deployed in radio transmitters. AT&T and General Electric were two dominating companies that produced highly functional, compact, handheld vacuum tubes.

For decades, radio improvements relied on incremental enhancements in vacuum tube design – tubes that required less current, lasted longer, and could run on household electrical current, instead of storage batteries. It was only with a deeper understanding of solid state physics that it was made possible to refine oscillating crystals into practical "transfer resistors" (transistors). These transistors allowed lightweight, flashlight-battery-powered radios to become a reality in the 1950s. The development of these transistors caused quite a stir. In October, 1948, *QST* magazine reviewed the transistors development by Bell Laboratories. This issue reviewed the construction of a simple superheterodyne receiver, noting that "these clever little devices are well worth keeping an eye on."

Another major communications advancement, the frequency-modulated (FM) radio, was developed between World Wars I and II, in the late 1920s and early 1930s by Edwin H. Armstrong. After serving in the US Army Signal Corps during World War I, Major Armstrong realized that a device was needed to reduce the effect of ignition and other noises surrounding vehicle radios. The FM radio was first adapted for military use by the US Army.

The USA entered World War II with highly developed radio communication systems and many electronic navigational aids in development. The US Navy deployed telephone systems and loud-speaking voice amplifiers on naval vessels. The US Air Force deployed radio communication to link their bases and landing fields, and developed airborne long-range, medium-range, and short-range radio equipment for air-to-ground and air-to-air communication.

The World War II battle theater was more technologically equipped than that of World War I. The air, infantry, artillery, and armored teams created new requirements for split-second radio communication among all members. Portable radio sets were provided as far down in the military echelons as the platoon. In every tank there was at least one radio and in some command tanks as many as three. High-powered mobile radio sets became common at division and regimental levels. With these sets, mobile telegraph communication could be conducted at distances greater than 100 miles. With all these developments, major telephone switchboards, of much greater capacity, were desperately needed. With great need comes innovation, and these switchboards were manufactured. They were then issued for use in all tactical headquarters to satisfy the high traffic telephone channels needed to coordinate the movements of highly mobile field units.

Radio relay, born out of the necessity for mobility, became the outstanding communication development of World War II. Sets employing frequency modulation and carrier techniques were developed, as were radio relay sets that used radar pulse transmission, reception techniques, and multiplex time division to enable multiple voice channels on one radio carrier. Radio relay telephone and teletypewriter circuits spanned the English Channel for the Normandy landing and later furnished communication services for General George S. Patton, following his breakout from the Normandy beachhead.

The need for communication between the CONUS and many far-flung theaters of war gave rise to improved long-range overseas communication systems. A system of radioteletypewriter relaying was devised, by which a radioteletypewriter operator in Washington, London, or any other capital could transmit directly, by teleprinter, to the commander in any theater of war. A system of torn-tape relay centers was also established so that tributaries could forward messages through the major centers and retransmit quickly, by transferring a perforated tape message from the receiving position to the transmitting positions. In addition, a holding teletypewriter conference system was developed, called "telecons." This enabled a commander (or his staff) at each end to view the incoming teletypewriter messages, as fast as the characters were received, on a screen. Questions and answers could be passed rapidly back and forth over the thousands of miles separating the Pentagon in Washington, DC, for example, from the Supreme Allied Headquarters in Europe or General Douglas MacArthur's headquarters in the Far East.

During the latter years of the war, many new and improved communication and electronic devices came forth from collaboration between government and industry. One of these innovations was Loran, a new long-range electronic navigation device that was used for both naval vessels and aircraft. Loran was developed in conjunction with a short-range navigational system, Shoran. The joint deployment of radar and communications perfected the ability to land an aircraft in zero visibility. One such integrated system was the GCA, or ground-controlled approach system. The ground controlled intercept aircraft (GCI) was a hybrid system, composed of radio direction-finding radar and communications systems. Radio-controlled guidance of falling bombs enabled an operator in a bomber to easily direct a bomb to the target. While communications and navigation systems made this progress, electronic countermeasures made their appearance in the form of jamming transmitters. They jammed radio channels, radar, navigation, and other military electronics.

The military services learned the importance of scientific research and development in all fields, including communication electronics. Advances were made in the communication capacity of wired and radio relay systems and in improved electronic aids for navigation. Measures to provide more comprehensive and more reliable communication and electronic equipment continued to be stressed by powers in all military branches.

The Vietnam War era of radios used the very high frequency (VHF) communication band. That band was divided as follows: armor/infantry/artillery. VHF was designed that way for a reason; the infantry radio frequency had to overlap both sides to call in support. High frequency (HF) communication was designed for long-distance communications (over the horizon), and to enable communications from ground to air, a separate link had to be established. Ultra high frequency (UHF), for air-to-air and air-to-ground communications, was only possible with the development of a completely different radio. Satellite communications and the development of transistors and solid state technology altered the Vietnam era communication paradigm.

As jamming technology advanced, the USA moved to spread-spectrum radios, such as EPLRS, SINCGARS, and Link-16.

The battlefield communications capabilities will keep evolving and one day the war theater may have very little need for human presence; unmanned devices with artificial intelligence will determine who is superior in future wars. Maybe, one day, the same military technology that gave the human race this lethal destruction capability, will lead the human race to engage in war, but in such a way that no human life is lost.

Bibliography

1. Young, C.D. (1995) A Unifying Dynamic Distributed Multichannel TDMA Slot Assignment Protocol – CSD WP95-1001 – 02 November 1995.
2. Young, C.D. (1996) USAP: a unifying dynamic distributed multichannel TDMA slot assignment protocol. Proceedings of the IEEE Milcom 1996.
3. Young, C.D. (1999) USAP multiple access: dynamic resource allocation for mobile multihop multichannel wireless networks. Proceedings of the IEEE Milcom 1999.
4. Young, C.D. (2000) USAP multiple broadcast access: transmitter and receiver directed dynamic resource allocation for mobile, multihop, multichannel, wireless networking. Proceedings of the IEEE Milcom 2000.
5. Young, C.D. (2006) The Mobile Data Link (MDL) of the joint tactical radio systems wideband networking waveform. Proceedings of the IEEE Milcom 2006.
6. Young, C.D. and Amis, A. (2011) UCDS: unifying connected dominating set with low message complexity, fault tolerance, and flexible dominating factor. Proceedings of the IEEE Milcom 2011.

7. Mission Need Statement (MNS) for the Joint Tactical Radio (JTR) of 21 Aug 97.
8. Future Combat System Challenges and Prospects for Success, US Government Accountability Office, March 2005.
9. Kimura, B., Carden, C., and North, R. (2008) Joint tactical radio systems – empowering the warfighter for joint vision 2020. Proceedings of IEEE Military Communications Conference 2008, RST2.1.
10. (2004) Joint Requirements Oversight Council Memorandum (JROCM) 095-04, 14 June 2004.
11. (2007) Tactical Wireless Joint Network (TWJN) Concept of Operations (CONOPS) Version 1.0 of 30 April 2007.
12. Alberts, D.S., Garstka, J.J., and Stein, F.P. (1999) Network Centric Warfare.
13. (2000) Joint Chiefs of Staff J5, Joint Vision 2020, June 2000.
14. (2006) Joint Chiefs of Staff, Chairman of the Joint Chiefs of Staff Instruction (CJCSI) 6212.01D, March 2006.
15. Dastangoo, S. (2009) Performance analysis of distributed time division multiple access protocols in mobile ad hoc environments. Proceedings of the IEEE Milcom 2009.
16. Wu, J and Li, H. (1999) On calculating connected dominating set for efficient routing in ad hoc wireless networks. Proceedings of the 3rd International Workshop on Discrete Algorithms and Methods for Mobile Computing and Communications, ACM, pp. 7–14.
17. Wu, J., Dai, F., Gao, M., and Stojmenovic, I. (2002) On calculating power-aware connected dominating sets for efficient routing in ad hoc wireless networks. *Journal of Communications and Networks*, **4** (1), 1–12.
18. Raghavan, V., Ranganath, A., Bharath, R. *et al.* (2007) Simple and efficient backbone algorithm for calculating connected dominating set in wireless ad hoc networks. *World Academy of Science, Engineering and Technology*, **33**.
19. Bhattacharjee, S., Tripathi, J., Mistry, O. *et al.* (2006) Distributed Algorithm for Power Aware Connected Dominating Set for Efficient Routing in Mobile Ad Hoc Networks, IEEE 1-4244- 0731-1/06.
20. Wu, J. and Wu, B. (2003) A Transmission Range Reduction Scheme for Power-Aware Broadcasting in Ad Hoc Networks Using Connected Dominating Sets, IEEE 0-7803-7954-3/03.
21. Stojmenovic, I., Seddigh, M., and Zunic, J. (2002) Dominating sets and neighbor elimination-based broadcasting algorithms in wireless networks. *IEEE Transactions on Parallel and Distributed Systems*, **13** (1).
22. Ingelrest, F., Simplot-Ryl, D., and Stojmenovic, I. (2007) Smaller Connected Dominating Sets in Ad Hoc and Sensor Networks Based on Coverage by Two-Hop Neighbors, IEEE 2007 1-4244-0614-5/07.
23. Wu, J., Lou, W., and Dai, F. (2002) Extended multipoint relays to determine connected dominating sets in MANETs. *IEEE Transactions on Computers*, **55** (3).
24. Acharya, T. and Roy, R. (2005) Distributed algorithm for power aware minimum connected dominating set for routing in wireless ad hoc network. IEEE International Conference on Parallel Processing Workshops 2005.
25. Kim, D., Wu, Y., Li, Y., Zou, F., and Du, D. (2009) Constructing minimum connected dominating sets with bounded diameters in wireless networks. *IEEE Transactions on Parallel and Distributed Systems*, **20** (2).
26. Gupta, A. (2005) A distributed self-stabilizing algorithm for finding a connected dominating set in a graph. IEEE Proceedings of the 6th International Conference on Parallel and Distributed Computing, Applications and Technologies 0-7695-2405-2/05 2005.
27. Pond, L. and Li, V. (1989) A distributed timeslot assignment protocol for mobile multihop broadcast packet radio networks. Proceedings of the IEEE MILCOM 1989, vol. 1.
28. Prohaska, C. (1989) Decoupling link scheduling constraints in multihop packet radio networks. *IEEE Transactions on Computers*, **38** (3).
29. Chlamtac, I., Farago, A., and Anh, H.Y. (1994) A topology transparent link activation protocol for mobile CDMA radio networks. *IEEE Journal on Selected Areas in Communications*, **12** (8).
30. Arikan, E. (1984) Some complexity results about packet radio networks. *IEEE Transaction on Information Theory*, **IT-30**.
31. Alzoubi, K., Wan, P.-J., and Frieder, O. (2002) Message-optimal connected dominating sets in mobile ad hoc networks. Proceedings of the 3rd ACM International Symposium on Mobile Ad Hoc Networking and Computing (MobiHoc '02), June 2002, pp. 157–164.
32. Cardei, M., Cheng, X., Cheng, X., and Du, D. (2002) Connected domination in multihop ad hoc wireless networks. 6th International Conference on Computer Science and Infomatics (CS&I 2002), March 2002.
33. Chen, Y.P. and Liestman, A.L. (2003) A zonal algorithm for clustering ad hoc networks. *International Journal on Foundations of Computer Science*, **14** (2), 305–322.
34. Das, B. and Bharghavan, V. (1997) Routing in ad-hoc networks using minimum connected dominating sets. IEEE International Conference on Communications (ICC '97), June 1997, pp. 376–380.

35. Dubhashi, D., Mei, A., Panconesi, A., Radhakrishnan, J., and Srinivasan, A. (2003) Fast distributed algorithms for (weakly) connected dominating sets and linear-size skeletons. Proceedings of ACM-SIAM Symposium on Discrete Algorithms (SODA), pp. 717–724.

36. Sivakumar, R., Das, B., and Bharghavan, V. (1998) The clade vertebrata: spines and routing in ad hoc networks. Proceedings of the IEEE Symposium on Computers and Communications.

37. Wan, P.J., Alzoubi, K.M., and Frieder, O. (2002) Distributed construction of connected dominating set in wireless ad hoc networks. IEEE INFOCOM, June 2002.

38. Lin, Y., Chennikara-Varghese, J., Gannett, J., *et al.* Cross layer QoS design: overcoming QoS unaware slot allocation. Proceedings of IEEE Military Communications Conference 2008, NPM 4.1.

39. Ying, L., Srikant, R., and Greenberg, A.G. (1997) Computational techniques for accurate performance evaluation of multirate, multihop communications networks. *IEEE Journal on Selected Areas in Communications*, **5** (2), 266–277.

40. Yang, Y. and Weerackody, V. (2010) Estimation of link carrier-to-noise ratio in satellite communications. Proceedings of IEEE Military Communications Conference 2010.

41. MIL-STD-188-EEE (2007) Interoperability and Performance Standard for Full-mesh MF-TDMA SHF SAT-COM Communications, March 2007.

42. Wiss, J. and Gupta, R. (2007) The WIN-T MF-TDMA Mesh network centric waveform. Proceedings of IEEE Military Communications Conference 2007.

43. International Telecommunications Union (2004) Maximum Permissible Levels of Off-Axis e.i.r.p. Density from Earth Stations in Geostationary-Satellite Orbit Networks Operating in the Fixed-Satellite Service Transmitting in the 6 GHz, 13 GHz, 14 GHz, and 30 GHz Frequency Bands, Recommendation ITU-R S.524–528, September 2004.

44. L3 Communications Linkabit Division (2007) MPM-1000 IP Modem Software defined MF-TDMA DAMA/FDMA Modem and Network Controller Data Sheet, 23 March 2007, p. 2258.

45. Bennett, B., Meyer, C., and Raghunath, J. (2008) Defense information systems network satellite service. Proceedings of IEEE Military Communications Conference 2008, SAC 8-1.

46. Wang, L. and Jezek, B. (2008) OFDM modulation schemes for military satellite communications. Proceedings of IEEE Military Communications Conference 2008, SAC 8-5.

7

Cognitive Radios

In today's world, one cannot write a book about tactical wireless communications without dedicating a chapter to cognitive radios (CRs). This topic applies to both tactical and commercial wireless communications.

From the tactical wireless communications perspective, we have covered two types of radios: the Link-16 conventional radio, introduced in Chapter 5, and the wideband networking waveform (WNW), software defined radio (SDR), or software programmable radio (SPR),[1] introduced in Chapter 6. In this chapter, cognitive radios[2] are another generation of radios, as shown in Figure 7.1. If one thinks of SPR as conventional radio with added software architecture, re-configurability, easy-to-upgrade design, and so on, then one can think of a CR as an SPR with intelligence, awareness, learning, and observations.

7.1 Cognitive Radios and Spectrum Regulations

The tactical wireless arena has addressed spectrum scarcity in waveforms, such as the high-band networking waveform (HNW) previously mentioned, by using directional antennas. Unlike omnidirectional antennas, which dissipate spectrum in all directions, directional antennas' focused beams allow for more efficient spectrum reuse. In commercial cellular wireless, base stations use a similar concept, with sector/carrier planning, to aid spectrum reuse. Even with such techniques, spectrum reuse without CRs is inefficient. Studies completed by the US Federal Communications Commission (FCC) Spectrum Policy Task Force, in rural and major metropolitan[3] areas, showed that the overall spectrum usage ranges from 15 to 85 percent. The FCC questioned its regulatory approach and started seeking solutions to solve this spectrum bottleneck challenge, without touching the allocations controlled by major industries (primary users such as radio and TV broadcast) that might oppose an overhaul of the current system. One possible solution is to create secondary users which can be controlled by the primary users; however, this could result in a logistical nightmare for the FCC. Another approach, referred to as "opportunistic spectrum sharing," would allow secondary users to access primary users' spectrum (without primary user regulation) while

---

[1] In this book, we use the terms software programmable radio and software defined radio interchangeably.
[2] Cognitive radios are still evolving at the time of writing this book.
[3] You can find a chart of the FCC spectrum allocation at: http://www.ntia.doc.gov/files/ntia/publications/2003-allochrt.pdf.

Tactical Wireless Communications and Networks: Design Concepts and Challenges, First Edition. George F. Elmasry.
© 2012 John Wiley & Sons, Ltd. Published 2012 by John Wiley & Sons, Ltd.

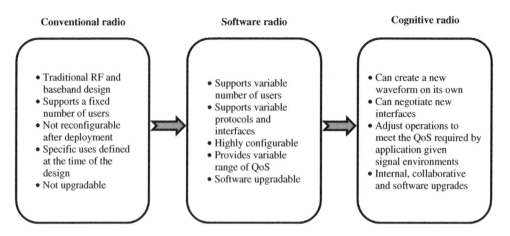

Figure 7.1   Evolution of tactical radios.

the spectrum is not in use. As long as secondary users do not interfere with primary users in specific frequency bands, secondary users can operate in these bands. One clear example is the spectrum used by a TV station which does not operate from midnight to six in the morning. A secondary user may use that spectrum during these 6 hours without interfering with the TV broadcast.

In 2002, the FCC approved some opportunistic spectrum sharing in the ultra wideband (UWB) spectra, with server restrictions on the transmission power level of the secondary user. However, this approval did not solve the known problem of "white space" (unused spectrum) in spectrum utilization. The FCC identified CRs as the platform to implement more aggressive opportunistic spectrum sharing. The FCC considered allowing a radio to transmit at higher power if it can reliably identify unused spaces in the spectra and prevent interference to primary users. CR engineers and scientists argue that an order of magnitude improvement in spectrum capacity can be achieved with the advances in CR protocols and FCC regulation. The gigahertz wideband spectra are utilized below 10 percent, which gives CRs the chance to approach a 10-fold improvement in spectrum utilization. It is argued that it is possible for *every* cell phone and WiFi user to have real-time video download at all times in ubiquitous wireless infrastructure architecture.

Dynamic utilization of bandwidth, in reference to NCW and WNW, was discussed in the previous chapter; the TDMA resource allocation can vary depending on traffic demand at each node within a single network. Dynamic utilization of spectra takes this concept a step further by dynamically allocating spectrum between competing networks as opposed to competing nodes within a network. For example, services that may require large amounts of bandwidth during short periods of time, such as public safety and emergency relief, can be given the ability to temporarily occupy large bandwidth with high priority access and no interference if they can look for available (unused) bandwidth on their prospective area of deployment.

While researchers and engineers define a CR as a node able to reason about external resources and build learning and reasoning capabilities, the FCC definition is narrowed to spectrum access:

A cognitive radio (CR) is a radio that can change its transmitter parameters based on interaction with the environment in which it operates. The majority of cognitive radios will probably be SDRs, but neither having software nor being field programmable are requirements of a cognitive radio.

The bottom line for spectrum regulation is that secondary users have to relinquish spectrum when it is requested by the primary user. The primary users may detect the existence of secondary users and communicate with them, or the primary users may claim their spectrum and force the secondary users to search for different frequency bands. In either case, the secondary users are still required to continuously sense the frequencies they use and detect the appearance of the primary user.

CR techniques must minimize the impact of both primary users and secondary users on each other. Secondary users can spread their subcarriers over a wide range of the spectrum such that when a primary user appears, its impact on reduced physical layer resources is minimized. Primary users can use beaconing to ensure that secondary users sense the upcoming spectrum use and free the spectrum. Secondary users can continuously sense the spectrum to ensure no interference with the primary user. Nevertheless, one needs to consider the effect of noise on sensing time since the secondary user senses the signal energy. At low SNR, secondary user sensing time can increase to accommodate for noise variance.

Researchers looked into spectrum pooling as a means to have CRs sense the spectrum based on the application needs. These pools can be defined as: very low band (26.9–399.9 MHz), low band (404–960 MHz), mid band (1390–2483 MHz), and high band (2483–5900 MHz). Factors such as application requirements for bandwidth, propagation distance, and traffic characteristics can decide which of these pools are most suitable.

Researchers also considered the existence of management authorities that manage the unused spectrum and charge for its use with varying prices, depending on bandwidth, location, interference level, and so on. Secondary users could then get cost-effective spectrum access through this management authority.

Example: An OFDMA based (orthogonal frequency division multiple access) centralized spectrum pooling architecture was studied. This approach envisions an 802.11-like access-point system consisting of a CR base station and CR mobile users. OFDMA was selected since this modulation approach assumes that when a primary user appears, it will not request the entire pool. OFDMA can create subcarriers of the spectrum pool, ensure no interference between subcarriers and enable the monitoring of the spectral activity. In order to detect primary users, the base station periodically broadcasts *detection frames*, which are frames containing no data. During that broadcast period, all CR users sense the spectrum. All the sensing data is sent to the base station. Sensing data is generated by the CR users which modulate a complex symbol at maximum power by using only those subcarriers where no primary user is detected. All CRs execute this modulation simultaneously. The base station will receive a reply to its detection frames, in the form of an amplified signal on all subcarriers which have been deemed "not occupied" by each CR. This excessive use of spectrum happens in a very short period, so as to minimize any possible interference with the primary user. The concept of disabling adjacent subcarriers, next to an active primary user, for the aim of reducing interference, was also studied and it was shown that this decreases the bandwidth available for secondary users.

This example proves the applicability of spectrum sharing between primary and secondary users, but it also shows that there are many possibilities and ambiguities yet to be addressed.

7.2   Conceptualizing Cognitive Radios

Please refer to Section 6.2.1 to review cross layer signaling (CLS) with the WNW radio. You can see how the channel parameters play a major role in ensuring that the higher layer protocols adapt to the changes in the physical layer parameters. We will refer to this approach as the CLS and compare it to the simplest concept of cognitive radio settings (CRS).

In the WNW waveform, CLS communicates physical layer information to the mobile data link (MDL) layer and the MDL layer makes the decision to change the radio parameters (power level, modulation mode, error control coding level, etc.). CRs have similar CLS as shown in Figure 7.2. In addition to CLS from the physical layer to the cognitive engine, the radio metrics are also sent to the cognitive engine (e.g., spectrum sensing metrics). The cognitive engine generates optimum parameters for the radio, which include changing the actual software modules (e.g., change the modulation type) and possibly creating new software components based on knowledge of the surrounding environments. Please be aware that some CR references use the term "radio" loosely. The physical layer is sometimes referred to as the radio. References that focus on cognitive algorithms conceptualize a CR as having three main components: a radio or physical layer, a MAC layer with a cognitive engine as its focus and operating environments (OEs)[4] as indicated on the right-hand side of Figure 7.2.

7.2.1   Cognitive Radio Setting (CRS) Parameters

With CLS, one can see a myriad of parameters being used by the radio to adapt to the dynamic environment. With CRS, the cognitive engine deals with even more parameters from the physical layer, MAC layer, and the radio OEs to include:

- From the physical layer, the radio metrics can include received signal parameters such as power, angle of arrival, delay spreading, Doppler spreading, and fading patterns. The radio metrics can also include noise parameters and interference power.

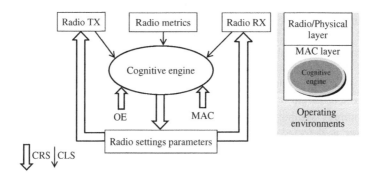

Figure 7.2   Cross layer signaling (CLS) and cognitive radio settings (CRS).

---

[4] The term operating environments will become clear as you go through the software communications architecture explained later in this chapter.

- From the MAC layer, the cognitive engine deals with issues such as frame error rate, transmission data rate, multiple-access options, and channel slot allocation. The cognitive engine addresses how the MAC layer selects frame types, changes frame size, and implements compression and encryption. It also influences how the MAC layer performs error control coding and interleaving.
- From the radio OE, the cognitive engine may deal with its own node transmitting power, power consumption rate, spreading type, spreading code, and modulation type (including symbol rate, carrier frequency, antenna diversity, dynamic range, etc.) in order to adapt the node to the changes in its surroundings. The cognitive engine also considers its own node computational power, battery life, and CPU allocation for ensuring proper use of the node internal resources.

7.2.2 The Cognitive Engine

Now we come to the more complex part of making CRs a reality. The challenge becomes how to architect *true* CRs with cognitive engines or "intelligent agents" that perform cognition tasks. These agents must be intelligent, in the sense that they can make accurate, consistent decisions. The agents must be able to collect information from surrounding environments and process this information with low overhead computation and without excessive battery consumption. These cognitive engines need to be adaptable to the dynamic environment and capable of making the optimum decisions to meet the applications needs, given the available spectrum resources.

There have been initiatives to base the cognitive engine design on approaches that range from machine learning, such as neural networks, to generic statistical based algorithms. The use of game theory for interactive decision making, between the different cognitive engines in a network, has been studied extensively in the literature. The concept of using some form of centralized control in a cognitive network, where nodal interest can be overridden by the network interest, is more applicable to the tactical use of CRs. The use of policy based network management (PBNM), or a policy engine to control the behavior of a cognitive network is also of great interest to the tactical applications of CRs. The software architecture demonstrated in Figure 7.3 (proposed by Virginia Tech) shows the cognitive engine with different modules and different interfaces to other radio components. In Figure 7.3, note the four main components: radio, radio interface, user interface, and a cognitive system module or cognitive engine. Researchers in the area of artificial intelligence (AI) who work on CRs abstract the radio capabilities and make the cognitive engine central to the radio design. Communications engineers, on the other hand, focus on the spectrum efficiency and see the cognitive engine as part of the MAC layer.

Notice how the cognitive engine has the following components:

- **Resource monitor component**: Continuously feeds the radio metrics through the cognitive engine-radio interface API (application programming interface).
- **Wireless system genetic algorithm (WSGA) component**: Implements the approach used to adapt the radio to the changing environment. As mentioned above, many approaches are proposed in the literature. Genetic based algorithms implement one type of approach.
- **Evolver component**: Enables the cognitive engine to control the search space by limiting the number of generations, crossover rates, mutation rates, fitness evaluations,

and so on, that the cognitive engine cannot be left to evolve without boundary. The evolution process is controlled to ensure legal and regulatory compliance.

- **Decision maker component**: Feeds the WSGA with initial settings, WSGA parameters, objectives, and weights. The settings are referred to by genetic algorithm developers as the initial chromosomes, where every radio parameter is represented by a gene.
- **Knowledge base component**: Contains capabilities such as short-term memory, long-term memory, WSGA parameter set, and regulatory information.

Figure 7.4 illustrates the biological metaphor of radio parameters. The genes of the radio chromosomes represent the traits of the radio (error correction, frequency bands, modulation, spreading code, payload, signal power, etc.). The WSGA analyzes the information fed to it from the cognitive system module in order to make the radio adapt to the changes in its environment. Changing the radio parameters (genes) is the equivalent to the creation of a new radio chromosome.

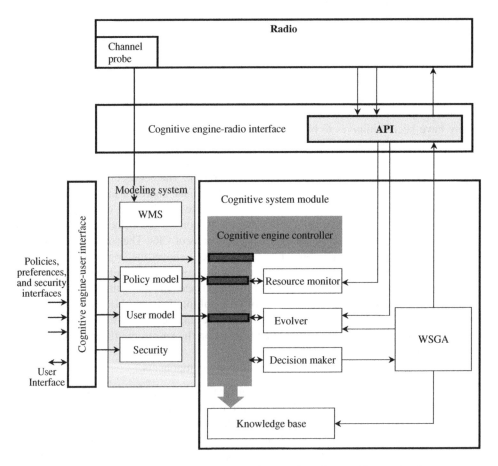

Figure 7.3 Software architecture of a cognitive radio with cognitive engine focus.

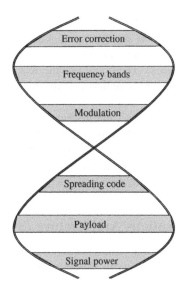

Figure 7.4 The biological metaphor of cognitive radio parameters.

7.3   Cognitive Radios in Tactical Environments

The Defense Advanced Research Projects Agency (DARPA) generated its own CR initiative, the XG initiative, to define an architectural framework where radios can be controlled using policy rules. A CR can be deployed anywhere in the world with only a change in policy rules. This radio can sense for spectrum opportunities, use the policies to decide which channels to use among a domain of CRs (in the DARPA initiative, this is referred to as the "XG domain"). A collection of centralized and distributed protocols are used to achieve the three steps necessary to have a functioning radio: spectrum awareness, spectrum allocation, and spectrum use.

The DARPA XG initiative is technically not limited to the tactical use of CRs, although the concept of universal deployment is certainly applicable to military deployment. Tactical CRs are thought of as machines that sense their environment (the radio spectrum) and respond intelligently to it. They seek other radios with which they want to communicate while avoiding or outwitting enemies (interfering radios). They conform to the etiquette defined by the FCC and deliver services to the user. They can deal with entirely new situations and learn from such experiences.

At the time of writing this book, it is hard to predict the magnitude of the impact that CRs will bring to tactical communications. The US Department of Defense realizes the drawback of SPR. It also realizes the time and fiscal investment it took to realize SPR.[5] CRs still need significant research; we are not even at the point where we can accurately model the decision processes, the learning processes, and so on. We also have yet to know

---

[5] SPR is a US government initiative from the JTRS program. Some commercial radios have begun to follow this model.

what hardware support CRs require. Although the FCC took some steps toward regulating CRs, we still have regulatory concerns since we are not dealing with just the FCC (there are multiple organizations around the world who manage spectrum, and in wartime there is no guarantee that regulators will agree). There remains the fear of losing control of the radio, and a deployed group of radios may have undesirable adaptations. There are initiatives, being brought forth, on having a defined architecture for more controllable tactical CRs. One needs to start from the software communications architecture (SCA), which is the core of the JTRS radio architecture, to see the path toward CR development. Please refer to the web site below for understanding the US government rights to the SCA data. Here, we will briefly cover the highlights of the SCA. You can access the SCA details after agreeing to the US government terms and conditions. http://www.public.navy.mil/jpeojtrs/sca/Pages/default.aspx.

7.4   Software Communications Architecture (SCA)

SCA's intended goal is to provide architecture for the deployment, management, interconnection, and intercommunication of software components in embedded, distributed-computing communication systems. SPR developers, following this approach, should maximize software application portability, reusability, and scalability. There are many definitions of a SPR; the JTRS program defines it as:

> A radio or a communication system whose output signal is determined by software and the output is entirely reconfigurable at any given time, within the limits of the radio or system hardware capabilities (e.g. processing elements, power amplifiers, antennas, etc.) by loading new software as required by the user.

This new software is referred to as "waveform software" or simply as the "waveform." Thus, a radio is capable of multiple mode operations (including variable signal formatting, data rates, and bandwidths) within a single hardware configuration. Multichannel configuration can thus provide simultaneous multimode operation. Achieving this SPR capability is dependent on the ability to select and configure the appropriate hardware that supports the software required for a specific system. The selection of hardware elements is not solely based on the input/output (I/O) devices of the communication system (analog-to-digital converters (ADCs), power amplifiers, etc.), but also on the processing capabilities of the elements. An element can be a general purpose processors (GPPs), digital signal processors (DSPs), or field-programmable gate arrays (FPGAs), and so on. A waveform may require many components of different elements, based on performance requirements. A variable collection of hardware elements needs to be connected together, based on the specific software loaded onto the system to form a communication pathway. SCA provides a common infrastructure for managing the software and hardware elements in a system, thus ensuring the produced waveform meets requirements and capability standards. SCA accomplishes this by defining a set of interfaces, the core framework, that isolate the system applications from the underlying hardware. SCA also provides the ability to ensure that software components deployed on a system are able to execute and communicate with the other hardware and software elements in the system.

7.4.1   The SCA Core Framework

The core framework abstracts the system software and hardware elements and emphasizes the open application-layer interfaces and services. These interfaces are based on the existence of the components that provide an abstraction for the SCA compliant products. These components are:

- **base application components**: provide the management and control interfaces for all system software components.
- **base device components**: allow the management and control of hardware devices within the system through their software interface.
- **framework control components**: control the instantiation, management, and removal of software from the system.
- **framework services components**: provide additional support functions and services.

The core framework interfaces are:

- base application interfaces;
- base device interfaces;
- framework control interfaces (FCIs);
- framework services interfaces (FSIs).

7.4.2   SCA Definitions

SCA differentiates between waveform or application software and the software that provides the capabilities for waveforms to execute and access the systems hardware resources. Waveform software essentially manipulates input data and determines the output of the system. Waveform software has to adhere to the base application interfaces mentioned above. The software components that provide access to the system hardware resources are referred to as SCA "devices." These devices are purely software and have to adhere to the base device interfaces mentioned above. Resources provided by the system for use by the applications are "services." Keep in mind that SCA standardizes the component interfaces and does not have any implementation requirements on the software.

The software components which provide for the management and execution of the applications and devices are referred to as the operating environment. The OE is essentially the following components:

- operating system (OS);
- framework control interfaces;
- framework service interfaces;
- transfer mechanism.

The transfer mechanism role will be clarified and further discussed in the next section.

7.4.3 SCA Components

In the SCA, an application consists of multiple software components that are loaded onto the appropriate processing resource. As demonstrated by Figure 7.5, note the following properties:

- The application components communicate with each other (to execute the implemented protocol) and with the services and devices provided by the system through the base application interface defined above. The base application interface is native to the specific systems.
- The application components access the services and devices through the "System Components" as shown in Figure 7.5. The APIs to the services and devices are standardized for a given domain. This way, the system component can communicate between the OE and the application component uniformly across multiple systems.
- The FCIs manages the system components according to the base device interfaces.
- The application components communicate to the FSIs through the transfer mechanism.
- The SCA calls for these interfaces but does not standardize them.

Any application can access OS functionality according to a defined profile. Portable operating system interface (POSIX) specifications are developed to define this access.

SCA provides for component deployment, management, and interconnection. The management hierarchy of these entities is shown in Figure 7.6. The root of this hierarchy is the domain manager, which is made aware of all implementations installed or loaded onto the system. The domain manager also has references to all file systems (through the file manager), device managers, and all applications (and their resources). Each device manager contains complete knowledge of a set of devices and/or services. A system may have multiple device managers. Each device manager must register with the domain manager giving the domain manager a full view of the system. A device manager may have an associated file system or file manager (if the device manager needs to be supported by multiple file systems). Applications and their components are instantiated on the systems and are accessed by the *application* interface.

The SCA defines a set of files named the domain profile, which describe the characteristics and attributes of the services, devices, and applications installed on the system. The domain profile is a hierarchical collection of descriptor files that define the properties of all software components in the system. Each software element in the system is described by a software package descriptor (SPD) and a software component descriptor (SCD) file.

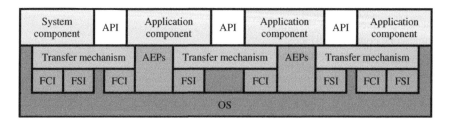

Figure 7.5 SCA layer diagram.

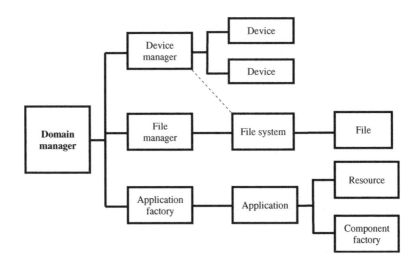

Figure 7.6   SCA management hierarchy at instantiation.

The SPD provides:

- identification of the software (title, author, etc.), the name of the code file (executable, library, or driver), and the implementation details (language, OS, etc.);
- configuration properties including initialization properties (contained in a properties file);
- dependencies to other SPDs and devices;
- references to a SCD.

The SCD defines interfaces supported and used by each specific component.

Since applications are composed of multiple software components, a software assembly descriptor (SAD) file is defined to determine the composition and configuration of the application. The SAD contains the following:

- references to all SPDs and SADs needed for the application;
- definitions of the connections required between application components;
- definitions of the needed connections to devices and services;
- additional information on locating the needed devices and services;
- definitions to any co-location (deployment) dependencies, and identification of component(s) within the application as the assembly controller.

An SCA application consists of one or more software modules that, when loaded and executed, create one or more components. These components access the platform devices and services. The software profile for an application consists of one SAD file that references (directly or indirectly) one or more SAD and SPD files. An SPD file details the software modules that are to be loaded and executed for each application. The SPD file does so by referencing the SCD and other files known as the property files. The SPD specifies the *device* implementation requirements for loading dependencies (processor kind, etc.) and processing capacities (e.g., memory, process) for an application software module.

A SAD may also reference an application deployment descriptor (ADD) that defines the channel deployment precedence order for the application.

Similar to the application SAD, a device manager has an associated device configuration descriptor (DCD) file. The DCD identifies all devices and services associated with this device manager, by referencing the associated SPDs. The DCD also defines the properties of the specific device manager, enumerates the needed connections to services (e.g., file systems), and provides additional information on how to locate the domain manager. In addition to an SPD, a device may have a device package descriptor (DPD) file which provides a description of the hardware associated with this (logical) device including description, model, manufacturer, and so on.

7.4.4 SCA and Security Architecture

SCA, in a nutshell, uses an object-oriented approach where a component is "an autonomous unit within a system or subsystem which provides one or more interfaces for user access." Components are defined as stereotypes that represent the bridge between the interface definitions and the products that will be built in accordance with the SCA. The component definitions specify any required behaviors, constraints, or associations that must be adhered to when the corresponding SCA products are built. Within this specification, the unified modeling language (UML) can be used to graphically represent interfaces and component definitions, while the interface definition language (IDL) contains the textual representation.

Figure 7.7 demonstrates a view of SCA considering security architecture and using common object request broker architecture (CORBA) as the transfer mechanism and IDL to represent the APIs. Security architecture forces the use of separate hardware (GPP, DSP, and FPGA) on the plain text (red) side from the hardware (GPP, DSP, and FPGA) used at the cipher text (black) side. Security architecture relies on the security processor JTRS AIM (advanced InfoSec machine) to provide a secure hardware platform on which software based cryptographic algorithms and higher-level crypto equipment applications can execute. JTRS AIM is totally programmable, enabling the SPR in which it is embedded to be modified or upgraded with a download of software. The JTRS AIM design includes three independent cryptographic processors that make it possible for a JTRS radio to have three security enclaves (plain text subnets): a sensitive compartmented information (SCI) subnet, a secret subnet, and a top-secret subnet. Any exchange of information between the red (plain text side) and the black (cipher text side) has to go through the AIM and must be approved by NSA.

The SCA OE adheres to the NSA separation of the plain text hardware from that of the cipher text hardware, as well as the separation of security classification enclaves' hardware from each other.

Some defense industries from other nations have adapted the SCA approach, indicating that the trend is gaining momentum in defense applications. With commercial applications, SCA is being considered by some systems developers and is being resisted by others. The fact is that SCA has a way to go before some commercial communications systems fully buy into it.

With technologies such as cell phones, due to market share and the number of units produced, developers tend to use specialized hardware (e.g., wireless phones have specialized

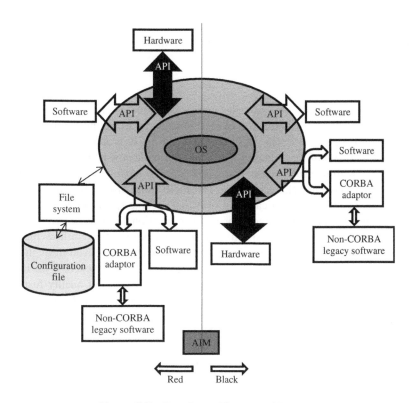

Figure 7.7   Security architecture with SCA.

hardware to reduce size). These developers view SCA as GPP-centric, which can increase end user product size.[6] These commercial applications find some of the SCA capabilities unnecessary and have no need for waveform portability. Also, commercial producers of such communication systems maintain an interest in static products to keep market demand high.

On the other hand, other types of commercial applications, such as base stations, see much promise in SCA and the development of multistandard base stations (e.g., 3G/4G and Worldwide Interoperability for Microwave Access (WiMAX))[7] has promoted the use of SCA. The developers of commercial technologies such as base stations expect US government funding to finance the evolution of SCA capabilities, thus avoiding any financial burden to themselves.

In the end, reduction of cost will be the driving force behind the popularization of SCA. If the commercial industry sees promising cost savings, it will surely adapt. As the number of protocols supported by SCA increases and software reusability increases, cost cutting will naturally follow. The advancement in hardware capabilities are on the side of SCA. Low-cost standard ADCs, flexible RF modules, and powerful GPP are becoming an available reality. Since low-cost hardware has been developed, software modularity and standard interfaces

---

[6] Advances in GPP may combat this concern.
[7] The field of neural networks was researched for decades before its concepts became applicable. Likewise, it will likely take decades from the time of this book until we know how advanced the radio cognitive engine can become.

have become common practice today in many other non-communications applications. These trends are helping advance SCA and paving the way for the CRs of the future.

7.5 Spectrum Sensing

Spectrum sensing is the only area of radio communications where we understand how to couple communications with CRs. We have a long ways to go before we can develop deployable cognitive engines.[7] The FCC definition of a CR is mainly geared toward the autonomous exploitation of local unused spectrum to provide new paths for spectrum access.

In order for a CR to measure, sense, learn, and adapt, it must be aware of the parameters related to the radio channel characteristics. The radio must also know the availability of spectrum and the level of power it can emit over that specific spectrum. One can argue that spectrum sensing is the most important aspect of CR functionality since it provides the radio with awareness regarding the local spectrum usage and information on how the spectrum is being utilized by primary users. Spectrum awareness can be achieved through the usage of geo-location information, beacons, and/or sensing by CRs. With beacons, the transmitted information can include occupancy of a spectrum and other channel quality features. This section of the chapter addresses spectrum sensing performed by the CR for the purpose of obtaining spectrum usage characteristics covering time, space, and frequency bands. A CR will also learn the types of signals occupying the spectrum, including modulation techniques, waveforms, bandwidth, carrier frequency, and so on. This type of sensing requires advanced signal analysis techniques with high computational complexity.

7.5.1 Multidimensional Spectrum Awareness

The conventional definition of spectrum opportunity is "a band of frequencies that are not being used by the primary user of that band at a particular time in a particular geographic area." This definition covers: frequency, time, and space. One other dimension that can be considered is the spreading code in spread spectrum modulation (see Section 2.6). Another dimension is the hopping pattern mentioned in Section 2.6.2. Although these dimensions can bring about new opportunities for spectrum usage, they can also create new challenges. With directional antennas and beam forming, the beam angle can be another dimension to be exploited for spectrum reuse. One can see how additional dimensions of spectral space can increase spectrum sharing. Naturally the CR would need more advanced sensing techniques.

Now with CR, one can think of spectrum sensing as a multidimensional problem that may be defined as "a hyperspace occupied by radio signals, which has dimensions of location, angle of arrival, frequency, time, and other dimensions." This hyperspace is called many names in the literature, including electrospace, transmission hyperspace, radio spectrum space, and spectrum space. The definition of space essentially describes how the spectrum can be shared among primary and/or secondary systems. It is essential to know that each dimension has its own parameters that should be sensed for complete spectrum awareness. A CR defines such an n-dimensional space for spectrum sensing. The radio must have the capabilities for spectrum sensing in all dimensions of the spectrum space and have techniques for finding spectrum *space* holes. Before CRs become a reality, advanced spectrum sensing techniques that offer awareness in multiple dimensions of the spectrum space should be developed. Let us discuss these dimensions briefly.

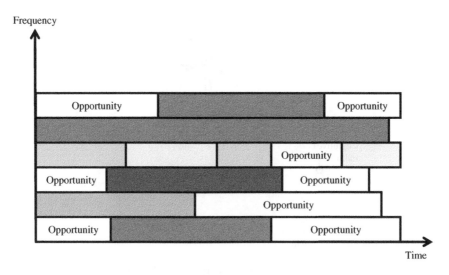

Figure 7.8   Spectral opportunity in terms of time and frequency.

7.5.1.1   Time and Frequency

Spectrum availability in the frequency domain is divided into chunks of narrow bands. When all the bands are not simultaneously used, an opportunity for usage appears. This opportunity is defined for a specific band in time as shown in Figure 7.8. For each frequency band, there will be times when it is available for opportunistic usage. One can see that a frequency band can be occupied for a given time, but it might also be empty at another time. As such, the temporal dimension goes hand in hand with the frequency dimension. Some researchers exploit the idle periods between bursty transmissions of wireless LAN signals, by finding spectral holes for very short durations of time.

A primary user must be able to claim his frequency bands at any time, which requires CRs to be able to identify the presence of primary users at any moment and relinquish the frequency band immediately. This requirement creates a challenge for CR developers. There is a tradeoff between the speed (sensing time) and reliability of sensing. If the primary users' occupation of a spectrum changes slowly, sensing frequency requirements can be relaxed. Consider the case of the TV broadcast, which (in a given geographical area) does not change frequently. The IEEE 802.22 draft standard, covered later in this chapter, defines the sensing period as 30 seconds. Interference tolerance of primary license owners is also an important factor. An opposing example of the TV broadcast case could be a CR exploiting opportunities in public safety bands, where sensing must be as frequent as possible to prevent interfering with sudden urgent use of the spectrum. Thus, sensing times can greatly affect the performance of secondary users.

7.5.1.2   Geographical Space

In the graphical space dimension, a CR location is defined by its latitude, longitude, elevation, and distance from the primary users. At any given time, spectrum opportunity may

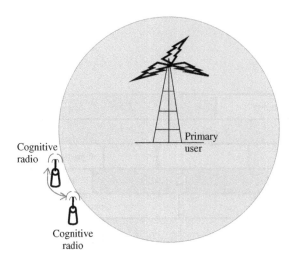

Figure 7.9 Geographical separation between a primary user and cognitive radios communicating at the same frequency.

be available in some parts of a geographical area while being fully occupied in other parts of that same area. The graphical space dimension helps the CR make use of propagation loss (path loss) in space. A CR can simply look at the interference level with its signal. If it detects no interference, this means no primary user transmission in a given local area. Figure 7.9 shows an example of geographical separation between a primary user and CRs communicating at the same frequency.

Nevertheless, there remains the potential of interfering with a hidden primary user. Please refer to Section 4.1 to see how collision occurs with carrier sense multiple access (CSMA) MAC protocol and note that the hidden primary user issue is similar. This issue can occur if there is severe multipath fading or shadowing of the RF signal as it is emitted from the primary user while the secondary user is scanning for the primary users' transmissions.

Figure 7.10 illustrates a hidden node problem between a primary user and a CR, where the circles represent the area of coverage (AOC) of the primary user. One can see how the CR device close to the primary user transmitter AOC causes unwanted interference with the primary user's receiver since the transmitting signal from the primary user could not be detected. The literature discusses a technique called cooperative sensing which handles hidden primary user problems as discussed later in this chapter.

7.5.1.3 Spreading Code

This dimension considers the spreading code, time hopping (TH) or frequency hopping (FH) sequences used by the primary users. Making use of this dimension requires the CR to be aware of the timing information so that it can synchronize its transmissions with the primary user. In some cases, the CR can use a long spreading code (to transmit at a very low bit rate over a wide spectrum) causing minimal interference with the primary user. Spectrum use over a wideband might be allowed at a given time through spread spectrum or FH. The primary user could be aware of a secondary user, who is using a long spreading code.

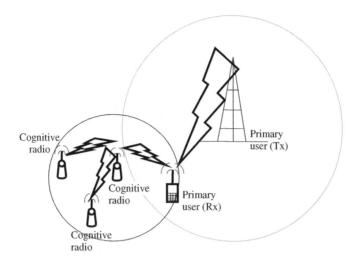

Figure 7.10   Hidden node problem.

Secondary users may continue with their communications without affecting the primary user communications capabilities.[8] Simultaneous transmission without interfering with primary users is possible with orthogonal spreading codes. This requires the CR to figure out the spreading code of the primary user.[9]

7.5.1.4   Angle of the RF Beam

When primary users use directional antennas (such as microwave links), the locations of the primary users and the directions of their RF beam (which include the azimuth and elevation angle) can be leveraged by the CR to create opportunistic spectrum. The CR's spectrum sensing techniques will estimate the location/position of the primary users and the direction of their beam. With a directional antenna, if a primary user is transmitting in a specific direction, the secondary user can transmit in the opposite direction without creating interference as shown in Figure 7.11.

7.5.2   Complexity of Spectrum Sensing

CRs rely on advanced ADCs, which can sample a large dynamic range of spectrum at high resolution, and powerful signal processors. It is important for CRs to have optimum techniques for noise variance estimation in order to contain its spectrum emission. Optimum implementation of modulation and coding is needed to help optimize spectrum emission

---

[8] In rural areas, a cognitive radio with a very small spectrum footprint (as in a home) can use this technique over some broadcast frequency band. As long as it never affects other subscribers on this broadcast band, it has no negative impact.

[9] The implementation of spreading codes in systems like 3G follows certain standards that can be reverse engineered to determine the spreading code of the base station and the spreading codes assigned to end users. Cognitive radio spectrum sensing techniques can make use of this vulnerability and create opportunities for spectrum reuse.

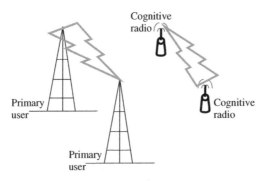

Figure 7.11 Cognitive radios leveraging of the primary user beam angle to create spectral opportunity.

power. For a CR subnet, physical layer resource allocation techniques need to have power control capabilities.

In conventional radios, the RF receivers are tuned to a specific bandwidth, thus making estimating noise and interference from other spectrum emitters easier. Receivers can have narrowband and baseband signal processing techniques that do not require high processing power. As we move to the CR, the complexity of noise and interference estimation problem increases drastically. With CRs, the RF receiver must process signals over much wider frequency bands, searching for opportunities. The RF components (antennas, power amplifiers, etc.) of CRs must operate over a wide range of frequencies and have high-speed DSPs and FPGAs to satisfy the computational demand of the wide frequency bands without excessive processing delay.

CR researchers study two different categories of spectrum sensing: single radio spectrum sensing and dual radio spectrum sensing, as explained below.

Single Radio Spectrum Sensing
With this approach, a specific timeslot is allocated for spectrum sensing. This will limit the spectrum sensing duration and accuracy of the sensing process. Also, the efficiency of spectrum use is compromised because a portion of the available timeslot is utilized by the spectrum sensing process instead of transmitting data.

One can see the obvious advantages of single radio spectrum sensing as being simplicity and lower cost.

Dual Radio Spectrum Sensing
In this approach, the RF signal is split into two chains (paths). One RF path is dedicated to data transmission and reception, while the other path is dedicated to spectrum sensing. With advanced signal processing techniques, only one antenna is needed for both chains. While this approach ensures that data transmission utilizes the entire available spectrum (after all CR concepts revolve around optimal spectrum use), the obvious drawbacks of such an approach are the increased complexity, power consumption, and hardware cost.

The literature is full of different proposals for spectrum sensing that are mutations of these two spectrum sensing categories.

7.5.3 Implementation of Spectrum Sensing

Different implementation methods are proposed for identifying the presence of signal transmissions. A CR would need to first detect the characteristics of the identified transmission. Some of the studied techniques are to include the following.

7.5.3.1 Energy Detector Based Spectrum Sensing

This technique is also known as radiometry or periodogram. It is a generic approach that has low computational and implementation complexities since receivers do not need knowledge of the primary user's signal. The primary user signal is simply detected by comparing the output of the energy detector with a threshold which depends on the estimated noise. Naturally, selection of the threshold can be a challenge where interference and noise can mislead the CR. Also, this technique cannot work with a spread spectrum signal that has low energy.

Let us assess energy detector based spectrum sensing. Please refer to Section 2.2 for a review of signal detection and the effect of noise on the probability of the bit error rate for refreshing. Here, let us assume that the CR received a signal with additive white Gaussian noise (AWGN) that can be expressed as:

$$y(n) = s(n) + w(n), \tag{7.1}$$

where $s(n)$ is the signal to be detected, $w(n)$ is the AWGN, and n is the sampling index.

Note that $s(n) = 0$, if the primary user is not transmitting. The energy detector in the CR senses the signal energy as a vector of multiple sampling points that can be expressed as:

$$M = \sum_{n=0}^{N} |y(n)|^2, \tag{7.2}$$

where N is the size of the observation vector.

Now the energy detector needs to compare the decision metric M against a fixed threshold λ_E. The detector needs to distinguish between two hypotheses, one for the presence and one for the absence of the primary user signal. These two hypotheses are:

$$H_0 : y(n) = w(n), \tag{7.3}$$

$$H_1 : y(n) = s(n) + w(n). \tag{7.4}$$

The detection algorithm can successfully detect the frequency with probability P_D and the noise can cause a false alarm with a probability of P_F. The detection problem can be expressed as:

$$P_D = \text{Pr}(M > \lambda_E | H_1), \tag{7.5}$$

$$P_F = \text{Pr}(M > \lambda_E | H_0), \tag{7.6}$$

where P_F is the probability that the energy detector falsely decides that the opportunity frequency band is occupied when it is not. Naturally, the CR desires to minimize P_F, in order to maximize spectrum utilization. If the energy detector has knowledge of the signal and noise power, maximum likelihood decisions can be applied as discussed in Section 2.2. Practically, the decision threshold λ_E is selected to create a balance between P_D and P_F.

Signal detection techniques similar to that discussed in Section 2.2 can be used to estimate the noise power. Estimating the signal power is difficult since it can change due to transmission characteristics, and the distance between the CR and the primary user. The threshold λ_E can be chosen to meet a given false alarm rate that can be acceptable. This makes it sufficient to know the noise variance, assuming a zero-mean Gaussian random variable with variance σ_w^2. With this assumption, we can express the noise as $w(n) = (0, \sigma_w^2)$. We can also simplify how we model the signal $s(n)$ in order to make this analysis possible. We can express the signal as a zero-mean Gaussian variable[10] as well as making $s(n) = (0, \sigma_s^2)$.

This assumption allows us to express the decision metric M as a chi-square distribution with $2N$ degree of freedom (χ_{2N}^2) giving,

$$M = \begin{cases} \dfrac{\sigma_w^2}{2}\chi_{2N}^2 & H_0, \\[2mm] \dfrac{\sigma_w^2 + \sigma_s^2}{2}\chi_{2N}^2 & H_1. \end{cases} \tag{7.7}$$

Thus, the energy detector can calculate P_D and P_F as:

$$P_F = 1 - \Gamma\left(L_f L_t, \frac{\lambda_E}{\sigma_w^2}\right), \tag{7.8}$$

$$P_D = 1 - \Gamma\left(L_f L_t, \frac{\lambda_E}{\sigma_w^2 + \sigma_s^2}\right), \tag{7.9}$$

where $\Gamma(a, x)$ is the incomplete gamma function.[11]

The energy detector can use receiver operating characteristics (ROCs) to compare the performances for different threshold values. ROCs are a set of curves that explore the relationship between the probability of detection and the probability of a false alarm for a variety of different thresholds. Based on given requirements, an optimal threshold can be reached. Figure 7.12 exemplifies different ROC curves for different SNR values, using $N = 15$ in Equation 7.2 SNR is defined as the ratio of the primary user's signal power to noise power $\frac{\sigma_s^2}{\sigma_w^2}$.

One can see from Figure 7.12 that as SNR increases, we can achieve higher P_D at lower P_F. It is critical to understand that this process relies on estimating the noise variance. Noise power estimation error can cause significant performance loss. Noise level can be estimated dynamically by separating the noise and signal subspaces. With this technique, noise variance is obtained as the smallest eigenvalue of the incoming signal's autocorrelation. This estimated noise variance can then be used to find the decision threshold λ_E that satisfies the requirements for a given false alarm rate. This algorithm is applied iteratively, where N can be a moving average window that normalizes the noise power.

---

[10] This approximation is hard to justify in the presence of fading.
[11] Please refer to mathematical references such as reference [1].

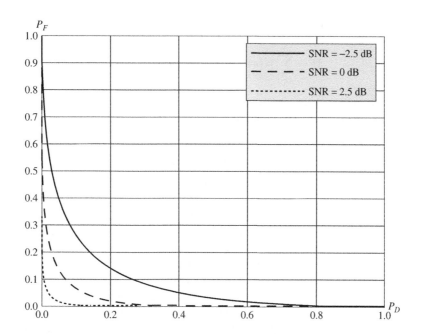

Figure 7.12   An example of different ROC curves for different SNR (not to scale).

7.5.3.2   Waveform Based Spectrum Sensing

This approach relies on preknowledge of the signal to be sensed. Some wireless signals use known synchronization patterns to align the receiving node processing to the received signal. These patterns can be exploited by CRs. Synchronization patterns can use any of the following techniques: preambles, mid-ambles, regularly transmitted pilot patterns, spreading sequences, and so on. These known patterns make it possible to perform sensing by correlating the received signal with a known copy of itself (it is a form of coherent detection similar to that discussed in Chapter 2). Because of the correlation of the primary user received signal with a noise-free version, waveform based sensing outperforms energy detector based sensing. The reliability of the correlation process increases with increasing the known signal length, resulting in further enhancement of the CR's primary user sensing. This approach applies to known signals such as that of the IEEE 802.11x waveforms.

Let us use the same AWGN model discussed in the energy detector based sensing to analyze waveform based sensing. The sensing metric here can be expressed as:

$$M = Re\left[\sum_{n=1}^{N} y(n)s^*(n)\right],\tag{7.10}$$

where $*$ represents the conjugation operation and $y(n)$ is given by Equation 7.1.

In the absence of the primary user, the sensing metric becomes,

$$M = Re\left[\sum_{n=1}^{N} w(n)s^*(n)\right].\tag{7.11}$$

In the presence of the primary user, the sensing metric becomes,

$$M = \sum_{n=1}^{N} |s(n)|^2 + Re\left[\sum_{n=1}^{N} w(n)s^*(n)\right]. \tag{7.12}$$

Similar to the energy detector based sensing, the presence of a primary user signal is determined by comparing the decision metric M against a fixed threshold λ_E.

7.5.3.3 Cyclostationarity Based Spectrum Sensing

In some cases, OFDMA waveforms are altered by the transmitter to add signatures in the form of cycle frequencies at certain frequencies. These signatures are needed to increase the robustness against multipath fading. CRs can make use of these features for primary user signal sensing. In these cases, the received signal has statistical mean and autocorrelation features that can be exploited. The signal itself can also have periodicity features. The introduced cyclic frequencies and the periodicity features make the signal cyclostationary. The cyclostationary characteristics of the signal follow a spectral density (cyclic spectral density function – CSDF) that can be leveraged for the detection process that differentiates noise from primary users' signals. While the modulated signals are cyclostationary, the noise has wide-sense-stationary characteristics with no correlation. Cyclostationarity can also be used for distinguishing among different types of primary users.

The CSDF of the received signal y can be calculated as[12]:

$$S(f, \alpha) = \sum_{\tau=-\infty}^{\infty} R_y^{\alpha}(\tau)e^{-j2\pi f\tau}, \tag{7.13}$$

where f is the signal frequency, α is the cyclic frequency, and $R_y^{\alpha}(\tau)$ is the cyclic auto-correlation function (CAF), which can be expressed as:

$$R_y^{\alpha}(\tau) = E[y(n+\tau)y^*(n-\tau)e^{j2\pi\alpha n}]. \tag{7.14}$$

The CR can be informed about α for specific primary users in the area or the radio can extract α from sensed signals and use it to identify signals.

7.5.3.4 Radio Identification Based Spectrum Sensing

Knowing the specific technology used by the primary user can inform the CR about the dimensions of the spectrum characteristics of the primary user signal with high precision. The CR can extract features from the received signal and select the most probable primary user technology. The amount of energy detected and the distribution of the signal energy across the spectrum can be used to map the signal to a specific technology (e.g., Bluetooth has very specific distribution of signal energy). Channel bandwidth and signal shape can be an excellent discriminating parameter. Energy detectors can identify the operation bandwidth and center frequency. Standard deviation of the instantaneous frequency and the maximum

---

[12] Please look at probability and random signal analysis references for more details about types of density functions that apply to RF signals.

duration of a signal can be extracted using time–frequency analysis. The CR literature has shown that neural networks concepts can be used to classify the primary user signals, and studies that compare neural network concepts to statistical based models are available.

7.5.3.5 Matched Filtering Based Spectrum Sensing

Matched filtering is the most efficient method that a CR can use when all aspects of the primary user signal are known (bandwidth, operating frequency, modulation type and order, pulse shaping, and frame format). For a predefined probability of false alarm, the CR can sense the primary user signal with high accuracy in a short period of time. Sensing is completed very quickly, far exceeding the speed of other methods mentioned above. The disadvantages of this sensing method include:

- the CR must demodulate the received signals;
- the implementation complexity of the sensing unit in the CR is not practical since the CR would need receivers for all signal types;
- large power consumption is needed to execute the various receiver algorithms.

7.5.3.6 Other Sensing Methods

There are many sensing techniques proposed in the literature to including:

- multi-taper spectral estimation – this is a maximum likelihood estimator of power spectral density and works well for wideband signals.
- random Hough transform – as a transform domain detection approach, this suits signals with periodic patterns. It exploits the statistical covariance of noise and signal. This method is shown to be effective at detecting digital television (DTV) signals.
- wavelet transform based estimation – this is the method that allows wavelets to be used for detecting edges in the power spectral density (transition from the occupied band to the empty band and vice versa) of a wideband channel (see Section 2.5). This detection allows the frequency spectrum to be labeled occupied or unoccupied in a binary fashion. Analog implementation of wavelet transform based sensing was also studied, which yields low power consumption and can be implemented in real time.

7.5.4 Cooperative Spectrum Sensing

The idea behind this approach is for the CRs to share information among themselves and combine results from various measurements. The optimum information fusion rule for combining sensing information is referred to as the Chair-Varshney rule and is based on log-likelihood ratio tests. Cooperative sensing has the following advantages:

1. reduction of noise uncertainty;
2. can overcome shadowing of the primary signal;
3. solves the hidden primary user problem;
4. decreases sensing time;
5. provides spectrum reuse capabilities better than local sensing.

The downside of cooperative sensing is the need to develop mechanisms for information sharing, which increases the complexity of implementing CRs. The control channel used for sharing channel allocation information can also be used for sharing spectrum sensing results among CR.

7.5.5 Spectrum Sensing in Current Wireless Standards

At the time of writing, some commercial standards have started developing cognitive features. It is a challenge to base wireless standards on wideband spectrum sensing and opportunistic exploitation of the spectrum. Wireless technologies refer to these cognitive features as spectrum sensing for dynamic frequency access (DFA). This leads to the development of advanced receivers with adaptive interference cancellation.

7.5.5.1 IEEE 802.11k

The 802.11k is an extension of the IEEE 802.11 specification which defines the noise histogram report and channel load report. The noise histogram report provides methods for measuring interference levels from all non-802.11 energy at the 802.11 mobile units. The access point collects channel information from each mobile unit in addition to its own measurements. The access point uses this data to regulate channel access. The channel load report is used to improve traffic distribution. Consider a scenario where we have multiple 802.11 access points. A mobile unit connects to the access point with the strongest signal at its location. In some cases, one access point can become overloaded (increased collision) while other access points are underused. When one access point becomes overloaded, a new mobile unit can be assigned to one of the underused access points, even if the received signal at the new mobile unit from the assigned access point is weaker than the signal received from the oversubscribed access point. The end result is an improved overall system throughput and more efficient utilization of spectrum resources.

7.5.5.2 Bluetooth

Bluetooth standards added a new feature called adaptive frequency hopping (AFH) to reduce interference between wireless technologies sharing the 2.4 GHz unlicensed radio spectrum. This spectrum is shared between the IEEE 802.11b/g devices, cordless telephones, and some microwave ovens. AFH identifies transmissions in bands of this spectrum and avoids their frequencies. AFH employs a sensing algorithm to determine the presence of other devices, to avoid colliding with them. This sensing algorithm collects statistics per frequency band on packet error rate, bit error rate, received signal strength, and so on, to classify an HF as *good*, *bad*, or *unknown*.

7.5.5.3 IEEE 802.22

IEEE 802.22 standard is in the development phase at the time of writing, and is truly a *CR standard* because of the cognitive features it contains. The IEEE 802.22 standard has well-defined spectrum sensing requirements. The IEEE 802.22 wireless regional area network

(WRAN) devices sense TV channels and identify transmission opportunities. The sensing is based on fast and fine sensing. The fast sensing stage uses energy detectors. There are many proposed techniques for the fine sensing stage to include waveform based sensing, cyclostationary detection, and matched filtering. The base station acts as a centralized collaborative sensing entity and coordinates the sensing process with the subscriber stations.

7.6 Security in Cognitive Radios

Security for CRs has yet to be thoroughly studied. The same factors that give CRs outstanding potential can be used maliciously to attack CR networks. Consider a malicious user who can modify his own air interface to mimic a primary user. This user could mislead the spectrum sensing performed by legitimate primary users. This attack type is referred to as the primary user emulation (PUE) attack. Even if CRs or primary users can identify a PUE, there are challenges in developing effective countermeasures. One can see how primary users have to change the way they work to accommodate CRs. A form of public key encryption is necessary to identify legitimate primary users and prevent PUE attacks. Primary users using digital modulation would be required to transmit a signature along with their transmissions. This signature would be used for validating the primary user. This is just one example of CR vulnerabilities. Every developed and deployed technology has been misused one way or another. As CRs are deployed, attackers will find ways to compromise the network, and radio developers will have to develop techniques to combat such problems.

7.7 Concluding Remarks

7.7.1 Development of Cognitive Radios

The field of producing hardware for CRs is still evolving. There are some available hardware and software platforms as mentioned below.

GNU[13] radio is a free software signal processing toolkit that helps engineers and scientists experiment with building and deploying SPR systems. GNU is rich with capabilities that allow its users to experiment with manipulating and using the electromagnetic spectrum. One of the basic design concepts of GNU radio is to allow its users to use single generic radio hardware. This hardware can be used to feed complex CR formations into the powerful GNU signal processing software.

The GNU radio toolkit also utilizes the universal software radio peripheral (USRP) which is a computer based transceiver with powerful ADC and DAC converters with circuitry to interface with a host computer. USRP interfaces with several transmitters and receivers covering frequency bands from 0 to 5.9 GHz.

Based on the DARPA XG project, an XG experimental radio was shown to reuse the 802.16 spectrum.

One can anticipate that CR hardware and software platforms will become more sophisticated and that tactical use of CRs will have more security focus and restrict rules for the radio evolution.

---

[13] GNU started as project at MIT for free software which initiated an operating system called GNU ("GNU" is a recursive acronym which stands for "GNU's Not Unix"). GNU radio is a toolkit of GNU.

7.7.2 Modeling and Simulation of Cognitive Radios

Modeling and simulation of communications systems has undergone major overhauling as technology has advanced over the past few decades. Until the early 1980s, protocols were simple and could be developed according to specifications, and tested for acceptance. As communication system complexity increases, engineers will not be able to create specifications for development without the help of modeling and detailed simulators. Discrete event simulation tools have allowed the development of protocols in the simulation environment. This provided insight to the engineers who designed communication specifications. Scientists also used modeling and simulation tools to develop new protocols. There was a trend from the US military to encourage the use of modeling and simulation as a design aid for developing communications systems. As tactical communications systems became ever more complex, government sponsors began identifying waste in the modeling and simulation process. They declared that resources used for building models are a waste since the same protocols are developed twice, once in the modeling and simulation environments and once in the actual product. The modeling and simulation trend then transitioned to portable software. This meant that modeling and simulation software became the actual product software. New capabilities for the development of communication systems were produced since modeling, simulation, development, and testing all relied on a single environment.

The trend of using portable code in the modeling and simulation environments may be of great interest to CR developers. Creating specifications for a radio that morphs and changes applications may not be possible. If CRs take off – and we are at the point where we are developing AI in communications systems – we would need a modeling and simulation environment where hardware specifications in the modeling environment are identical to the product. The GNU radio toolkit took the first step in this direction. In addition, the fear of having a deployed CR network morph into an undesired state must be addressed in this development/modeling and simulation/testing platform. With the advancement in CRs, a new generation of development platforms will need to be created. These platforms will need to encompass the technology development, modeling and simulation, laboratory testing, production testing, and operational testing needs of today's world.

7.7.3 Historical Perspective

The term "cognitive radio" appeared in the literature in the 1990s as an extension of SPRs with "radio etiquettes" that moderate the use of spectrum patterns. Within a decade, standardization began, although with limited focus on spectrum reusability. Only time will determine to what degree CRs will become an integral part of our society. The tactical world's interest in CRs, and government funding, will help to advance the technology, although one can expect a more contained form of CRs for tactical applications (since the commander's intent has to override the CR engine).

Bibliography

1. Abramowitz, M. and Stegun, I. (eds) (1970) *Handbook of Mathematical Functions With Formulas, Graphs, and Mathematical Tables*, 9th edn, National Bureau of Standards.
2. Mitola, J. (2007) in *Cognitive Radio Architecture, in Cognitive Networks: Towards Self-Aware Networks* (ed. Q.H. Mahmoud), John Wiley & Sons, Ltd, Chichester, doi: 10.1002/9780470515143

3. Mitola, J. (1999) Cognitive radio for flexible mobile multimedia communications. Proceedings of the IEEE International Workshop on Mobile Multimedia Communications.

4. Mitola, J. (2000) Cognitive radio an integrated agent architecture for software defined radio. PhD thesis. KTH Royal Institute of Technology, Stockholm, Sweden.

5. Haykins, S. (2005) Cognitive radio: brain empowered wireless communications. *IEEE Journal on Selected Areas in Communications*, **23** (2) 201–220.

6. Mitola, J. and Maguire, G. (1999) Cognitive radio: making software radio more personal. *IEEE Personal Communications*, **6** (4), 13–18.

7. Mitola, J. (1992) Software radios-survey, critical evaluation and future directions. Proceedings of IEEE National Telesystems Conference, New York, pp. 13/15–13/23.

8. Federal Communications Commission (2005) Notice of Proposed Rulemaking and Order: Facilitating Opportunities for Flexible, Efficient, and Reliable Spectrum Use Employing Cognitive Radio Technologies. ET Docket No. 03-108, February 2005.

9. Yucek, T. and Arslan, H. (2009) A survey of spectrum sensing algorithms for cognitive radio applications. *IEEE Communications Surveys and Tutorials*, **11** (1), 116–130 (First Quarter).

10. Hu, W., Willkomm, D., Abusubaih, M. *et al.* (2007) Dynamic frequency hopping communities for efficient IEEE 802.22 operation. *IEEE Communications Magazine*, **45** (5), 80–87.

11. Cabric, D.B. (2007) *Cognitive Radios: System Design Perspective*, University of California, Berkeley, UMI Microform 3306077.

12. Lundén, J., Koivunen, V., Huttunen, A., and Poor, H.V. (2007) Spectrum sensing in cognitive radios based on multiple cyclic frequencies. Proceedings of the IEEE International Conference on Cognitive Radio Oriented Wireless Networks and Communications (CROWNCOM), Orlando, FL, July/August 2007.

13. Papadimitratos, P., Sankaranarayanan, S., and Mishra, A. (2005) A bandwidth sharing approach to improve licensed spectrum utilization. *IEEE Communications Magazine*, **43** (12), 10–14.

14. Digham, F., Alouini, M., and Simon, M. (2003) On the energy detection of unknown signals over fading channels. Proceedings of IEEE International Conference on Communication, Seattle, WA, May 2003, vol. 5, pp. 3575–3579.

15. Cordeiro, C., Challapali, K., and Birru, D. (2006) IEEE 802.22: an introduction to the first wireless standard based on cognitive radios. *Journal of Communications*, **1** (1) pp. 3 8–47.

16. Ghasemi, A. and Sousa, E.S. (2007) Asymptotic performance of collaborative spectrum sensing under correlated log-normal shadowing. *IEEE Communications Letter*, **11** (1), 34–36.

17. D'Amour, C., Life, R., Elmasry, G.F., and Welsh, R. (2004) Determining network topology using governing dynamics based on nodal and network requirements. Proceedings of MILCOM 2004, Monterey, CA, November 2004, U076.

18. Holma, H. and Toskala, A. (2000) *WCDMA for UMTS, Radio Access for Third Generation Mobile Communications*, John Wiley & Sons, Ltd, September 2000.

19. Mathur, C.N. and Subbalakshmi, K.P. (2007) Digital signatures for centralized DSA networks. First IEEE Workshop on Cognitive Radio Networks, Las Vegas, NV, January 2007, pp. 1037–1041.

20. Lehtomäki, J., Vartiainen, J., Juntti, M., and Saarnisaari, H. (2006) Spectrum sensing with forward methods. Proceedings of the IEEE Military Communications Conference, Washington, DC, October 2006.

21. Marcus, M. (2005) Unlicensed cognitive sharing of TV spectrum: the controversy at the federal communications commission. *IEEE Communications Magazine*, **43** (5), 24–25.

22. Geirhofer, S., Tong, L., and Sadler, B. (2006) A measurement-based model for dynamic spectrum access in WLAN channels. Proceedings of the IEEE Military Communications Conference, Washington, DC, October 2006.

23. Cabric, D., Tkachenko, A., and Brodersen, R. (2006) Spectrum sensing measurements of pilot, energy, and collaborative detection. Proceedings of the IEEE Military Communications Conference, Washington, DC, October 2006.

24. Pawełczak, P., Janssen, G.J., and Prasad, R.V. (2006) Performance measures of dynamic spectrum access networks. Proceedings of the IEEE Global Communication Conference (Globecom), San Francisco, CA, November/December 2006.

25. Guzelgoz, S., Celebi, H., and Arslan, H. (2011) Statistical characterization of the paths in multipath PLC channels. *Proceedings of the IEEE Transactions on Power Delivery*, **26** (1), 181–187.

26. Gandetto, M. and Regazzoni, C. (2007) Spectrum sensing: a distributed approach for cognitive terminals. *IEEE Journal on Selected Areas of Communications*, **25** (3), 546–557.

27. Geirhofer, S., Tong, L., and Sadler, B. (2007) Dynamic spectrum access in the time domain: modeling and exploiting white space. *IEEE Communications Magazine*, **45** (5), 66–72.

28. http://jpeojtrs.mil/sca/pages/default.aspx.

29. Blossom, E. (2004) GNU radio: tools for exploring the radio frequency spectrum. *Linux Journal*, **2004** (122).

30. Ettus, M. Universal Software Radio Peripheral. [Online]. www.ettus.com.

31. McHenry, M., Livsics, E., Nguyen, T., and Majumdar, N. (2007) XG dynamic spectrum sharing field test results. Proceedings of the IEEE International Symposium on New Frontiers in Dynamic Spectrum Access Networks, Dublin, Ireland, April 2007, pp. 676–684.

32. Leu, A., McHenry, M., and Mark, B. (2006) Modeling and analysis of interference in listen-before-talk spectrum access schemes. *International Journal of Network Management*, **16**, 131–147.

Part Three

The Open Architecture Model

8

Open Architecture in Tactical Networks

Up to the 1980s, the defense industry believed that military networks should be based on proprietary protocols. With the increased complexity of tactical networks, swelled cost of design, development, and deployment, and the success of IP, the concept of an open architecture started to take root. This idea held fast since it would contain costs and leverage commercial capabilities. As time went by, tactical networks started to adapt open architecture concepts. One such integrated network is the software communications architecture (SCA) presented in Section 7.4 which promotes the use of APIs and modular software. In this chapter we will discuss the open architecture applicability to tactical networks from a wider perspective.

The success of open architecture within commercial wireless networks (specifically those with fixed infrastructure, such as 3G/4G/LTE[1]) have significantly improved service and resulted in drastically reducing cost. Today, one can purchase a low-cost smart phone with multimedia services, travel internationally, then utilize said device for wireless service at nearly any other location. Even developing countries, which missed out on the wired, microwave, and satellite eras of telephony, now have excellent wireless infrastructure. The monumental success of commercial wireless networks is evident from the steep price drops of high quality services, prevalent connectivity, and compact end-user equipment with innovative features.

Compared to the leaps of commercial wireless capabilities, the latest tactical wireless networks are a modest improvement over the legacy networks which were in use in the early 1990s when the US Congress decided to adopt the goal of having light, agile forces for fast military deployment; thus starting the Joint Tactical Radio Systems (JTRS) program. Similar initiatives were adapted by coalition forces. This wave of tactical wireless networks began alongside the latest wave of commercial wireless networks. Deployment of full IP-based MANET waveforms began facing many delays and substantial increases in cost. This prompted high-ranking military/government officials to ask: Why is it possible to have low-cost and high-quality services from commercial vendors but not from tactical

[1] Third generation, fourth generation, and long term evolution (LTE) are the benchmark for cellular technology.

Tactical Wireless Communications and Networks: Design Concepts and Challenges, First Edition. George F. Elmasry.
© 2012 John Wiley & Sons, Ltd. Published 2012 by John Wiley & Sons, Ltd.

wireless networks? Engineers and scientists started looking into ways to utilize commercial capabilities, while ensuring that tactical needs were still met. This provided even more reasons to consider open architecture features for tactical networks.

Tactical networks differ from commercial networks in their security constraints, anti-jamming needs, mobile/dynamic topology, scarcity of bandwidth, excessive delay, and so on. One can argue that the plethora of major differences can justify any lag in achieving the goals set for tactical wireless networks. One also can argue that there are noteworthy lessons to be learned from the phenomenal success of the commercial model. This chapter attempts to construct an encompassing open architecture tactical wireless model that applies to all IP-based tactical waveforms, as well as non-MANET nodes. This open architecture would meet tactical requirements and leave ample room for advancement. It would also make it possible to build and deploy the desired tactical wireless infrastructure, at a lower cost, with a variety of secure tactical radios and commercial capabilities.

The more we shy away from deploying open architecture in tactical wireless communications and networking, the harder it becomes to progress, reduce costs, and continue development. By contrast, open architecture lends itself to fast advances, cost reduction, and seamless utilization of the rich capabilities within commercial wireless communications and networking.

The model presented in this chapter does not exclude SCA. Instead of looking at a single radio and creating operating environments with plug-in software modules as with SCA, the model presented here approaches open architecture from the open system interface (OSI) model and defines the stack layers as entities with well-defined interface control documents (ICDs) between them.[2] The goal is to have the technology developers focusing on a single protocol stack layer, based on defined ICDs, to eliminate the need for proprietary techniques. Instead of having differing plain text IP layer implementation in different radios (which makes interoperability a nightmare), the defined ICDs will ensure seamless communications between the different technologies, while also meeting security and information assurance constraints and adapting to the dynamics of tactical networks. This will ultimately provide the warfighter with rich applications and excellent communications capabilities, while allowing waveform and platform integrators to acquire protocol stack layers (software for Generation Partnership Project (GPP) or full functioning hardware) from different developers at low cost.

8.1   Commercial Cellular Wireless Open Architecture Model

This section assesses the general architecture of commercial cellular wireless networks, highlighting their successful development in terms of excellent coverage and delivery of affordable, yet reliable, services. A typical commercial service provider has mobile end users and fixed network towers with base stations. Multiple base stations could be connected to a base station controller, and base station controllers are connected to a core network through a gateway.

Details regarding the functionality of commercial wireless protocols are beyond the scope of this book. Here, we focus on the open architecture adopted after regulations were established through standardization committees, such as the Third Generation Partnership Project

---

[2] The next chapter details some of these ICDs with a focus on the plain text IP layer capabilities that are not available in commercial networks, thus further making the case for the concept presented in this chapter.

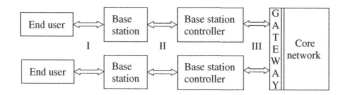

Figure 8.1   Commercial wireless definition of entities and well-defined ICDs.

(3GPP). These standardizations contributed to the enormous success of commercial cellular wireless networks. An important aspect of this technology is the definition of open standard ICDs. Consider a typical message flow (end user to end user), where packets flow from an end user to a base station, to a base station controller, and then to the core network and onward to a base station controller, then to another base station and finally to the other end user. Figure 8.1 shows the four main entities of this flow, with three main ICDs needed to define the interfaces. ICD I defines the end user to base station air interface, ICD II defines the base station to base station controller interface, while ICD III defines the base station controller to core network interface. Note that there also exist more detailed ICDs, such as those defining end user handover between two different base stations. One can see that an open architecture with well-defined entities and interfaces would provide the following advantages:

- Containment: Organizations that develop the technologies (i.e., vendors) need not be concerned about defining interfaces. By implementing an open architecture, vendors must adhere to predefined interfaces and focus on functionalities within the entity (user equipment, base station, base station controller, core network, gateway, etc.) instead of focusing on defining interfaces.
- Competition: To build a network, a given service provider can purchase some base stations from vendors A and B, while purchasing some base station controllers from vendors C and D. Said service provider can then purchase a core network gateway from vendor E, then interconnect these devices and build a reliable, seamless network. This creates competition and reduces cost.
- Specialization: With this approach, some vendors can specialize in building user equipment and successfully market it at a low price, while others can specialize in base stations, and so on. Specialization leads to innovation and drastic cost reduction.
- Innovation: With standardization committees defining the entities and alleviating concerns about interoperability, the vendors can focus their efforts on important issues, such as traffic engineering, resource management, quality of service (QoS), call admission control (CAC), and so on. Vendors can then create innovative techniques in such areas, as long as they adhere to the ICDs.

The above case is made for 3G/4G/LTE wireless architecture. The same case can be made with other commercial wireless technology, such as WiFi, where there is a standard air interface between the end user equipment and the wireless access point. The interface between the wireless access point and the service provider core network is standardized. As a result, the cost of creating a hotspot, which relies on WiFi, is low. Also, the installation and maintenance of the hotspot can be done by a non-expert.

To a large extent, one can attribute the affordability and increased quality of commercial wireless to the above advantages. One must also recall the role of a huge market size, which consists of millions of end users. Vendors were able to invest in developing the best competitive entities, and service providers were able to invest in the commercial wireless infrastructure. These investments were based on the projected return on investments that would be realized through the ever-growing wireless userbase. Obviously, there is no such market size for tactical networks. However, the lessons learned from the commercial wireless model can, and should, be adopted for the benefit of the tactical world. On this note, the use of an open architecture approach, that allows us to incorporate the mature commercial wireless technology into tactical communication capabilities, was a praiseworthy move.

8.2   Tactical Wireless Open Architecture Model

Before we present the open architecture model for tactical networks, let us review where the JTRS program stands at the time of writing. Monumental successes and failures have come from the JTRS and other programs that influence the path to a comprehensive open architecture model.

The JTRS program is the first serious attempt to attack tactical MANET challenges and, in spite of delays, it has achieved monumental successes, such as:

- Tactical IP MANET became reality. For the first time, an IP-based, true MANET waveform was built, despite an abundance of architectural challenges.
- Cross layer signaling became a reality. The JTRS model implemented cross layer signaling, thus showing the importance of deviating from the decades old OSI model. When cross layer signaling came into effect, it became apparent that ICDs between tactical network protocol stack layers are needed, because the commercial definitions are insufficient. These ICDs are not needed in wired networks.
- New capabilities at different layers of the protocol stack were implemented, such as the wideband networking waveform (WNW) cipher text IP layer utilization of radio open shortest path first (ROSPF). The issue of scalability of a MANET subnet, where control traffic does not eat up the scarce wireless bandwidth, was finally resolved. Real data, showing the amount of bandwidth saving, could now be measured. This allowed the tactical wireless communications community to realize that protocol stack layers, in tactical networks, need different implementation, and that a well-defined open architecture model must be taken into consideration.
- SCA was developed, thus creating the first true modularity of software components, with tactical radios defining APIs.

The JTRS program has achieved the ability to create some form of tactical MANET, but the non-open architecture model has some pitfalls that delayed deployment. Some of these pitfalls include:

1. *No separation of entities*: A JTRS waveform (soldier radio waveform SRW or WNW) has been defined as a single entity. The WNW waveform has many layers and sub-layers (e.g., mobile internet – MI and mobile data link – MDL) specific only to this waveform. The non-open architecture JTRS approach defines a complex waveform as a single program of record, which covers all the protocol stack layers from the plain

text side transport layer to the physical layer. This increases the problem scope of the waveform and does not allow for commonality between peer layers, from different waveforms, that need some commonalities (e.g., plain text IP layer).[3]

2. *No room for competition*: Deployment of a tactical network is left to a single integrator who faces many overwhelming challenges, especially with interoperability. Integrators tend to only trust the in-house solutions, which can lead to cost overruns before deploying a waveform.

3. *No room for contributions from specialized organizations*: If the layers below the cipher text IP layer (refer to Figure 6.3 for the WNW case and Figure 6.14 for the SRW case) were defined as separate entities (radio entity), with a well-defined ICD to the cipher text IP layer, firms specializing in this type of technology would have room to produce low-cost waveforms (with no IP layer, just adhering to the ICD to the cipher text IP layer).

4. *Limited room for innovation*: The vendor responsible for producing a radio has address many challenges in order to produce a working radio. They must develop software components that interface properly, port the software into the hardware, test and evaluate the radio, and so on. As a result, this vendor has very little, if any, resources left to focus on innovation.

A true open architecture in software defined radios (SDRs) would mean different vendors producing different layers of the protocol stack, such as plain text IP layer, encryption, cipher text IP layer, and *radio layers* that adhere to a given standard definition. One can then use a plain text IP layer from vendor x and a plain text IP layer from vendor y, and both layers communicate seamlessly either on the same waveform or on different waveforms. Both of these plain text IP layers will have tactical capabilities above what the commercial IP layer has, as will become clear in the next chapter.

Adhering to a form of open architecture is essential for the evolution of tactical wireless communications and networking. This would make good use of existing commercial technologies by getting the most capabilities into the hands of the warfighter with contained cost. The challenge here is to create an open architecture model that does not create any vulnerability or jeopardize security.

It is worthwhile noting that the line between tactical and commercial wireless communication and networking blurs in some areas. Researchers in both fields see the value of cross layer signaling, cognitive spectrum management, cognitive radios, network coding, and so on. However, the security needs for tactical wireless communications and networking will always lead to an unavoidable deviation. Nevertheless, distinction should not negate the need for an open architecture.

8.3 Open Architecture Tactical Protocol Stack Model

Figure 8.2 introduces a tactical wireless networking open architecture model. The figure shows the stack layers for two peer nodes. The layers above the plain text IP layer (application layer, session layer, and transport layer) are not shown. We assume that COTS

---

[3] The JTRS concept is that each waveform has an intended use. Hence, each waveform will have a different implementation at all stack layers. The only peering between nodes that are not from the same subnet is at the IP layer using standard IP protocols.

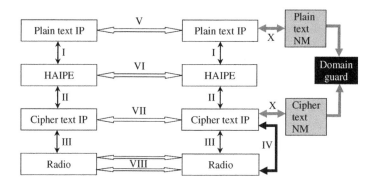

Figure 8.2   Tactical wireless model with well-defined entities and ICDs.

applications or non-COTS applications will be used with standard IP ports that use standard TCP over IP or UDP over IP. Here, "applications" refer to the application layer, the session layer, and the transport layer in Figure 1.1. The radio layer represents all the layers below the cipher text IP layer (or the DLL, MAC, and physical layers in Figure 1.1). Each radio can have its own implementation of these layers (or a different mutation of these layers as with the WNW and SRW). The peer-to-peer relationship of the radio layer is expressed by the horizontal arrows, VII in Figure 8.2, to indicate that within the same subnet, the different radio layers can peer with each other.

The IV arrow in Figure 8.2 indicates the cross layer signaling between the radio and the cipher text IP layers. Cross layer signaling between the cipher text and plain text IP layers can be allowed, if approved by NSA (in the USA), and could be processed through network management (NM) entities that relay this information through a domain guard.[4] Other implementations that are not approved by NSA can allow the encryption layer to leak cross layer signaling between the cipher text and plain text IP layers. This cross layer signaling is represented by the arrow labeled X. Although only one side of the peer nodes shows this cross signaling, it can exist in every node. Cross layer signaling between the radio layers (DLL, MAC, and physical layers) is not shown here and is assumed to exist, as you have seen with the WNW waveform layers in Chapter 6.

Many ongoing attempts at standardizing tactical networks can fall under the approach presented here. For example, the attempts to standardize the interface between modulation entities at the physical layer and the remaining radio components can be an extension of this approach – see reference [1]. Another example would be the attempts to standardize the hardware components of the radio, in order to create hardware modularity similar to the SCA software modularity. The take-away concept is that we can create standardization by using a top-down approach. Any of the entities presented here can be further broken down to modules, and interfaces between these modules can be defined.

The conventional OSI model, developed for wired networks, does not include cross layer signaling. The conventional OSI model assumes that layers are independent of each other. Engineers can develop a conventional OSI layer, with upward and downward traffic flow that also signals to the peer layers without dependencies on adjacent layers. Although cross layer signaling goes against the grain of the conventional OSI model, we have seen

---

[4] The next chapter will show how explicit congestion notification bits (which were approved for use by NSA) can be used to create cross layer signaling between the plain text IP layer and the cipher-text IP layer.

the importance of cross layer signaling for tactical MANET from the WNW waveform introduced in Chapter 6.

Figure 8.2 introduces concepts that could lead to the development of an open architecture model, for tactical wireless nodes, that can be adapted to a C2 node and lightweight soldier radios (please refer to the node notional architecture defined in Figure 6.2). The open architecture model presented in Figure 8.2, attempts to separate the entities and use well-defined ICDs (vertical arrows). Much like the OSI model, this open architecture pays great attention to security constraints and requires added functionality to meet tactical needs. However, one must note that this architecture requires cross layer signaling. The model creates entities (plain text IP layer, High Assurance Internet Protocol Encryption (HAIPE), cipher test IP layer, and radio) that are explained in detail below.

The relationship between the cipher text IP layer and the radio layers is one-to-many. A cipher text IP layer can interface to multiple radios, where each radio has its own implementation of ICD III and cross layer signaling IV. The cipher text IP layer can also interface to a wired medium (e.g., standard Ethernet). Having multiple interface capabilities is necessary in order to route to different waveforms or links. Each link can have its own version of cross layer signaling to the IP layer.

On the other hand, the relationship between HAIPE and the plain text IP layer is one to one. HAIPE architecture has a single plain text port and a single cipher text port. However, the relationship between the cipher text IP layer and HAIPE can be one to one (for a single security enclave nodal architecture) or one to many for a multisecurity enclave nodal architecture (please refer to Figure 6.21).

8.3.1   Tactical Wireless Open Architecture Model Entities

The model presented in Figure 8.2 is limited to the following four main layers.

8.3.1.1   Plain Text IP Layer

This layer is a fully capable IP layer with added functionality to address the existence of HAIPE and the unique characteristics of the cipher text core network. These add-ons optimize the performance of protocols, such as multicast over the HAIPE encrypted network core. It is important to note that with HAIPE, the entire encrypted network core looks like a black box to the plain text IP layer. The plain text IP layer has to pay attention to the performance of multicast protocols over the HAIPE tunnels. Other functions, such as CAC over HAIPE, flow control, and delivery assurance over HAIPE can be additional plain text IP layer functionalities that distinctly define this layer from conventional IP. Mobility and jamming can cause high bit error rates, beyond the radio DLL correction capacity, thus resulting in a large percentage of erroneous packets. Since HAIPE drops any packet that has even a single bit error, the plain text IP layer can face a huge percentage of packet loss. The combination of packet loss, limitation of available bandwidth over some core network paths, and the red/black[5] separation issue leaves the plain text IP layer with many challenges regarding admission control, multicast, and functionality of protocols such as TCP.

---

[5] HAIPE constraints prevent the passing of information between the plain text (red) IP layer and the cipher text (black) IP layer; this is known as red/black separation.

With the open architecture model, a common software solution for the plain text IP layer could be defined across all tactical nodes. This software solution can be deployed as a part of an SDR, or as a LAN router on stand-alone hardware for large-scale nodes, such as C2 nodes. This layer will be detailed in Chapter 9.

In the absence of an open architecture model for tactical wireless networks, there is no common agreement on how different plain text IP layers will communicate across different technologies deployed within the US forces or across joint forces (i.e., there will be no definition of ICD V in Figure 8.2). Different technologies can have different solutions for CAC over HAIPE, flow control over HAIPE, delivery assurance over HAIPE, and so on. However, more often than not, these solutions are not compatible. Defining ICD V in Figure 8.2 is necessary in order to create commonality between peer plain text IP layers. It will also make engineers and scientists focus on optimizing these add-on techniques instead of being bogged down in the interoperability details. Chapter 9 explains some plain text IP layer capabilities that can be common across all tactical network nodes for the US, joint and coalition forces.

8.3.1.2   HAIPE Entity

Although HAIPE standards are very well defined and are backward compatible, commonality is still advantageous. If different technologies are deployed with different versions of HAIPE, this would make the plain text IP layer even more complex. Consider, for example, CAC over HAIPE; a solution based on HAIPE version 3.0 or earlier cannot utilize the explicit congestion notification (ECN) bits, while a solution based on version 3.01 and later can utilize the ECN bits and can give a more powerful CAC capability. With the lack of HAIPE commonality, the plain text IP layer will have to implement different CAC techniques and communicate to different remote nodes with different CAC algorithms. The same argument applies for multicast over HAIPE: if a HAIPE version implements multicast internet protocol security (IPSec) tunnels over the cipher text core network, and an earlier version does not support multicast IPSec tunneling over the cipher text core, then the plain text IP layer implementation of multicast has to handle both cases. Chapter 9 will further delve into the discussion of these issues.

8.3.1.3   Cipher Text IP Layer Entity

The cipher text IP layer entity is the most complex entity within tactical wireless networks. Using COTS IP routers, tactical wireless cipher text IP layer ran into many complex problems. These problems included: how to create load balancing over different radio links; how to adapt to changes in the radio bandwidth (how to make packet service rate of the cipher text IP layer for a specific port change in real time as the throughput of the radio layer beneath changes)[6]; how to contain open shortest path first (OSPF) protocol traffic in scaled subnets (ROSPF mentioned in Section 6.2.1.1 is native to the WNW); how to accommodate and standardized protocols like point-to-point protocol over Ethernet (PPPoE) used between some satellite terminals and COTS routers; how to address dynamic area ID (to allow

---

[6] Notice that the WNW cross layer signaling is native to the WNW only and is not standardized across the different cipher text IP layers within the GIG. Bandwidth fluctuation can happen in many other cases such as the NCW satellite explained in Section 6.2.4.

nodes to roam between different IP areas); and how to overcome intermediate blockage of the physical layer through the use of techniques such as network coding.

One can see three areas to be considered for improving the cipher text IP layer:

1. How the user traffic flows to the radios (ICD III in Figure 8.2 defining the user plane). ICD III can address packet prioritization based on real-time and survivability needs and the radio throughput at any given moment.
2. Cross layer signaling with the radios (ICD IV) where protocols like ROSPF need to be implemented to contain the increase in link state advertisement (LSA) traffic. This ICD can also include standardization of protocols, like PPPoE, to create commonality on how to synchronize the cipher text IP layer's service rate to the radio's fluctuation in bandwidth.
3. Peer interface with the peer cipher text IP layers (in remote nodes) to address issues like load balancing (ICD VII).

If you refer back to Figure 6.21, you will see the repetition of the cipher text IP layer in the platforms with multiple waveform capabilities. While the radios have a fully capable cipher text IP layer, the COTS router also has a fully capable IP layer. When the WNW radio throughput changes, there is no standardized way to make the COTS router adapt to this dynamic change without flooding the backbone network with link state update (LSU). The WNW GIG port runs standard OSPF and the COTS gateway router, in Figure 6.21, can route traffic over the WNW subnet, with a fixed interface speed, without consideration of the fluctuation in the WNW subnet throughput. On the other hand, having a single cipher text IP layer, with well-defined ICDs,[7] will simplify radios (the WNW waveform will require the developer to focus on the MI, MDL, and SIS layers). This standardization can enhance performance and reduce the challenges we face in deploying mixed technologies. Note that the WNW design solves the scalability challenge within the WNW subnet only. Only standardization can make use of such a solution across all multiple access waveforms and point-to-point links of the tactical GIG.

8.3.1.4 Radio Entity

With an open architecture model, well-defined interfaces between the cipher text IP layer, and the radio layer, can be created. One can then confine radios to their specific DLL, MAC, and physical layers. Radio vendors will not have to build a full implementation of the cipher text IP layer; instead they can focus on innovative development of different types of radios. With this model, there is even room to adopt well-matured, commercial, technologies, while interfaces III and IV in Figure 8.2 are adhered to. Organizations specializing in this line of technology can have the opportunity to introduce cutting-edge waveform innovations such as waveforms with small areas of coverage that use omnidirectional antennas (adapting the IEEE 802.11x standards) to long-range waveforms (such as WiMAX) to point-to-point links or satellite links that will all adhere to ICDs III and IV. Integrators can then build and deploy tactical networks with a mix of commercial and tactical radios that are communicating seamlessly at the cipher text IP layer.

---

[7] Some researchers refer to this enhanced cipher text IP layer as the "tactical router."

8.3.2 Open Architecture Tactical Wireless Model ICDs

Before introducing the tactical wireless model ICDs, it is worth mentioning that the best definitions of ICDs are those which only define the minimum interoperability needs, omit the implementer details and verify that the interface is working seamlessly. With tactical wireless networking, the open architecture model would allow integrators to acquire protocol stack layers from different vendors and have them interoperate in a plug-and-play manner.[8] The ICDs within the tactical wireless model, shown in Figure 8.2, can be defined as follows:

8.3.2.1 ICD I: Plain Text IP to HAIPE

This ICD covers the user traffic flow and addresses issues defining the maximum size allowed for a plain text IP packet entering the HAIPE layer. This packet size can be chosen to avoid fragmentation over the cipher text IP layer (which can reduce throughput efficiency). Note that because HAIPE adds extra headers and padding to the plain text IP packet, one cannot allow the maximum Ethernet frame size to flow from the plain text IP layer down to the encryption layer. To consider the overhead, introduced by HAIPE, the maximum size for the packets entering HAIPE should be less than the Ethernet frame size. This ICD can explicitly define the maximum plain text IP packet size flowing down to the HAIPE layer.[9] Further details of this ICD are in Chapter 9.

8.3.2.2 ICD II: HAIPE to Cipher Text IP

Since the HAIPE standards are well-defined, in an open architecture model, one should aim to adhere to a single version of HAIPE (or limited number of versions) to create commonality with the least burden on the plain text IP layer. Chapter 9 discusses how HAIPE tunnels are formed above the cipher text IP layer. This exemplifies the extent of what can be defined within this ICD.

8.3.2.3 ICD III: Cipher Text IP Layer to Radio – User Traffic

This ICD covers user traffic flow and can include standardization of issues like QoS prioritization (i.e., which packet is given precedence in service when there is contention for bandwidth). Chapter 9 details how a combination of integrated services (IntServ) QoS, at the plain text enclaves, and differentiated services (DiffServ) QoS, at the encrypted core, ties the cipher text IP layer implementation of QoS to the plain text IP layer implementation of QoS. This then creates a coherent end-to-end solution for QoS across tactical networks. Chapter 9 will illustrate an example of DSCP mapping of the different types of traffic. This mapping can be used by the cipher text IP layer to prioritize serving packets.

---

[8] Each of these protocol stack layers or entities can be software based, ported on GPP, or can come on specialized hardware (compact board) with ports implementing the ICDs, thus allowing the integrator to create nodes with minimal cost.
[9] Packet fragmentation reduces the encrypted core throughput.

8.3.2.4   ICD IV: Cipher Text IP to Radio – Control Plane

This ICD covers all cross layer signaling issues between the radio and the cipher text IP layer, like bandwidth fluctuations and how link states can be reported from the radio to the cipher text IP layer. Queue depth at the cipher text IP layer (reflecting traffic demand) can be reported from the cipher text IP layer to the radio (radio terminals, with dynamic resource management between multiple terminals that share the same physical layer resources using this to dynamically reallocate resources based on the traffic demand as explained with WNW and network centric waveform (NCW)[10]), and so on. It requires a large effort to identify and utilize all details of this ICD since many challenges arise as one tries to develop a comprehensive cipher text tactical router. These challenges include the complexity of the cipher text IP layer (as mentioned in Section 8.3.1.3), the one-to-many relationship between the cipher text IP layer and the radio layer, the existence of different radio technologies with different ICD requirements, and the dynamics of the radio links and multiple access waveforms.

8.3.2.5   ICD V: Plain Text IP Peer-to-Peer

ICD V defines the peer relationship between the plain text IP layers in different nodes. This ICD is essential in order to have a common solution for issues like multicast over the encrypted core. This ICD could address common solutions of CAC over HAIPE, multicast over HAIPE, resource reservation protocol (RSVP) over HAIPE, and so on. Some of these techniques are covered in Chapter 9 and address:

- how to proxy user traffic such as TCP over HAIPE – there are many implementations of TCP proxy and not all of them are suitable for tactical MANET, and they are not compatible. This ICD can define how to proxy TCP over HAIPE such that different vendors can create different implementations of TCP proxy as part of the plain text IP layer while having all different implementations interoperate seamlessly.
- how to proxy VoIP traffic over HAIPE – there are different implementations of VoIP codec for tactical wireless MANET that are not compatible. A proxy technique that is interoperable and optimizes VoIP performance over the HAIPE encrypted core can be addressed in this interface.
- how to proxy video traffic over HAIPE – video streaming and interactive video requires a proxy technique that is different from that of voice and data. This interface can create commonality and optimize video transmission over HAIPE.
- how to proxy control signaling such as RSVP over HAIPE – because RSVP packets are encrypted by HAIPE, the cipher text routers do not make use of them. Thus one can proxy RSVP with an RSVP aggregate (RSVP-AGG) protocol that reduces this unused control signaling over the encrypted core, thus saving scarce wireless bandwidth for user traffic.

---

[10] Notice how these waveforms require a fully functional cipher text layer in order to perform dynamic resource allocation based on traffic demand. The existence of this ICD can make these waveforms simpler without IP layer implementation, while also getting traffic demand information through the defined ICD. Also, if there is a second IP layer (a router) feeding into the radio IP layer, this second IP layer will have to adapt to the radio IP layer that has to adapt to the waveform resource fluctuation, thus creating unnecessary hierarchy from the same layer.

8.3.2.6   ICD VI: HAIPE Peer-to-Peer

As mentioned for ICD II in Section 8.3.2.2, if HAIPE standards are well-defined and adhere to one version, they can help create common solutions across the different nodes of the tactical GIG.

8.3.2.7   ICD VII: Cipher Text IP Layer Peer-to-Peer

This ICD will define issues including load balancing over different paths (e.g., sending voice over terrestrial links and data over satellite links), and the implementation of dynamic area IDs where (in the case of multiple areas at the encrypted core) a mobile node can detach itself from one area and join another seamlessly.

8.3.2.8   ICD VIII: Radio Peer-to-Peer

This ICD is shown in Figure 8.2 with multiple arrows since each radio implementation can have its mutation of the DLL, MAC, or physical layers. Notice that this ICD is contained within a specific radio (a radio will form a subnet). If a cipher text IP layer interfaces to different radios, with each radio forming a subnet, then each radio will then have its own implementation of ICD VIII.

8.3.2.9   ICD X: Cipher Text to Plain Text (Control Signaling)

This ICD should be agreed on by the NSA and should be limited to the minimum flow necessary for protocols, such as multicast, to work. Abstraction of cipher text information communicated to the plain text IP layer is essential. For example, to implement multicast over HAIPE, at the plain text IP layer, one needs some topology information from the cipher text IP layer. One cannot pass the encrypted core entire topology to the plain text IP layer because the more information passed from the cipher text IP layer to the plain text IP layer, the more vulnerable the network users become to malicious attacks. Rather, one would need to find a way to minimize and abstract this information in order to protect the end users, at the plain text subnets, from malicious attacks. As an example, with the WNW waveform, the subnet mask of the waveform's encrypted core subnet is passed from the cipher text IP layer to the plain text IP layer. This subnet mask is used by the multicast algorithm at the plain text IP layer to figure out which nodes in a multicast address can be reached through the WNW waveform (notice that the WNW waveform is a multiple access waveform, which means that a single packet broadcasted over the radio can reach all nodes in the subnet while destinations drop packets that are not intended for them). Since the plain text IP layer multicast algorithm now knows those destinations in a multicast group that are reachable over a single broadcast WNW subnet, the algorithm will then be able to optimize multicast.

The GIG initiative could have started from an open architecture model as the model presented here. There are many advantages for adapting an open architecture. In addition to creating smaller problem spaces, the ICD definitions encourage commonality, containment, and competition while also simultaneously allowing ample room for innovation in the specialization areas. An open architecture model can be achieved by involving the stakeholders and relying on research and development (R&D) expertise. Such experts can create ICDs and flush out the interoperability and inefficiency at a low cost since the R&D community already has conducted many studies regarding the use of open architecture in tactical networks. With the establishment of this model, many organizations would have invested their own resources, and focused on creating software based solutions (for entities like the plain text IP, radio, etc.) while adhering to the predefined ICDs. More complex entities, such as the cipher text IP layer, could have become focused programs on their own. Although the cipher text IP layer could be the most complex entity in this model, it remains much simpler than the JTRS concept that covers all the stack layers.

An open architecture model does not require the integrators of tactical networks to define interfaces or conduct R&D in order to learn how layers peer with each other. Rather, this action is left for R&D organizations ahead of the development phases. With this model, the integrators will only need to harness tested technologies and build the systems (much like commercial service providers). Different organizations can produce plenty of waveforms (DLL, MAC, and physical layers), with a variety of capabilities, that adhere to the defined ICDs and can be utilized as plug-and-play with the cipher text IP layer.

The pillars of an open architecture model are simple engineering principles such as "do not attempt to solve a problem before you define it," and "keep it simple and straight." Although the military communications community is well aware of the value of modularity, they were so rushed to deploy an IP-based tactical MANET that properly defining entities or interfaces was overlooked. At the time of writing, 2012, the military communications organizations have come to realize the challenges with the non-standardized approach. Many industry and government organizations in the defense arena are currently creating consortiums to address standardization issues within tactical networks.

8.4   The Tactical Edge

The GIG architecture calls for the HAIPE-based encryption to be the common architecture of the tactical theatre. One important reason for HAIPE encryption is to ensure that every IP packet that goes over the air is securely encrypted. This level of encryption ensures that enemy technology that sniffs packets from over the air, will fail to get any useful information, since HAIPE encryption is the most complex known encryption technique.

8.4.1   Tactical Edge Definition

The diverse possibilities of how the plain text IP layer can exist over the HAIPE encrypted core network, with the GIG architecture, creates what we will refer to as the tactical edge. The tactical edge, in this book, refers to the access points to the HAIPE encrypted network core and specifically the plain text IP layer access points as shown in Figure 8.3.

In Figure 8.3, there could be many users (with their applications, sessions, and transport layers) within the IP wired subnet. Also, the gateway (referred to in Figure 8.3 as "GW"), between the legacy non-IP subnet and the plain text IP layer, can map each legacy node to an IP address. This creates many IP instantiations that access the GIG core through the tactical edge point. Since HAIPE has a single plain text port, one will have a single plain text IP access point to HAIPE. This single access point, or the GIG tactical edge, plays a major role in tactical wireless communications and networking. Optimization of

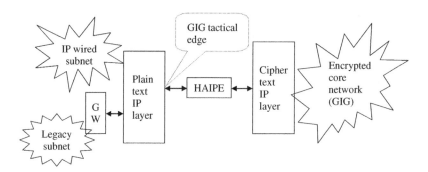

Figure 8.3   The tactical edge as the plain text IP layer access point to the GIG encrypted core.

the encrypted core resources will rely on some resource management and QoS techniques implemented at the tactical edge.

8.4.2   Tactical Edge Analysis

The tactical edge, being a single point of aggregation for various sources of traffic over the GIG encrypted core, opens the door for using resource management techniques that exploit statistical multiplexing properties of traffic flow. Let us take, for example, a transport layer protocol that creates redundancy packets to overcome high packet loss over the GIG encrypted core network as discussed in Chapter 3 and Appendix 3.A. This transport layer protocol relies on the collection of real-time statistics for the path over the network core to estimate the number of redundancy packets needed at a given time. This transport layer protocol can be implemented at the end user terminal, device transport layer or at the tactical edge.

If the transport layer is implemented at the end user terminal or at the device transport layer, then the protocol is collecting statistics for a single session. If packet flow for this session is n packets per second, then the statistics are collected as n samples per second.[11] If instead it is implemented at the tactical edge, then statistics are collected for multiple sessions. If we have m sessions going in parallel between two tactical edge points, then the statistics are collected for the aggregated traffic ($m \times n$ samples). This scenario provides a more accurate measurement of the GIG core path. In the case of a soldier radio, where the radio has only a single application, the two cases are identical. However, for a large platform, like a C2 node, this distinction is very important since there may be multiple sessions in parallel over the same GIG core path. Also, one would rather implement these techniques at a single point (tactical edge) than require the costly alternative of having every COTS application to change its transport layer.

If we were to take the measuring of end-to-end packet delay to estimate congestion and were to assume that each packet delay is independent from the previous and future packets,[12]

---

[11] Measuring packet loss in networks with limited bandwidth, like tactical networks, cannot use propping. Instead, the actual user traffic is used to collect path statistics.

[12] This assumption is needed for ease of analysis. Similar assumptions of packet arrival independency are made with the M/M/1 queue model in Chapter 4. Also, since we are talking about end-to-end delay, packets have gone through many hops, where flows are split and merged, thus reducing the dependencies of packet inter-arrival time and making this assumption valid.

then packet delay becomes a random variable that deviates from the average or expected delay. In order to keep up with the encrypted core dynamics, measurements are collected over a window of time. Let us assume that we collect packet delay between source S and destination D over a window of time, where we have m sessions going in parallel between S and D. For a single session, if the number of packets received, during the window time τ where the delay is measured, is n, then we have the measured average delay as a random variable. Let's refer to the average delay measured as $E[T]$ where,

$$E[T] = \frac{1}{n} \sum_{j=1}^{n} d_j, \tag{8.1}$$

where d_j is the measured delay of packet j

If we can collect the delay measurement over a very large window of time, then the expected value of the average delay measure should approach the actual mean delay μ over the path. Or,

$$E\left[E[T]\right] = E\left[\frac{1}{2} \sum_{j=1}^{n} d_j\right] = \frac{1}{n} \sum_{j=1}^{n} E[T] = \mu. \tag{8.2}$$

Since the time frame and number of samples are limited, the measured average delay $E[T]$ is a random variable that deviates from the mean. Let us compare the tactical edge and transport layer delay measurements.

In the transport layer case, where we have n samples, we can measure the estimated error in the measured delay as:

$$E[(E[T] - \mu)^2] = E\left[(E[T] - [E[T]])^2\right] = VAR[E[T]]. \tag{8.3}$$

The variance of the measured delay is:

$$VAR[E[T]] = \frac{1}{n^2} VAR[d_j] = \frac{n\sigma^2}{n^2} = \frac{\sigma^2}{n}. \tag{8.4}$$

For the tactical edge with m parallel sessions, one can calculate the estimated error based on $m \times n$ samples, which yields,

$$VAR[E[T]] = \frac{1}{(n, m)^2} VAR[d_j] = \frac{\sigma^2}{n.m}. \tag{8.5}$$

For Equation 8.5, as the number of parallel sessions over a single path increases, the error in the estimated average delay approaches zero.

Similar analysis can be made for packet loss considering that it is also a random variable. Thus, the larger statistical sample will yield a more accurate measurement of packet loss. It is worth emphasizing that the existence of the tactical edge aggregated access point suggests that some of the transport layer techniques have been shifted to the tactical edge. These techniques will be performed at the aggregated plain text IP point instead for the following reasons:

- The larger statistical sample yields smaller error in measurement-based QoS techniques.
- Since there are not any requirements that need to be imposed on the transport layer of the end user in order to make it applicable in tactical networks, we can use COTS

applications. The costly alternative is developing a specific transport layer for each application.

- Legacy net users do not have a transport layer (refer to Link-16 in Chapter 5). The only way to optimize the performance of their traffic, and ensure they have the necessary QoS over the GIG encrypted core, is through the tactical edge resource management techniques.

In the next chapter, we will cover how the tactical edge can implement:

- a TCP proxy that imposes no requirement on COTS applications using standard TCP/IP.
- a VoIP proxy that allows VoIP to work seamlessly over the tactical GIG core with limited bandwidth and high packet loss.
- a video proxy that allows video transmission over limited bandwidth and high packet loss.
- a proxy of some control signaling such as RSVP, where HAIPE encryption alters its performance.
- a proxy of multicast over HAIPE to optimize its performance.

You can see now why Figure 8.2 did not cover the layers above the plain text IP layer. We had assumed that applications are COTS and that the plain text IP layer in Figure 8.2 is the tactical edge aggregated IP access point.

8.5 Historical Perspective

It is interesting to know that the concept of cellular phones is rooted in military communications. One can argue that the first experimentation of using radio telephony can be accredited to Reginald Fessenden's ship-to-shore demonstration in 1906. World War II should be credited with giving birth to the cellular phone since during the war, radio telephony links were used on a very large scale. Radio telephones were also manually linked to the wired telephone network. By the late 1940s, automobile phones were available in the USA, although they were very bulky and consumed a lot of power. These radio networks were able to support a few calls at a time. In 1946, the St Louis mobile telephone service (MTS) system could only carry three parallel calls. It was not until 1964 that a system called improved mobile telephone service (IMTS) was introduced, which could carry more voice channels and link automatically to the public telephone network. Nevertheless, public demand for these vehicle-mounted phones continued to surpass the available channels.

Prior to the availability of the automobile phones, in the 1930s, telephone customers in the USA could place a call to a passenger on a liner in the Atlantic Ocean that modulates in the VHF band. The air time rate for these calls was very high: $7 per minute in the 1930s is equivalent to about $100 per minute in 2012. Even today, in areas where marine VHF radios are available, an operator can be reached from the public telephone network that can manually link the customer to a ship.

Motorola can be credited with a major milestone in cellular technology with its development of a backpacked two-way radio (the first walkie-talkie). Motorola also developed a large handheld, battery powered two-way radio for the US military which was about the size of a man's forearm.[13] Both these devices were vacuum tube based, and were used in World War II.

In conjunction with adapting the IMTS system, the USA also adapted the radio common carrier (RCC) system from the 1960s to the 1980s. RCCs handled telephone calls and were operated by private parties. RCCs used paired UHF 454/459 MHz and VHF 152/158 MHz frequencies, near those used by IMTS. RCC did not support roaming because there was no centralized billing availability. It is interesting to know that RCC was *not* developed according to standards. RCC equipment was basically a conglomerated system that used different signaling techniques. Some radio equipment was half-duplex, push-to-talk equipment such as Motorola handheld or RCA 700-series conventional two-way radios. Other vehicular equipment had telephone handsets, rotary or pushbutton dials, and operated full duplex like a conventional wired telephone. A few users even had full-duplex briefcase telephones.

By the end of the RCC era, the industry associations were working on a technical standard that would allow roaming and multiple decoders for some mobile users. The latter would enable the operation with more than one of the common signaling formats. This effort came a bit too late since a new generation of standardized cellular phones was already being introduced into the market.[14]

In December 1971, AT&T submitted a proposal for cellular service to the FCC. After years of hearings, the FCC approved the proposal in 1982 for the advanced mobile phone system (AMPS). Analog AMPS was eventually superseded by digital AMPS in 1990. This was the era when generations of cellular telephone networks started to be rolled out every few years, but AMPS can be considered as the first.

In the 1990s, the second generation (2G) mobile phone systems emerged, using the global system for mobile communications (GSM) standard. This differed from the previous generation since 2G used digital transmission and fast phone-to-network signaling. Following the introduction of 2G, mobile phone usage exploded, and 2G introduced a new form of communication, called SMS or text messaging. It was initially available only on GSM networks but eventually spread to all digital networks. The first machine-generated SMS message was sent in the UK on 3 December 1992, followed in 1993 by the first person-to-person SMS sent in Finland.

Demand for wireless access to the internet led to the newer generations of cellular technology, since 2G could not keep up. Thus 3G was born and used a packet switched design instead of the digital circuit switched GSM. This standardization focused on requirements more than technology.

During the development of 3G systems, 2.5G systems such as CDMA2000 1x and general packet radio service (GPRS) were developed as extensions to existing 2G networks. These provided some 3G features without fulfilling the promised high data rates or full ranges of multimedia services.

In the mid 2000s enhancements to the 3G systems began under the names 3.5G and 3G+. By the end of 2007 there were 295 million subscribers on 3G networks worldwide. The 3G services generated over 120 billion dollars of revenue during 2007 alone.

---

[13] The same Motorola organization is now owned by General Dynamics and can be credited with the deployment of the first JTRS radio that carried the SRW waveform, the Rifleman Radio.

[14] At the time of writing, we are at a similar junction with tactical wireless radios. A clean slate approach is certainly better than band-aid fixes.

By 2009, it had become clear that 3G networks could not handle the growth of bandwidth-intensive applications such as streaming media. The industry started to look into the data-optimized fourth-generation standards which promised a drastic improvement on speed up to 10-fold over 3G standards.[15] The first two commercially available technologies billed as 4G were the WiMAX standard (offered in the USA) and the LTE standard, first offered in Scandinavia. 4G, which is all IP-based, ushered in the treatment of voice calls comparable to any other type of streamed audio media.

In the meantime, the defense industry was struggling with the decision of whether or not to define standards. With the introduction of HAIPE encryption in tactical networks at the turn of the millennium, many challenges emerged due to HAIPE constraints. As you have seen in this chapter, the defense industry is now ready to learn a lesson from the commercial world regarding the same technology that the defense industry pioneered decades ago.

Bibliography

1. http://www.opengroup.org/tech/direcnet-task-force/.
2. http://www.3gpp.org/.
3. http://www.umts-forum.org/.
4. http://www.globalsecurity.org/military/library/report/2003/Future_Force_Black_Book_26_Aug_03_Final.pdf.
5. http://jpeojtrs.mil/.
6. Elmasry, G.F. (2010) A comparative review of commercial vs. tactical wireless networks. *IEEE Communications Magazine*, **48** (10), 54–59.
7. North, R., Browne, N., and Schiavone, L. (2006) Joint tactical radio system – connecting the GIG to the tactical edge. Proceedings of MILCOM 2006, Washington, DC, 23–25 October 2006.
8. Stine, J. (2007) A cautionary tale on testing and evaluating tactical wireless mobile ad hoc networks. *Proceedings of the ITEA Journal*, Mar/Apr 2007, 53–62.
9. (2003) Soldier-Level Integrated Communications Environment (SLICE) Soldier Radio Waveform (SRW) Functional Description Document (FDD), Version 1.3, November 2003.
10. (2003) Wideband Networking Waveform (WNW) System Segment Specification, Specification Number AJ01120, Boeing, 12 February 2003.
11. http://peoc3t.monmouth.army.mil/win_t/win_t.html.
12. http://www.govcomm.harris.com/solutions/products/000056.asp.
13. Wiss, J. and Gupta, R. (2007) The WIN-T MF-TDMA mesh network centric waveform. Proceedings of MILCOM 2007, Orlando, FL, 29–31 October 2007.
14. United States National Security Agency (2006) High Assurance Internet Protocol Encryptor Interoperability Specification, Version 3.1.0, 31 December 2006.
15. Trabelsi, C., Liu, H., and Elmasry, G.F. (2008) A method and system for channel element allocation for wireless systems. *Bell Labs Technical Journal*, Special Issue on Next-Generation Wireless Access Networks, **13** (1), 263–272.

---

[15] The 4G standardization included an LTE plan. In this book we use the term 3G/4G/LTE as the benchmark for the cellular technology standards.

9

Open Architecture Details

The open architecture model for tactical networks, introduced in the previous chapter, requires added features and interface control document (ICD) definitions. In this chapter we will detail these ICDs and address the needs of the plain text IP layer for additional features not defined within the standard IP protocols. Figure 9.1 shows the three relevant ICDs to the plain text IP layer taken from Figure 8.2. These ICDs are:

- ICD I, which defines the interface adhered to by user traffic flow between the plain text IP layer and HAIPE (see Section 8.3.2.1).
- ICD V, which defines the peer relationship between the plain text IP layer of different nodes (see Section 8.3.2.5).
- ICD X, which defines cross layer signaling between the plain text IP layer and cipher text IP layer through the HAIPE domain guard (see Section 8.3.2.9).

The reason for such an emphasis on the plain text IP layer will become clear as we go through this chapter. We will define the plain text IP layer as it relates to the tactical edge. The cipher text IP layer and its ICDs form another critical component of the open architecture approach. Many of the needed cipher text IP layer capabilities were discussed with WNW in Chapter 6. We will address the cipher text IP layer, in this chapter, in regard to how it relates to the plain text IP layer in areas such as quality of service (QoS). This is in order to achieve a comprehensive end-to-end QoS solution.

9.1 The Plain Text IP Layer and the Tactical Edge

It is important to emphasize the definition of the plain text IP layer as it relates to the tactical edge. The goal of defining the tactical edge is to obtain the open architecture presented in Figure 8.2, regardless of the protocol stack position within a node or platform. There are three cases where the plain text IP layer can exist.

The first case is a simple architecture, such as that used to deploy a soldier's communications device.[1] Here we can have a single application built on top of the plain text IP layer.

---

[1] Throughout the remainder of this chapter, the term "radio" will be used for the layers below the cipher text IP layer (MAC, DLL, and physical). A soldier's communications device is composed of the radio, cipher text IP layer, HAIPE, plain text IP layer, transport layer, session layer, and application layer.

Tactical Wireless Communications and Networks: Design Concepts and Challenges, First Edition. George F. Elmasry.
© 2012 John Wiley & Sons, Ltd. Published 2012 by John Wiley & Sons, Ltd.

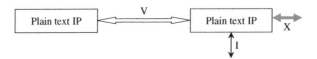

Figure 9.1 Three ICDs relevant to the plain text IP layer.

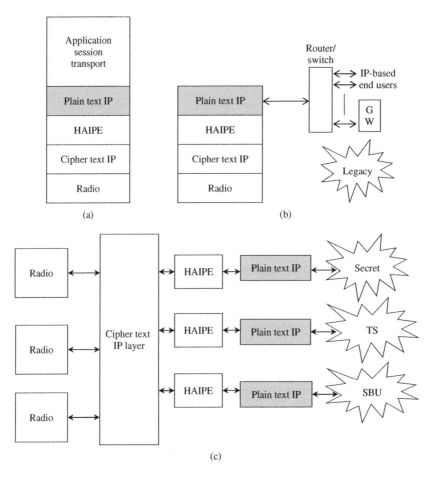

Figure 9.2 Three different possibilities for having an open architecture based plain text IP layer. (a) Single application device, (b) single radio vehicle or platform, and (c) multiple radio platform.

The soldier's communications device is composed of the radio (MAC, DLL, and physical layers), the cipher text IP layer, the HAIPE layer, the plain text IP layer, and the upper layers (transport, session, and application). This architecture is demonstrated in Figure 9.2a. Although all these layers are integrated in a single device, an open architecture would allow for the integration of this device, at a low cost and for the creation of peer-to-peer communication with other nodes, within or external to this soldier's radio subnet, as explained in Chapter 8.

The second case is for the communications architecture of a vehicle or a platform, with a single communications link. The plain text IP layer of the communications link can have an IP port connected to a router or switch in the vehicle, as shown in Figure 9.2a,b. An entire subnet (supporting multiple applications over the vehicle) can exist behind the plain text IP layer. Each end user in the subnet is COTS and uses standard TCP/IP or UDP protocols. A legacy waveform, with an IP gateway (marked "GW" in Figure 9.2b), can also be connected to the router. The plain text IP layer becomes the tactical edge. We are interested in the capabilities of the plain text IP layer (or tactical edge) that allows for seamless operation of the COTS end user applications, over HAIPE encrypted tactical networks with limited bandwidth and high packet loss, delay, and jitter.

The third case is that of a platform architecture with multiple waveform capabilities, as shown in Figure 9.2c. Here, the cipher text IP layer can use separate hardware (tactical router) connected to the different radios, according to the ICDs mentioned in Chapter 8. Multiple HAIPE devices can be connected to the cipher text IP layer to form different security enclaves (secret, top secret – TS, and sensitive but unclassified – SBU). Each of these security enclaves has its own plain text IP layer and each can have a subnet router similar to the single subnet in Figure 9.2b.

9.2   Measurement Based Resource Management

In the absence of any information flow from the cipher text side to the plain text IP layer, one needs to consider managing the resources of the encrypted core, at the plain text IP layer, solely based on measurements. We assume that if congestion, packet loss, or packet delay occur at the encrypted core, the cipher text IP layer cannot pass any information to the plain text side due to HAIPE constraints (we will cover how encrypted core information allowed through the HAIPE can be utilized for better resource management techniques). measurement based resource management (MBRM) relies on user packets to collect statistics about the state of the encrypted core. If the plain text IP layer at node a can measure packet loss, delay, delay variation, and so on. of the packet flow coming from node b, then node a knows the status of the encrypted core path from b to a. Node a can then send a feedback token[2] to node b so that node b can know the status of the path from node a to node b and from node b to node a. The collected measurements can be used for protocols such as:

- **Call admission control (CAC)** – a technique known as measurement based admission control (MBAC).
- **TCP proxy flow control** – throttles back or forth a data flow from a TCP session at the plain text enclave through a TCP proxy technique as explained in this chapter.
- **Voice proxy state** – but VoIP should have a fixed rate flow; this chapter presents a technique to control the flow in terms of how much bandwidth a VoIP call can use, over the encrypted core, as well as robustness techniques to overcome packet loss.
- **Video proxy state** – similar to VoIP, and can control bandwidth consumption of video and create robustness capabilities for the video stream.
- **Multicast optimization** – the ability to tie multicast over HAIPE to admission control.

---

[2] Closed-loop feedback is a common approach between peer protocol stack layers. Recall that the WNW link adaptation in the MDL layer creates closed loop metrics. Also, the NCW MAC/DLL layer uses the FOW, ROW, and DCOM closed-loop measurements.

9.2.1   Advantages and Challenges of MBRM

MBRM relies on measuring the traffic characteristics based on packet flow and uses these measurements to create admission control, flow control, and proxy policies. The technique comes with some advantages and challenges. The advantages include:

- suitability for partitioned environments, such as the partition created by HAIPE.
- a closed-loop feedback mechanism that can make MBRM policies adaptable to the condition of the network, especially for radio links with variable bandwidth and high packet loss.
- real-time measurements of the source to destination path condition.
- path condition measurements for each class of service (CoS).

On the other hand, the challenges include:

- **Overhead traffic** – if the implementation technique creates probing packets to collect measurements, it would increase traffic and may overload the network. With tactical networks, one can have up to 64 different differentiated services code points (DSCPs). To collect large enough statistical samples for all active DSCPs, at useful intervals, it would require a huge amount of probing traffic to measure the status of the encrypted core network. Thus, probing should be avoided, and measurements ought to be piggy-backed on user traffic instead. Also, the feedback mechanism that reports the congestion of a path from a destination node to a source node should not burden the network with extra traffic.
- **Ripple effect** – if all the nodes, sharing a congested link, detect congestion, they might simultaneously throttle back the flow of traffic (the admission control and flow control functions) and then discover that congestion is relieved and throttle forward. This ripple effect will result in an oscillating, poorly utilized, unstable encrypted core network.
- **Fairness** – admission and flow control policies should be fair in the sense that two nodes sharing a congested link should throttle back traffic flow with the same precedence in the same way. This is essential with tactical networks, since survivability information should be preserved at the expense of information with lesser precedence. In other words, two nodes sharing the same congested link should preempt routine traffic, then priority traffic, and so on, using some form of synchronization.[3]
- **Effect of other entities in the network** – the implementation must consider the behavior of other entities in the network, such as how queues in the encrypted core network build up.[4] For example, there could be a lag time between initial congestion at an encrypted core hop and the time the MBRM technique senses this congestion.
- **Differentiation between physical layer and IP layer effect** – if the physical layer is dropping packets, due to foliage or RF signal blockage, one cannot interpret packet loss measurements at the plain text IP layer in the same manner as packet loss due to congestion (queue overflow) at the cipher text IP layer.

---

[3] Tactical IP networks have different precedence levels in the following ascending order: routine, priority, immediate, flash, and flash override. These precedence levels should not be confused with the classes of services (voice, video, streaming data, interactive data, etc.). Each CoS can have these five precedence levels. While a CoS indicates real-time needs, precedence level indicates survivability needs of the packet flow.

[4] See Chapter 4 for queue behavior.

The implementation of MBRM, at the plain text IP layer, should address the above challenges in order to create a robust, measurement based, resource management solution. This solution would function in the absence of any encrypted core network health information, due to HAIPE constraints.

9.2.2   Congestion Severity Level

Consider traffic flow from node X to node Y, over a single DSCP. At node X, before a downstream packet is encrypted, the plain text IP layer stamps the packet with the current network time GPS time synchronization can be used) and the packet sequence number with respect to the flow ID (flow ID is a unique identifier for a call/session). At node Y, after an upstream packet is decrypted, the plain text IP layer extracts the time stamp and the packet sequence number. For a window of received packets for a given DSCP, if the number of received packets is n, let us assume that we collect the following measurements:

- number of packets violating delay requirements of the DSCP class, n_d
- number of lost packets, n_l
- number of repeated packets,[5] n_r
- number of packets violating jitter requirements, n_j.

Congestion, of severity level S, can be defined as a function of measurements:

$$S = f\left(n, n_d, n_l, n_r, n_j\right), \tag{9.1}$$

where the function f compares each of the past window parameters with established DSCP threshold values and failure ratios.

The function defined in Equation 9.1 can decide whether the DSCP exceeds QoS requirements, conforms to QoS requirements, or violates QoS requirements.

At a given node, the above function is computed for all active DSCP values and for all remote nodes (nodes that actively communicate to the specific node). Based on the violation counters in the calculation window, the status of a DSCP may change (status exceeds QoS requirements, conforms to QoS requirements, or violates QoS requirements). The change in the QoS status of a DSCP will trigger a change in the calculated congestion severity indication. The change in the congestion severity indication is reported to the node that initiated the traffic in the form of a token.

As the network congestion persists, the calculated congestion severity level increases, which leads to more throttle-back of flow control. The proxy techniques explained in Section 9.4 adjust traffic flow based on the persistence of congestion. The MBAC technique will result in greater blockage of new lower precedence calls, and preemption of existing low precedence calls. This throttle-back will continue until congestion is resolved and all active DSCPs conform to their QoS requirements. Congestion is considered to be resolved when all active DSCPs traffic conforms to the QoS requirements.

With MBAC, calls with high precedence can be admitted (even if the measurements suggest congestion), while calls with low precedence are preempted. The next section analyzes MBAC as a distributed Markov chain. Notice that implementing MBRM techniques

---

[5] TCP retransmission of packets is an indication of packet loss and the reaction of the transport layer to packet loss and delay.

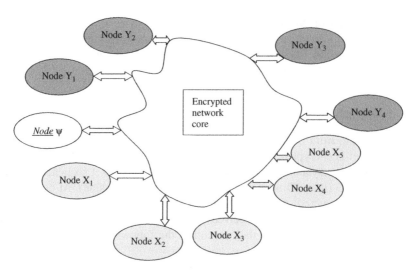

Figure 9.3 MBRM over the encrypted core network where node Ψ receives from nodes X_1, X_2, X_3, X_4, and X_5 and transmits to nodes Y_1, Y_2, Y_3, and Y_4.

at the plain text IP layer has to be distributed. This distribution needs to be such that for a local node, the upstream congestion level is calculated from the received traffic flow, and is reported via a token to the remote node where the traffic originates. The downstream congestion level is obtained from the token received from the remote node. Thus, the local node has information about the upstream and downstream path condition. Admission and preemption policies are made based on the condition of both the upstream and the downstream paths.

Now, let us consider a general case, as in Figure 9.3. Let us assume that we have multiple nodes communicating over an encrypted network core, as shown in the figure. Let us focus on node Ψ, which may be a source node, destination node, or both.[6] Nodes Y_1, Y_2, Y_3, and Y_4 are the nodes that node Ψ communicates to (here, Ψ initiates the calls/sessions). On the other hand, nodes X_1, X_2, X_3, X_4, and X_5 are the nodes that communicate to Ψ (here, these nodes initiate calls/sessions to Ψ). At node Ψ, MBRM stamps the traffic leaving to the encrypted core and receives a token from each of the nodes Y_1, Y_2, Y_3, and Y_4. Each token abstracts the congestion severity level of the corresponding path. Node Ψ also extracts the time stamp from all the received packets and performs severity level calculations for each source X_1, X_2, X_3, X_4, and X_5 independently. If the calculated congestion severity of a certain path changes, then Ψ sends a token to the corresponding node. Thus, node Ψ has information about the upstream and the downstream of each path it uses. Note how the calculation of the path condition of each pair of communicating nodes is independent and the path upstream and downstream calculations are independent.

MBAC is a critical part of MBRM since MBAC helps the plain text IP layer reach equilibrium of shared encrypted core network resources. The blockage and preemption of

---

[6] Descriptions relating to node Ψ apply to every node in the network. To make the technique more understandable, we are focusing on node Ψ.

lower precedence calls/session has greater effect than other MBRM function such as flow control when it comes to healing congestion. MBAC can lead to equilibrium between nodes contesting for the precious core network bandwidth. In the next section, we use a Markov chain to represent the MBAC states at each node.

9.2.3 Markov Chain Representation of MBAC

The goal of MBAC, much like many resource management protocols, is to reach a form of equilibrium while the network dynamics change. The MBAC goal is to have the admission control of the plain text side react to the changes in the encrypted core network. As explained above, destination nodes compute the impact of the admitted forward traffic on given network resources as a forward congestion severity level for the corresponding source node. The value of the forward congestion severity level is reported to the call/session source. Based on the received severity value and the locally calculated congestion severity level of the return traffic, the source node decides which calls/sessions need flow adjustment, blocking, or even preempting. Lower priority calls/sessions can be preempted in order to guarantee QoS for all classes of service.

Let us first focus on the forward traffic congestion severity level. Similar analysis applies to the return traffic. By computing a communication severity value, each destination node keeps track of the impact of traffic for each source sending messages to this destination. The severity is a summary of the traffic load for the various classes of traffic. Each time the severity value changes, the destination node sends an updated severity value to the message originator.

Let us assume that voice, video, and data traffic travels from source nodes $S = \{X_1, X_2, \ldots, X_r\}$ to destination nodes $D = \{Y_1, Y_2, \ldots, Y_s\}$. Let us also assume that we have the following DSCPs associated with classes of services:

- data class: $D_1, D_2, \ldots, D_{n1}$
- voice class: $V_1, V_2, \ldots, V_{n2}$
- video class: $VTC_1, VTC_2, \ldots, VTC_{n3}$

mapped to $\{C_1, C_2, \ldots, C_n\}$, where C_i corresponds to a DSCP and $n = n_1 + n_2 + n_3$.

For $\{C_1, C_2, \ldots, C_n\}$, there are corresponding QoS attributes $\{Q_1, Q_2, \ldots, Q_n\}$, where $Q_i = \langle R_i, T_i, E_i, ..\rangle$ with:

$R_i =$ Required fraction of packets of class C_i with end-to-end delay less than T_i seconds.
$E_i =$ Acceptable packet loss ratio.

At time t_{k+1}, if destination node Y_i receives a packet of DSCP C_p from source node X_j, then Y_i computes a cost function $S(t_{k+1})$, where S returns a value of $(-1, 0, \text{or } 1)$ based on end-to-end delay, packet loss ratio, and previous update time. Y_i updates the severity value V with:

$$V(t_{k+1}) = V(t_k) + S(t_{k+1}). \tag{9.2}$$

If V changes value, Y_i does the following:

1. updates its upstream severity value corresponding to destination node X_j;
2. returns the new value to X_j, which updates its downstream value for the corresponding destination Y_i.

Based on the updated value of congestion severity level V for forward traffic and the locally computed congestion severity level value for return traffic, X_j may change its admission policy.

One can assume that $S(t_{k+1})$ depends on the current utilization of the system and is independent of the value of V for time prior to t_k, for any sequence of times $t_0 < t_1 < \cdots < t_{m+1}$ and values $U_0, U_1, \ldots, U_{m+1}$ in the range of V. Based on this assumption, the following probability equality holds:

$$Pr[V(t_{m+1}) = U_{m+1} | V(t_m) = U_m, \ldots, V(t_0) = U_0]$$

$$= Pr[V(t_{m+1}) = U_{m+1} | V(t_m) = U_m]. \qquad (9.3)$$

Thus V is a discrete Markov chain, taking values in the range $\{1, 2, \ldots, N\}$ (see Figure 9.4). The cost function S is defined, such that the ripples in V are removed, that is, a certain delta time (measurement window duration) must transpire before S increments or decrements, which ensures that the congestion severity level cannot decrement or increment by more than one state. Please note that the cost function $S(t_{k+1})$ expresses persistence in congestion estimation. If congestion estimation persists, V is incremented by one. If congestion estimation is alleviated, V is decremented by one. At state zero, if the estimation of no congestion is unchanged, V stays zero. At state N, if congestion persists, V stays at state N (highest level of congestion). At all other states, a duration of time must pass before we can transition between states, to combat ripple effect. The transition probabilities of this Markov chain depend on the implementation details and are not defined here.

A similar Markov chain applies for the return traffic. The source node admission and flow control policies can be made based on the highest of the congestion severity levels, or a combination of the two. In Figure 9.4, V_i represents the severity value V at node X_i for all nodes sending traffic to X_i or receiving traffic from X_i.

A change of network topology or network traffic demand can dramatically impact the utilization of network resources and will be detected by V. Application of the cost function and the consequent change in severity value can engender blocking of class C_i calls/sessions.

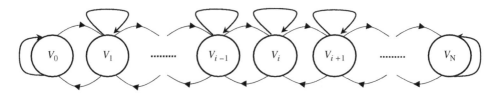

Figure 9.4 Congestion severity level V as an N-state Markov chain.

9.2.4 Regulating the Flow of Traffic between Two Nodes

Based on Figure 9.4, let us assume that at a certain congestion severity level V_γ, the call admission policy at the source node will block large calls/sessions of class C_i. Note that blocking could mean either dropping the call/session or caching non real-time messages for later delivery. When the congestion severity level is alleviated, blocked messages of class C_i can be readmitted. Note that for a given TCP session, preemption can be achieved by blocking the local TCP port associated with the session to be preempted, thus forcing TCP to abort the session.[7]

If M_i represents the number of messages of class C_i from Ψ to Y_j per unit time, say seconds, and P_i is the average message size for that class, then $M_i(t) * P_i$ represents the amount of traffic, of class C_i, which flows from source Ψ to destination Y_j in t seconds, and

$$T = M_1(t).P_1 + M_2(t).P_2 + \cdots + M_n(t).P_n, \tag{9.4}$$

represents the total traffic of all classes from node Ψ to node Y_j in t seconds.

Now let us assume that the forward traffic from node Ψ to node Y_j at time t_{k+1} causes the cost function $S(t_{k+1})$ to return $+1$. Then, an updated congestion severity level $V_\psi(t_{k+1})$ is sent from Y_j to Ψ. This update in congestion severity level causes Ψ to reduce the admitted traffic by lowering the traffic contribution from the lowest priority DSCP C_d, that is, lowers T in Equation 9.4 by lowering the traffic associated with the term $M_d(t).P_d$. Attenuation of the lowest priority traffic can be achieved by blocking large messages in the class C_d, that is, by attenuating traffic associated with P_d and M_d. If QoS is still violated, the cost function $S_\psi(t_{k+2})$ will cause the congestion severity level $V_\psi(t_{k+1})$ to be incremented and reported back to node Ψ. Based on the new congestion severity level, node Ψ may decide to completely eliminate the traffic flow of class C_d, and thus remove all traffic represented by $M_d(t).P_d$ in Equation 9.4

If the QoS is violated again, then all messages of class C_d will be blocked and large messages of class C_{d-1} will be blocked as well. These steps continue until QoS requirements for high precedence sessions/calls are met.

9.2.5 Regulating the Flow of Traffic for Multiple Nodes

The same general procedure as in the two-node case will take effect here, with the exception that each source node X_i will receive severity values from all possible destination nodes Y_j. MBAC handles this multiple congestion severity level in a distributed, independent fashion. Each pair of communicating nodes has its N-level Markov chain that controls the admission policy, independent of other pairs communicating over the same network. Although one may think that there may exist some sort of cross-correlation between the congestion severity level corresponding to the flow of traffic from X_i to Y_j and the congestion severity level corresponding to the flow of traffic from X_i to Y_k, this cross-correlation only occurs if the two paths share a network resource (a certain link) that is causing congestion. MBAC does

---

[7] We will see later in this chapter how MBAC can be used with an RSVP proxy in order to signal TCP sessions with RSVP deny. This gives the TCP at the end user the same effect as if the RSVP deny were coming from a core network cipher text router.

not attempt to exploit this cross-correlation in order to avoid any ripple effect associated with MBRM techniques.

9.2.6   Packet Loss from the Physical Layer

One major challenge that MBAC faces with the tactical network is the differentiation between packet loss due to congestion at the IP layer of the encrypted core and packet loss due to physical layer problems. The RF signal can suffer from periods of blockage, which can vary depending on the terrain (urban, suburban, foliage, etc.).[8] One RF blockage mitigation technique is to use a link layer ARQ protocol directly over the radio link suffering from blockage. The link layer ARQ protocol detects packets that are received in error and requests retransmission to ensure that all packets are communicated to the receiving end of the radio link in a reliable manner. One drawback of using a link layer ARQ is the tradeoff of loss for delay. That is, the TCP layer does not sense packet loss; instead it senses high delay and jitter. The TCP layer can interpret this delay as a sign of congestion. Excessive delay may cause the TCP protocol to time out.[9]

With MBAC, detecting RF signal blockage over a given path can be used to enhance flow control and CAC policies by separating RF signal blockage from congestion and thus not erroneously dropping or blocking lower precedence traffic. Similarly, RF signal blockage detection can also be used to direct flow control when using a TCP proxy (please refer to the adaptive TCP proxy in Section 9.4.1). MBAC can collect measurements for all traffic coming from all plain text IP layers communicating to the local plain text IP layer. Let us assume that the following measurements are collected:

- packet loss (UDP traffic only);
- packets retransmitted (TCP traffic only);
- end-to-end delay, which is the time that a packet takes to traverse the encrypted core
- memory jitter (J1), which is the delay difference between two consecutive packets;
- memoryless jitter (J2), which is a measure of delay variance over a window of packets.

When RF signal blockage occurs without a method of blockage detection, the above measurements can lead the MBAC function to erroneously assume that a path is congested. For example, when an RF signal becomes blocked, the UDP packet loss will increase and TCP retransmission will increase. In the case of link layer ARQ enabled over the link with RF signal blockage, each successive TCP packet will report high delay and jitter. In tactical networks, MBAC must differentiate between a congested path and a path that is suffering from RF signal blockage.

9.2.6.1   RF Signal Blocking Detection Technique

In tactical networks, RF signal blockage can vary depending on the terrain. In urban areas, blockage can result from high buildings that obstruct point-to-point radio links, and a satellite beam with a low elevation angle can suffer from blockage as well. Blockage periods in urban

---

[8] The presence of jammers can also create a period of RF blockage.
[9] TCP timeouts may also surface when excessive delay occurs due to the use of multiple satellite links over the same path of traffic. The flow control feature of the TCP proxy, explained in Section 9.4.1, remedies satellite delay.

areas can sometimes last for minutes (a vehicle moving through a tunnel). In suburban areas, small buildings can cause blockages that last for shorter periods. Foliage can have an effect on the quality of radio signals as well. MBAC can use the following levels of blockage in tactical network:

- **Level 0: no blockage** RF signal blockage is not detected or detected in very small intervals (in the order of microseconds) and will amount to less than 5 percent of the total studied time interval.
- **Level 1: minor blockage** RF signal blockage will happen in small intervals (in the order of milliseconds) and will amount to less than 20 percent of the total studied time interval.
- **Level 2: moderate blockage** RF signal blockage will happen in larger intervals (in the order of a fraction of a second) and will amount to between 20 and 33 percent of the total studied time interval.
- **Level 3: severe blockage** RF signal blockage will happen in long intervals (in the order of seconds) and will amount to more than 33 percent and less than 66 percent of the total studied time interval.
- **Level 4: continuous blockage** RF signal blockage will happen for very long intervals (tens of seconds) and will give the impression of a lost link.

At a higher level, the tree in Figure 9.5 explains the basis of the presented technique for measurement-based RF signal blockage detection. The idea is to calculate a blockage status by first checking if the DSCP carries UDP or TCP traffic, and if so, then checking whether the link layer ARQ is ON or OFF[10] and if the TCP proxy discussed later in this chapter is

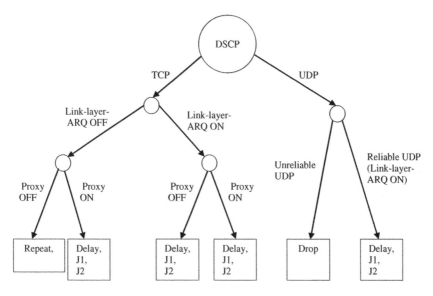

Figure 9.5 Violations considered for calculating blockage for different DSCPs (e.g., for unreliable UDP, use only the percentage of packet drop).

---

[10] The delay and loss pattern can be used to estimate whether or not the link layer ARQ is ON or OFF. With link layer ARQ, delay increases and packet loss decreases.

enabled or disabled (ON or OFF). This will make us reach a leaf in the tree. Based on this leaf, we only look at the corresponding parameters. For example, if the DSCP carries UDP and the link layer ARQ is OFF (or it is not reliable UDP), we only consider the packet loss ratio violation. Any other violation (delay or jitter) will be due to network congestion. Thus, following this tree leads us to choose which measurements are needed to determine RF blockage detection. Other measurements can be ignored or made to have minimal effect.

Note that the violations considered above are calculated over a given interval of time for each separate DSCP. The overall blockage level for the given destination can be calculated based on the estimated blockage level of all active DSCPs.

9.2.6.2   Blockage Detection as Matched Filters

The use of the five measurements mentioned above, for detecting a congested path in the network, can lead to considering a radio link blockage as congestion. The tree in Figure 9.5 tries to eliminate some violations from the RF signal blockage detection calculations. This can be represented as a matched filter approach, which can be expressed as follows.

Let P_i, for $1 \leq i \leq 6$, represent each of the communication protocols represented in the tree in Figure 9.5, in particular:

$P_1 = $ TCP + link layer ARQ OFF + proxy OFF
$P_2 = $ TCP + link layer ARQ OFF + proxy ON
$P_3 = $ TCP + link layer ARQ ON + proxy OFF
$P_4 = $ TCP + link layer ARQ ON + proxy ON
$P_5 = $ UDP + unreliable UDP
$P_6 = $ UDP + reliable UDP (link layer ARQ ON).

Let d_j, for $1 \leq j \leq 5$, represent each of the measured parameters over a window of time, in particular:

$d_i = $ percentage of packets violating delay requirements
$d_2 = $ percentage of dropped packets
$d_3 = $ percentage of repeated packets
$d_4 = $ percentage of packets violating memoryless jitter
$d_5 = $ percentage of packets violating memory jitter.

In the presented technique, blockage B and congestion C are calculated as linear functions of d_j. In particular, B can be represented as:

$$B = \begin{bmatrix} b_{11} & b_{12} & c_{13} & c_{14} & b_{15} \\ b_{21} & b_{22} & b_{23} & b_{24} & b_{25} \\ b_{31} & b_{32} & b_{33} & b_{34} & b_{35} \\ b_{41} & b_{42} & b_{43} & b_{44} & b_{45} \\ b_{51} & b_{52} & b_{53} & b_{54} & b_{55} \\ b_{61} & b_{62} & b_{63} & b_{64} & b_{65} \end{bmatrix}. \tag{9.5}$$

The blockage, $B(P_i)$, for the protocol P_i can be computed as:

$$B(P_i) = \sum_{j=1}^{5} b_{ij} * d_j. \qquad (9.6)$$

Finding the coefficients b_{ij} will be explained later in this section. Similarly, congestion $C(P_i)$ for P_i can be represented as:

$$C(P_i) = \sum_{j=1}^{5} c_{ij} * d_j. \qquad (9.7)$$

In Equation 9.7 c_{ij} is similar to b_{ij} and is given as:

$$C = \begin{bmatrix} c_{11} & c_{12} & c_{13} & c_{14} & c_{15} \\ c_{21} & c_{22} & c_{23} & c_{24} & c_{25} \\ c_{31} & c_{32} & c_{33} & c_{34} & c_{35} \\ c_{41} & c_{42} & c_{43} & c_{44} & c_{45} \\ c_{51} & c_{52} & c_{53} & c_{54} & c_{55} \\ c_{61} & c_{62} & c_{63} & c_{64} & c_{65} \end{bmatrix}. \qquad (9.8)$$

Further, blockage and congestion have been observed to be additive. So, in a system (path) which includes blockage and/or congestion, the total received interference T on P_i is:

$$T(P_i) = \sum_{j=1}^{5} (b_{ij} + c_{ij}) * d_j = \sum_{j=1}^{5} t_{ij} * d_j. \qquad (9.9)$$

Note that relying on all measurements generates an indication of congestion based on the $T(P_i)$ estimation. That is, when both RF signal blockage and congestion occur, both affect the estimated congestion severity. The RF signal blockage estimation considers the estimated RF signal blockage as a *signal* and the estimated congestion severity as a *signal plus noise* (when the noise is considered as the congestion part of the estimation above).

From the tree in Figure 9.5 above, the following can be seen:

P_1: *all but* t_{13} *are* 0
P_2: $t_{22} = 0$, $t_{23} = 0$
P_3: $t_{32} = 0$, $t_{33} = 0$
P_4: $t_{42} = 0$, $t_{43} = 0$
P_5: *all but* t_{52} *are* 0
P_6: $t_{62} = 0$, $t_{63} = 0$.

Thus, the first step to filtering the effect of congestion is to use the matched filter analogy and eliminate the measurements that reflect congestion. The second step is to find the values of the used coefficients of the filter.

Let us consider the case of a DSCP carrying unreliable UDP. The packet drop ratio is the only measurement that estimates the RF signal blockage. A percentage of the dropped packets could be due to actual blockage of the radio, while the remainder could be a result of queue overflow over a congested path. Since congestion can cause delay, jitter, and packet drops, one can use the delay and jitter measurements to estimate how many packets were dropped at the queues and not at the radio. Note the following:

- The CoS of the considered DSCP can be a factor in distinguishing between the effects of the radios versus the effects of the queues. For example, a DSCP carrying voice is expected to have some acceptable percentage of packet drop at the queue (since the queue depth could be short) while a DSCP carrying non-real-time traffic is expected to have a low drop ratio and high delay. Thus, the delay, jitter and packet drop ratio thresholds used at the plain text IP layer differ from one DSCP to another.
- The estimation of RF signal blockage for a certain CoS can be more accurate than another. For example, real-time UDP classes (e.g., voice) may give the most accurate estimation of RF signal blockage (since we use only one coefficient and the behavior of packet drop at the queues can easily be estimated). The confidence factor for one class over the other can be considered in the selection of the B coefficient of the matched filter above.

Now that we have seen how MBRM and MBAC work, let us delve into the ICDs mentioned in Chapter 8. We will examine which function each ICD can implement.

9.3   ICD I: Plain Text IP Layer to HAIPE

As mentioned in Chapter 8, this ICD covers the user traffic flow. ICD I addresses those capabilities performed by the plain text IP layer that do not require the peer plain text IP layer to have special protocol or software modules. An example of an ICD I capability is the selection of the size of the plain text IP layer packets sent to HAIPE for encryption.

The packet payload size can be large with data and video packets (the video packets contain the I-frame information). The maximum transmission unit (MTU) of Ethernet,[11] which was created to maximize the ratio payload size to header size, must be revisited with HAIPE encryption.[12] HAIPE encrypts the entire plain text IP packet, including its headers, and makes the encrypted packet the payload of the cipher text IP packet as shown in Figure 9.6 for the IP v4 case. HAIPE implementation of IP SEC ESP tunnel mode can add a considerable number of bytes[13] to the plain text IP packet size.

The use of the MTU size for standard IP, over Ethernet protocols, needs to be adjusted in the plain text IP layer to avoid fragmenting the cipher text packets.

Ethernet V2 defines the MTU as 1500 bytes. If we need 20 bytes for the Ethernet header, the plain text IP layer or the end user TCP layer creates a packet with the payload of 1440 bytes (considering plain text headers are 40 bytes and Ethernet headers of 20 bytes). Then, as this packet is encrypted by HAIPE, it will become $1500 + x$, where $x = (20 + 8 + 8 + (2 \rightarrow 257) + 12)$ bytes as can be seen in Figure 9.6. The cipher text IP layer of HAIPE will have to fragment this packet into two cipher text packets. The first packet can be 1500

---

[11] MTU size differs depending on the transmission media. For Ethernet V2, it is 1500 bytes.

[12] The path MTU discovery protocols cannot be used over HAIPE.

[13] No specific size is mentioned here since padding of the packet is required in some cases to make the packet size a multiple of the encryption key size. Padding can be as much as 257 bytes.

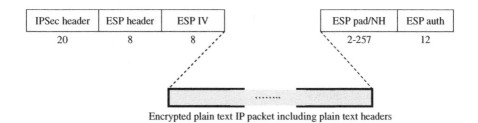

IPSec header	ESP header	ESP IV		ESP pad/NH	ESP auth
20	8	8		2-257	12

Encrypted plain text IP packet including plain text headers

Figure 9.6   HAIPE use of IPSec in tunnel mode increases the plain text packet size.

bytes (including the plain text IP header), while the second packet will have a payload of x and header of $x + \Delta$, where Δ is determined by the padding process. Clearly creating two packets, instead of one, will reduce the throughput efficiency of the cipher text IP layer.

ICD I can define specific sizes for the MTU that prevent fragmentation by the HAIPE cipher text IP layer.

9.4   ICD V: Plain Text IP Layer Peer-to-Peer

In this section, we cover a number of protocols that are needed for the plain text IP layer peer-to-peer communications. These protocols optimize the use of resources over the HAIPE encrypted core network. These protocols are explained in Sections 9.4.1–9.4.5.

9.4.1   TCP Proxy over HAIPE

Please review Appendix 3.A where a transport layer hybrid ARQ over HAIPE was presented. This hybrid ARQ over HAIPE compresses the packet payload and generates redundancy packets to overcome packet loss from the encrypted core path. A TCP proxy over HAIPE can

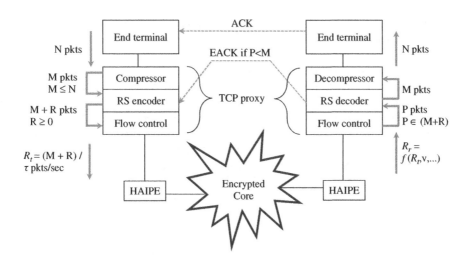

Figure 9.7   Plain text IP layer TCP proxy elements.

use this technique, as shown in Figure 9.7. In the figure, the TCP proxy over HAIPE has three main components: the compressor/decompressor, the Reed–Solomon (RS) encoder/decoder, and the flow control component. Standard TCP proxy[14] has only the flow control component. Let us consider the following two cases to conceptualize how this proxy works.

First, consider the case where retransmission is not necessary. Packet flow between end-terminals is conducted as follows:

1. Let N be the number of packets the transmitting terminal sends in a given window of packets.
2. Let M be the number of packets the compression technique produces. M represents the number of data packets that the RS encoder will encode (the encoder is designed to encode any number of packets as explained in Appendix 3.A). Keep in mind that N packets get compressed to output M packets and $M \leq N$. The MTU size can be implemented to ensure the compliance of the M packets to the MTU size over HAIPE as discussed above.
3. The RS encoder adds R redundant packets to the data stream, where R is chosen based on network conditions and policy as explained in Appendix 3.A. If the network conditions are good, R might be zero. $M + R$ packets are transmitted through the encrypted core network to the remote proxy.
4. The remote proxy receives P packets, and all of the packets are a subset of those that were transmitted. Because of packet loss, $P \leq (M + R)$. It is possible that $P > (M + R)$ because some packets might be duplicated, but the duplicates will be ignored by the receiver.
5. The RS decoder requires at least M packets to regenerate the original data. Those packets can be any M packets within the original set of $(M + R)$. Assuming that $P \geq M$, the RS decoder will output the original M packets.
6. The decompressor will decompress the M packets into N packets, and transmit them to the receiving end terminal.
7. The receiving end terminal will send a TCP ACK to the transmitting end terminal.

Now let's consider what happens when there is more packet loss, and the RS decoder realizes that it hasn't received a sufficient number of packets $P < M$:

1. The receiving proxy will realize that $P < M$ because of gaps in the sequence numbers.
2. The receiving proxy will send an EACK (explicit acknowledgment) to the sending proxy. The EACK will acknowledge all of the packets except the last $(M - P)$ missing packets. In other words, it can request the transmission of the last $(M - P)$ missing packets. For example, if $M = 100$ and $R = 20$, the sender will send 120 packets. If we receive only $P = 90$ packets and are missing packets 10–39, then the sender may not request all 30 packets, but will only request packets 30–39.
3. The sending proxy will send the $(M - P)$ packets that were requested, and also R' redundant packets, where R' is chosen based on the current network conditions and policy.

---

[14] Standard TCP proxies can have names such as TCP accelerator and performance enhancement proxy (PEP).

9.4.2   VoIP Proxy over HAIPE

The use of VoIP over tactical networks with HAIPE encryption is known to face many challenges such as:

1. **Excessive delay**. QoS techniques at the cipher text IP layer can help with this problem. Real-time packets can be expedited to ensure that packets from real-time applications (e.g., voice) are delivered to the destination within a specific end-to-end delay requirement. The drawback is that expedited forwarding (EF) queues tend to be short. Although the use of EF combats delay, it can also cause packet loss. Excessive delay can also occur when VoIP packets go over multiple hops, or satellite links can suffer from high accumulative delay.
2. **Jitter (delay variation)**. VoIP codecs are designed with a jitter buffer that trades jitter for delay (to absorb the variations of delay of the IP packet stream). However, this jitter buffer has a limitation on how much jitter can be absorbed. In some cases, tactical networks can introduce higher jitter than the VoIP codec jitter buffer can handle.
3. **Increase in bit error rate**. Degraded SNR can cause high bit error rates (BERs). Forward error correction (FEC) in the DLL of the tactical radio can help mitigate this problem. However, in many cases, the error correction fails to correct the error pattern. The use of CRC codes at the upper stack layers results in the drop of these packets, making the end user experience packet loss.
4. **Packet loss at the IP layer**. The IP layer can introduce more packet loss as a result of queue overflow (this can happen in both wired and wireless media). If the VoIP packet traverses multiple hops, the cumulative effect of packet drop can be significant.

Packet loss is one of the biggest problems facing the deployment of VoIP in tactical networks. Typical human beings can tolerate at most 10% drop ratio (provided the drop is random). As the packet drop ratio increases, the human perception of the voice quality becomes intolerable.

9.4.2.1   VoIP COTS Technique

Faced with these issues with VoIP, many techniques have been developed to overcome VoIP shortcomings. Some of the techniques are described below.

Interpolation

VoIP codecs widely use interpolation to smooth some errors within the received bit stream. Interpolation may mask unexpected patterns as the bit stream is turned into an analog signal. Interpolation is an excellent technique for smoothing the effects of bits or bytes of error (which correspond to a short period of time in the analog signal). However, as the error patterns become more sporadic (affecting a cascade of bytes), interpolation techniques become less effective and fail to provide full recovery from these errors (analog signal smoothing becomes less accurate as the time gap increases). In VoIP, packet loss creates an erasure of bytes. With G.729, the most widely used VoIP codec, a single lost packet

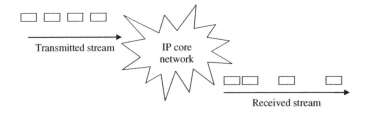

Figure 9.8   Introduction of jitter into the VoIP packet stream by the IP core network.

Figure 9.9   Absorption of VoIP jitter by the decoder buffer.

can create an erasure that causes a time gap as long as 20 ms. Because VoIP is a real-time application, techniques such as request for retransmission cannot be used. When the core network introduces high packet loss, VoIP codecs begin demonstrating poor performance. This has created the need for a VoIP proxy over tactical networks that can achieve packet recovery (in real-time) to overcome high packet loss situations.

Jitter Buffer Adaptation
VoIP packets can reach the decoder with variable delays. Figure 9.8 shows how the packets enter the core network in a more-or-less synchronized fashion (e.g., a G.729 encoder emits a packet every 20 ms), and reach the receiving end with delay variations that can be on the order of 100 ms.

A VoIP codec has a built-in jitter buffer as shown in Figure 9.9. This buffer trades jitter (delay variations) for constant delay to ensure that packets are decoded in a synchronized fashion. Many approaches have focused on jitter buffer adaptation where the jitter buffer can grow larger if the core network introduces a high jitter, and can get smaller if the core network introduces a low jitter. The reason for these approaches is the need to optimize the gap between the moment of speaking and the moment of hearing (VoIP users often complain about a perceived lag, resulting from jitter and jitter mitigation techniques).

FEC at the DLL
This is another important technique used to enhance the VoIP quality. As we have seen with the WNW waveform in Chapter 6, the waveform can react to increases and decreases in BER or SNR. These fluctuations result in a reduced packet loss ratio. Unfortunately, in some cases, SNR can be low and bandwidth can be limited. This results in a packet error rate that can be too high for VoIP.

Expedited Forwarding at the IP Layer
Another technique crucial to VoIP performance is EF at the IP layer. At each hop, the queuing techniques can use EF to ensure that time critical packets are served as soon as possible.

Without this expediting of VoIP packets, the VoIP traffic would encounter intolerable delays. Unfortunately, there are some complex issues that prevent relying on EF alone. Consider the following:

- Other traffic, such as control traffic, usually competes with the real-time traffic for EF privilege. Sometimes, this can reduce the speed of forwarding the VoIP traffic.
- Serving packets in real time results in a limited queue depth configuration. There is little reason to keep a packet at a hop if it will be served later and then discarded when it arrives late at the destination. This packet should instead be dropped at the delayed hop, in order to free resources for other packets. Referring to Figure 9.9, if a packet arrives at the destination and the jitter buffer is empty (all the packets preceding it have already been processed, and the synchronized server clock missed this packet), the packet will be discarded for late arrival. For this reason (a packet delayed too long at a hop is useless), queue depth for the real time traffic is very limited. At a given hop, when packets arrive in a burst (which is often the case with IP networks), the probability that a packet will arrive to a full queue increases, causing the packet to be dropped (this is referred to as tail drop).[15]
- There is oversubscription of the limited bandwidth. It is a well-known issue that the more bandwidth we add to networks, the greater the demand for bandwidth becomes. Bottlenecks are a major area of concern for all networks (wired or wireless – tactical or commercial). As oversubscription occurs, a hop can have a packet arrival rate that exceeds its service rate capacity (as determined by the available bandwidth). This then forces the hop to drop a large number of packets. The cumulative effect of packet drops below the IP layer – at the IP layer and along multiple hops – can cause an overall packet drop ratio well beyond the codec's ability to adequately process.

The techniques described above rely on processing the bit stream within the IP network. In addition to these techniques, a number of VoIP capabilities are widely used by the end user to increase VoIP robustness. Some of these techniques are as follows:

Reducing Coding Rate
There are many techniques used to reduce the coding rate of a voice codec that are already implemented and available in widely used products. As voice codecs generate a bit stream, they apply a quantization technique to turn the analog voice signal into a digital voice signal. Reducing the quantization rate (lossy compression) can reduce the produced bit stream. Another form of compression can be achieved by only sending the delta from the previous quantization value, instead of the entire new value.

A VoIP proxy at the plain text side needs to consider many factors, including the re-packetization of the bit stream into IP packet payloads, in order to efficiently ensure VoIP quality over the HAIPE encrypted network core while optimizing the scarce bandwidth resources and overcome high packet loss. Consider Figure 9.10, where the widely used G.729 codec payload is handled in IP tactical environments. The codec produces 20 bytes of payload per packet. With the addition of a UDP header, an IP header and an RTP (real-time

---

[15] Consider two queues in tandem, and the first queue is served by a TDMA controlled schedule. Packet arrival at the second queue will be bursty, and queue overflow at the second queue becomes more likely.

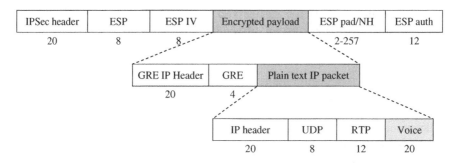

Figure 9.10 G.729 voice codec packets before encryption, with IP GRE and with IPSec in tunnel mode.

protocol) header, the packet becomes 60 bytes long. Some multicast implementations use generic routing encapsulation (GRE), which creates a tunnel (GRE tunnel) from the source to the destination subnets, communicating over the HAIPE encrypted core network. This GRE tunneling adds another 24 bytes of header information to the packet. HAIPE implementation of IPSec, as explained above, requires the addition of more header information. Thus, a payload of 20 bytes can require as many as 136 additional bytes per packet to transmit over the HAIPE encrypted core network.

With the use of VoIP over HAIPE encrypted core networks, attempting to reduce the coding bit rate makes little difference if the number of packets remains constant (due to the large ratio of header to information).

Adaptive Rate Codec

The adaptive rate codecs used in commercial technologies, such as 3GPP, was proven effective in making the best use of limited bandwidth resources. These codecs react to the changes in the network path (sensed through RTP header and RTCP – real-time transport control – packets) by reducing or increasing their coding rate. This reduction in coding rate is accepted by VoIP users since an end user would rather have a lower voice quality than lose their call altogether. An adaptive rate codec uses many of the techniques explained above to reduce the coding rate and relieve congestion. Although this is useful, one must keep in mind that with VoIP over HAIPE encrypted networks, the best gain is achieved by reducing the *packet rate*, not the *bit rate*.

Codec Conversion

Codec conversion is another method utilized in a multitude of currently available products in the commercial market. These conversion techniques simply allow the user to configure an edge (access point) in order to convert the packet stream from one codec to another. For example, one can convert a G.711[16] stream to a G.729 stream before going over an

---

[16] G.711 uses a much higher bit rate than G.729 and is commonly used with wired networks that have an abundance of bandwidth.

IP core network. As mentioned above, with HAIPE encrypted tactical networks, the main focus should be reducing the packet count and not the payload size.

Increasing Packet Flow

Approaches that add redundancy packets to overcome packet loss with VoIP are brute force techniques that increase the amount of consumed bandwidth. While some of these techniques repeat packets, others create a parallel codec stream (e.g., G.729 from a G.711), then send both streams and replace the lost G.711 packets with the information interpreted from the G.729 packets at the receiver. All of these techniques require an increase of bandwidth – the scarcest of all tactical wireless network resources.

9.4.2.2   How to Proxy VoIP over HAIPE

This plain text VoIP proxy technique would work at the IP packet stream. This proxy technique should be based on comprehension of how HAIPE encrypted VoIP packets have a small payload and a large header. Therefore it is effective to cross-correlate *only the payloads* of the IP packet streams by carrying the payload information of multiple consecutive packets in the same IP packet. If we can reconstruct the lost packet header at the receiving end, then we can use the payload information from the previous packet to reproduce the lost packet. This will make it possible to recover lost packets in near real time. This proxy technique should also ensure that reproduced packets are absorbed by the jitter buffer (and not dropped because they arrive too late). In a HAIPE encrypted tactical network with high packet loss, cross-correlating the payloads of the packet streams results in a small increase of the overall packets size, while the number of packets stays the same. The price paid for such an increased level of robustness is a minor increase in the payload size of the packets, but the hazard of adding any new packets, with large headers, is completely avoided.

The same technique used to reproduce lost packets can also be used to achieve VoIP compression. Within a tactical network core that has limited bandwidth and low packet loss, we can achieve compression by reducing the number of packets entering the IP network core (e.g., drop every other packet of the cross-correlated packet stream at the transmitting side).

The VoIP proxy at the plain text IP layer can address the main issues with VoIP (which lie in the IP area). In networks with high packet loss, lost packets can be recovered in near real time while the required increase in the bandwidth is minimal. In a network with low packet loss and limited bandwidth, shaping (dropping packets at the source) can decrease bandwidth usage over the encrypted core network. Figure 9.11 illustrates how two payloads of consecutive packets are cross-correlated to generate a VoIP packet stream. This stream

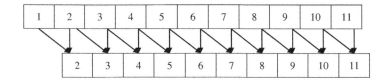

Figure 9.11   Cross-correlating the payload of two consecutive packets in the VoIP stream.

has the same number of packets, but has the ability to recover every other lost packet. Note that it is essential to make sure that the recovered packet will be considered by the jitter buffer.

This VoIP proxy can allow different levels of compression and different levels of robustness. It can react dynamically to any changes in the network path conditions, and analytically finds the "sweet point" between compression and robustness to optimize for the ambient conditions of the network path. It can sense the level of packet loss as well as the level of congestion, and adapts to the level of packet loss by creating an adaptive level of robustness. It also reacts to signs of congestion at the IP layer by creating a level of compression. Since the increase in compression can make the VoIP packet stream more sensitive to packet loss, this technique should analyze both of these conditions and find the best level of compression and the associated level of robustness – the "sweet point."

Example 9.1 Let us suppose we work with a G.729 codec that produces 50 packets per second and each packet has 20 bytes of payload. Let us assume that the cipher text packet is of size 136 bytes. Let us suppose that the cross-correlation technique in Figure 9.11 is used and the proxy technique creates 25 packets per second instead of 50 packets (every other packet is dropped at the transmitting side since it can be recovered at the receiving side). This was made possible since the given core network path has low packet loss and tight bandwidth. If every packet with the cross-correlated payload is then 156 bytes, instead of 136 bytes, the compression ratio could be calculated as $(136 \times 50)/(156 \times 25) = 1.74$. This compression ratio means we can put 1.74 voice calls over the same bandwidth instead of just one.

Example 9.2 Let us assume same parameters as above, but with a core network path condition that shows a high packet loss ratio. In this example, the packet loss probability is 0.25 and follows a random pattern. The VoIP proxy technique will then transmit all 50 packets per second and the compression ratio would be $(136 \times 50)/(156 \times 50) = 0.87$. That is, the bandwidth required increases. However, since the receiving end can recover lost packets, we have to lose two packets in a row before the proxy fails to send a packet to the VoIP decoder. The probability of losing two subsequent packets is $(0.25 \times 0.25) = 0.0625$. Although we would need to use slightly more bandwidth, the proxy technique would reduce the random probability of packet loss from 0.25 (which is intolerable to the human ear) to effectively 0.0625, which is quite tolerable to the human ear.

Example 9.3 Let us assume that we can cross-correlate three packets in a row. That is, in Figure 9.11, each packet on the bottom has information about three consecutive payloads from the top. For transmission over a low bandwidth reliable network core, we can shape (drop at the transmitting side) two out of every three packets and achieve a compression ratio of $(136 \times 50)/(176 \times 50/3) = 2.318$.

Example 9.4 Let us assume that we can cross-correlate three packets in a row as above. If we have a network core that has low bandwidth and high packet loss ratio, we can shape (drop at the transmitting side) one packet out of every three packets. This means we can have a compression ratio of $(136 \times 50)/[(176 \times 50 \times 2/3)] = 1.16$. In the meantime, we would have an enhancement in the effective packet loss ratio since we have to lose two subsequent packets before the codec can see a lost packet. That is, we can drop the effective packet loss ratio from 0.25 to 0.0625. By doing so, we can achieve both compression and robustness through the proxy technique.

9.4.3 Video Proxy over HAIPE

We have seen how TCP and VoIP can be in proxy over the HAIPE encrypted core. We have also seen how the plain text IP layer can ensure the use of COTS equipment within the plain text subnets, while optimizing the use of the HAIPE encrypted core resources through proxy protocols. Video proxy over HAIPE should use a different technique from VoIP and data since the video codecs were developed differently.

Video codecs were developed for COTS applications that do not communicate over HAIPE encrypted core networks. The codecs assume that a packet, with some bit or byte error, is forwarded up the stack where the decoder will then handle the error. Video codecs have error correction capabilities built into them. These error correction capabilities are designed with the assumption that the errors in the video stream will also reach the destination. Within HAIPE encrypted tactical networks, video packets are dropped. The video codec is faced with a packet loss ratio that can corrupt the displayed image.

Theoretically, one can build a video proxy that implements a new video coding approach that is designed to work over HAIPE encrypted networks. That is, for each video stream, the plain text IP layer, at the transmitting side, would decode the stream (to the analog signal), then re-encode it with a new technique that tolerates a high packet loss ratio. At the receiving side, the plain text IP layer would decode the received stream (to the analog signal) and re-encode it to match the end-user codec. This approach would require high processing power (to be able to work on each video stream in parallel) and may require special hardware. Instead, a simpler approach is presented here that relies on packet manipulation to proxy the video stream.

Video streams demand high bandwidth and reliable network paths; in many cases, this combination cannot be offered by tactical networks. The idea of video proxy is to conduct a tradeoff between bandwidth available for the video bit stream, and bandwidth needed for FEC. This tradeoff results in an optimal transmission bit rate. To clarify, let us consider the simplest mode of H.264 video standards. With this standard, a video stream consists of key frames (I-frames) and delta frames (P-frames), as shown in Figure 9.12. A frame is subdivided into slices, where each slice is carried in one RTP packet. I-frames consume around 10% of the video stream bandwidth (depending on the content).

Video is very sensitive to packet loss. At a packet loss ratio of 2%, the end user can perceive a corrupted image that lasts for a few noticeable seconds. This image distortion is a direct result of a corrupt I-frame. This I-frame would result in all subsequent P-frames being decoded erroneously. Also, in a sequence of P-frames, if one P-frame is corrupt, all subsequent P-frames leading up to the next I-frame can also be decoded erroneously. When

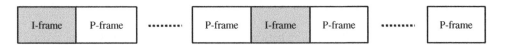

Figure 9.12 An H.264 video stream of I-frames and P-frames.

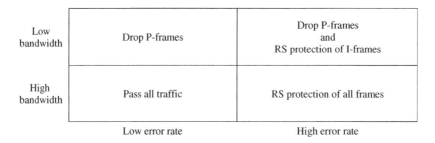

	Low error rate	High error rate
Low bandwidth	Drop P-frames	Drop P-frames and RS protection of I-frames
High bandwidth	Pass all traffic	RS protection of all frames

Figure 9.13   Video proxy using adaptive mode protection.

the core network introduces packet loss, a video proxy over HAIPE must then consider the integrity protection in order to prevent the corruption of the displayed image. The video proxy over HAIPE has to react to packet loss by freezing the last correct I-frame until a new clean image (I-frame) can be provided. That is I-slices are buffered and forwarded only if all packets belonging to the entire I-frame are received; otherwise, the I-frame is dropped. P-slices are then buffered and forwarded (or dropped) in the same manner as the I-slices. If an I-frame is dropped, all successive P-slices are dropped until the next correct I-frame is received.

Now let us assume that we have a HAIPE encrypted tactical network core with low bandwidth and high packet loss. Obviously, this network will not be able to support video transmission. If the proxy drops all P-frames at the transmitting side, we would be using roughly 10% of the bandwidth. However, the I-frame can suffer from error; to counteract this, we can use a portion of the remaining 90% of the bandwidth to utilize an RS code, for packet erasure, and thus protect the transmitted I-frames. This FEC is used to create and send redundant packets, much like the TCP proxy.

If we have abundant bandwidth and high packet loss, both I-frames and P-frames can be protected with an RS code. Adaptive algorithms that adjust the level of redundancy to network conditions can also be employed. Figure 9.13 demonstrates how the video proxy can adapt to bandwidth availability and network path conditions.

With this proxy technique, the dropping of P-frames would result in the receiving side displaying the video stream in a slideshow. As bandwidth availability increases, the proxy can just reduce the number of P-frames and result in a more fluid image. That is, the first half of the P-frames can be transmitted and protected.

9.4.4   RSVP Proxy over HAIPE

We have seen so far how the plain text IP layer can proxy user traffic (voice, video, and data) over HAIPE. This section will show how control traffic can also be in proxy over HAIPE. We will focus on Resource ReSerVation Protocol (RSVP) as an example for the plain text IP layer proxy of control traffic. This proxy uses Resource ReSerVation Protocol-Aggregate (RSVP-AGG), between the peer plain text IP layers, and standard RSVP signaling to the end user terminals.

As mentioned above, RSVP signaling between the plain text IP layer and the cipher text IP layer is not permitted. RSVP packets are encrypted and cannot be used by the cipher text IP layer. This makes encrypted RSVP control traffic a burden over the encrypted core with some low bandwidth links since the cipher text IP layer cannot utilize these packets. RSVP-AGG is associated with a single RSVP tunnel that exists between the two communicating plain text IP layers. Although the tunnel size can be based on some provisioned bandwidth, the MBAC technique mentioned above can be used in conjunction with provisioning. This is a result of the dynamics of the encrypted core since link bandwidths can fluctuate. If QoS violation is noticed before the provisioned bandwidth is reached, MBAC takes over as the default admission control. This proxy makes it possible for the end user to use COTS RSVP (which is designed for static networks) even if the core network is HAIPE encrypted and the mobility of the network nodes cause bandwidth fluctuation.

With this proxy, the encrypted core becomes an RSVP-AGG region. The reader can refer to RFCs 4804, 3175, and 4860 for details on how commercial networks defined RSVP-AGG. The presence of HAIPE creates unique circumstances. These circumstances define the manner by which the RSVP proxy behaves over HAIPE; often this behavior deviates from the RFCs defined in the commercial world.

With the RSVP proxy, individual RSVP flows that originate from the plain text subnet are aggregated. The aggregated flows are then propagated through the secure HAIPE IPSec tunnel. The bandwidth benefits (in bps) of using RSVP-AGG between peer plain text IP layers, will depend on the nature of traffic. With greater sessions aggregated, comes greater bandwidth savings. The ability to perform flow control and admission control, tying MBAC to RSVP signaling, is a crucial aspect of this proxy. The proxy reflects the estimated congestion through RSVP signaling (e.g., RSVP grant and RSVP deny) to the end user's terminal transport protocol, as shown in Figure 9.14.

Figure 9.14 demonstrates how COTS RSVP signaling can be used over the plain text subnet, while the RSVP proxy turns the RSVP signaling into RSVP-AGG over the HAIPE encrypted core. The figure also shows how RSVP-AGG is tied to the HAIPE IPSec tunnel. Over the core network, where RSVP signaling is encrypted, the QoS approach relies on DiffServ, while the QoS approach over the plain text enclaves is IntServ and uses COTS RSVP signaling.

9.4.4.1 Deviations from Existing RFCs

The deviations from existing RFCs focus mainly on the encrypted core side of the RSVP-AGG implementation, as the red enclaves fully adhere to COTS RSVP implementation.

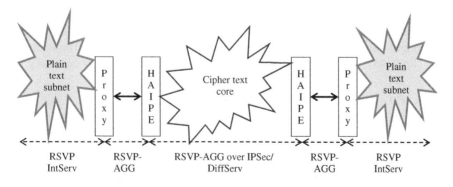

Figure 9.14 RSVP signaling with RSVP proxy.

There are two main points needed to be emphasized here. The first relates to the physical RSVP-AGG tunnel and the second to peer-to-peer association of the plain text IP layers.

No Physical RSVP-AGG Tunnel

The RSVP-AGG session is subject to the same limitations as RSVP sessions over HAIPE since it can be considered as a standard RSVP session that has a large reservation capable of accommodating all the associated RSVP sessions. In other words, the cipher text IP layer is not able to see any of the RSVP-AGG messages. Thus the creation of any physical reservation tunnels over the encrypted core remains pointless. As a result, the implementation of RSVP-AGG over HAIPE must deviate from the current RSVP-AGG protocol. The implementation of an RSVP-AGG session over HAIPE relies on the proxy protocol monitoring the connection state between all the HAIPE devices (it essentially ties the RSVP-AGG tunnel to the existing HAIPE IPSec tunnel). The MBAC component, mentioned above, plays a major role by informing the RSVP proxy that RSVP-AGG over the IPSec tunnel can accommodate a new session which is requesting a reservation. The RSVP proxy queries the MBAC component for the admission of the new RSVP session. The severity level of MBAC can also trigger the proxy to deny the reservation of an ongoing low precedence session when congestion is detected. By monitoring the connection state between the peer plain text IP layers, MBAC can notify the RSVP proxy when the peer remote plain text IP layer no longer has an active connection. This eliminates the need for implementing maintenance of the RSVP-AGG session. This implementation of RSVP-AGG over HAIPE reduces the bandwidth requirements by eliminating the session maintenance traffic.

In addition, the RSVP protocol does not need to be altered when sending the initial path and reservation messages. The cipher text IP layer is unable to see the RSVP messages, so they will never create reservations. However, the plain text IP layer RSVP proxies, at both ends, need the initial path and reservation messages in order to start the RSVP-AGG flow (must be known at both ends), allowing the RSVP sessions at the plain text IP layer to be mapped to the appropriate RSVP-AGG session over the encrypted core.

The fact that HAIPE eliminates the need to implement IP tunneling over the RSVP aggregation region is utilized in another way. Normally, RSVP-AGG would need to ensure that the packets for RSVP-AGG sessions all follow the same route as the initial path and reservation messages. In our case, the aggregation region does not contain any reservations due to HAIPE, so different routing at the encrypted core is not a concern at all. In other words, from the perspective of the plain text RSVP proxy, the encrypted core is a single hop, and packets will follow the same route over this single hop.

In summary, RSVP-AGG over HAIPE ties the RSVP-AGG tunnel to the IPSec tunnel. The initial path and reservation messages flow according to the COTS RFCs. The maintenance traffic is eliminated, and no IP tunneling over the HAIPE IPSec tunnel is necessary.

Plain Text IP Layer Peer-to-Peer Association

Plain text IP layer peer-to-peer association creates the most crucial deviations from commercial RSVP-AGG. Figure 9.15 shows how secure RSVP-AGG tunnels are established between peer plain text IP layers at their corresponding plain text subnets. Conversions between RSVP flows, in the plain text subnets and aggregated tunnels through the encrypted core, are performed by the RSVP proxy. The proxy creates an RSVP-AGG tunnel for each remote plain text subnet that the local plain text subnet communicates with. The interface

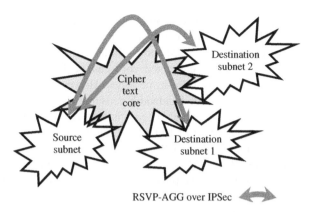

Figure 9.15 Peer-to-peer association of the RSVP proxy with an RSVP-AGG tunnel between a source subnet and each remote subnet.

between the RSVP proxy and the HAIPE device deviates from standard RSVP-AGG implementation over a router interface. The difference is that the RSVP proxy is able to create peer-to-peer association between the transmitting plain text IP layers and the receiving plain text IP layers, thus creating multiple RSVP-AGG tunnels at the proxy egress point.

The cipher text core is treated as one hop between the HAIPE devices. This makes the path between the RSVP proxy pair a natural fit for RSVP-AGG since there is a well-defined aggregation region. Each sender and receiver can be associated with a proxy pair. The proxy pair constitutes the aggregation region and can be associated with a set of RSVP-AGG sessions.

This RSVP proxy can be backward compatible and function despite a remote plain text IP layer failing to have proxy implementation. The standard RSVP protocol can be used between peer plain text IP layers when only one proxy is present. In fact, the RFCs explain how an RSVP session can be a mixture of standard RSVP and RSVP-AGG sessions. The RSVP sessions are determined by the destination, so the previous hop of an RSVP de-aggregator can be running standard RSVP for some senders and RSVP-AGG for other senders. The RSVP-AGG state can be stored for each sender to allow this flexibility. RSVP-AGG implementation over HAIPE thus allows backward computability.

9.4.4.2 Similarities with Existing RFCs

There are three major similarities between the RSVP-AGG protocol that is implemented over HAIPE and the existing RFCs. The first similarity is with respect to associating an RSVP session with an aggregate region with backward compatibility; if a remote plain text IP layer does not implement RSVP-AGG, the communications will still go on. The second similarity is with respect to adhering to the same acknowledgment scheme in order to create a robust aggregation region over the encrypted core. The third similarity is with respect to keeping the plain text subnets with COTS RSVP with no changes.

It is necessary to associate the RSVP sessions with an aggregation region. The sending and receiving enclaves are matched to an edge pair; this constitutes an aggregation region. The association (edge pairing) is determined by querying the MBRM component. It is important to perform a query to an admission control entity since some remote plain text IP layers might not use this RSVP proxy. In such a case, standard RSVP must be used (i.e., the implementation must preserve backward compatibility).

Note that messages sent over the encrypted core can be lost. Standard RSVP overcomes dropped messages by sending refresh messages. However, RSVP-AGG eliminates the refresh messages in order to reduce bandwidth consumption and lessen the burden on the routers. Therefore, an acknowledgment scheme is required for RSVP-AGG messages to ensure delivery. For example, a part of RFC 2961 can be used in this proxy to create a robust aggregation region, based on the RSVP-AGG message acknowledgment mechanism.

Senders, receivers and non-RSVP-AGG plain text IP layers still require RSVP refresh messages in order to avoid a timeout. The presented RSVP proxy sends path and reservation refresh messages. For RSVP-AGG sessions, the path and reservation messages are generated from the proxy itself and sent only over the plain text subnet. In this manner, the RSVP-AGG optimization is transparent to senders and receivers.

9.4.5 Multicast Proxy over HAIPE

The implementation of multicast over HAIPE is very difficult, and many challenges need to be addressed by the plain text IP layer. Some multicast over HAIPE approaches use GRE tunnels between the plain text IP layers. These GRE tunnels are in addition to the HAIPE IPSec tunnels and RSVP tunnels. Some encrypted core links have limited bandwidth, and cannot accommodate these multiple tunnel layers. To avoid this problem, a multicast proxy at the plain text side would be needed in order to eliminate the need for GRE tunneling. We have seen how the RSVP proxy in the previous section tied an RSVP-AGG tunnel to the IPSec tunnel. The existence of an IPSec tunnel can be leveraged by the multicast proxy to eliminate the need for GRE tunneling.

The multicast proxy interface to the plain text subnet must work with common multicast signaling such as Internet group management protocol (IGMP) and protocol-independent multicast (PIM). The multicast proxy should be compatible with both unicast and multicast HAIPE IPSec tunnels. Most importantly, the multicast proxy should understand the existence of multiple access waveforms at the encrypted core and leverage the multiple access capabilities. With multiple access waveforms, a packet sent over the air can be received by multiple receivers and if multicast over the plain text IP layer utilizes these multiple access capabilities, it can optimize bandwidth utilization by multicast traffic over the encrypted core.

Figure 9.16 provides a notional diagram explaining how the multicast proxy relates to other components in a network. In the figure we assume multicast packets from the source, S, being sent to the multicast address MC1 (top right-hand corner). H1, H2, H3, H4, and H5 are the grouping of the multicast address MC1. The multicast proxy will handle multicast over the cipher text core from the source to all hosts. If we assume that the IPSec tunnels between the HAIPE devices are unicast tunnels, the cipher text core will have three packets – one for H1 and H2, one for H3, and one for H4 and H5. Thus, by approaching this problem from the plain text side, the multicast proxy will eliminate the need for GRE tunneling – this provides a considerable bandwidth gain with small packets and small size cipher text

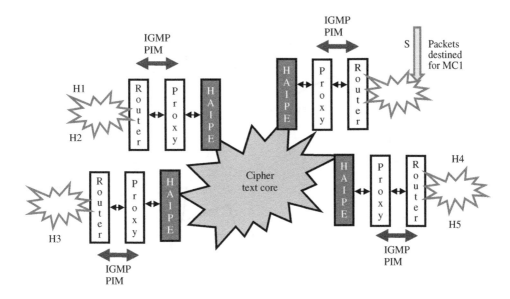

Figure 9.16 Multicast example as it relates to the multicast over HAIPE proxy.

core links. Also, assuming that the IPSec tunnels between the HAIPE devices are unicast tunnels, we can measure the optimization in terms of sending three packets instead of five packets over the cipher text network core.

Now let us see how the same multicast problem is seen at the cipher text IP layer. In Figure 9.17, we assume the existence of a multiple access waveform that connects the source node to the node of H3 and the node of (H4, H5) and a point-to-point link between the node of H3 and the node of (H1, H2). We also assume that HAIPE has multicast tunnels. In this case, we will have a multicast tunnel going over the cipher text IP of the source node S to the cipher text layer of the nodes of (H3) and (H4, H5). Optimum multicast implementation will have one packet travel over the air, instead of three packets. The cipher text IP layer of the node of H3 will forward a copy of the packet over the point-to-point link to the cipher text layer of (H1, H2).

The actual implementation of the multicast proxy will depend on how the HAIPE multicast tunnels are formed, regardless of the cipher text core topology. This implementation can rely on a protocol, HAIPE discovery protocol (HDP), briefly explained here. This protocol is very much tied to the deployment plan and the hierarchy of the nodes.

9.4.5.1 HAIPE's Role in Deployment Hierarchy

It is important to understand the role that HAIPE configuration plays in a typical deployment. In a deployment plan, not all HAIPE devices can talk to each other. If we have N nodes in the deployment, we will be forming $(N * (N - 1))$ tunnels in the network; this formation is impossible to maintain. Military hierarchy dictates which nodes can talk to other specified

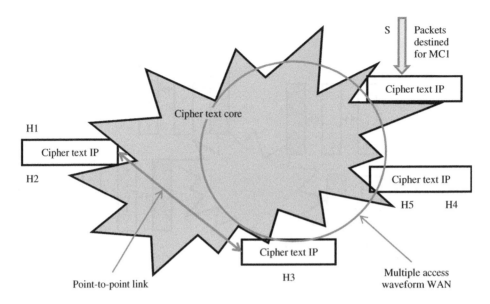

Figure 9.17   Multicast example as it relates to the cipher text IP layer.

nodes. This communication is achieved through the association of HAIPE cipher text IP addresses that are allowed to form IPSec tunnels.[17] For example, node i HAIPE can have the associations:

$$(B_i \leftrightarrow B_k) \, (B_i \leftrightarrow B_l).$$

This means that node i can only talk to nodes k and l. In other words, each HAIPE device knows which other devices it can form a tunnel with, in order to establish communications. In a typical deployment, the generic discovery server (GDS) configuration, explained below, defines all cipher text address associations.

9.4.5.2 HAIPE Discovery Protocol (HDP)

HDP relies on the existence of a GDS. In any given network, one deployed HAIPE device acts as the GDS, while the remaining HAIPE devices are generic discovery clients (GDCs). This client/server protocol is the foundation of HDP.

This client/server protocol is *only* a plain text side communications. Both the GDS and the GDC exist on the plain text side of HAIPE. No cipher text information is allowed in this protocol.

Each HAIPE device (client) can discover its own plain text enclave IP addresses by using standard IP discovery over the plain text port.[18] Practically, each HAIPE device is deployed with its unique cipher text side IP address. This unique cipher text IP address is part of a deployment plan that follows an IP addressing schema and ensures command hierarchy

---

[17] Military operational constraints are sometimes detrimental to network protocols. Many MANET protocols are developed without understanding of operational constraints and can be of no use in practical military applications.

[18] A HAIPE device (or chip) has only two ports. One is cipher text and the other is plain text.

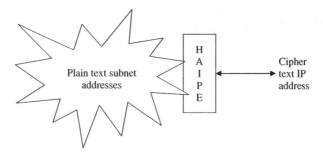

Figure 9.18 HAIPE associating its pre-configured cipher text IP address to its discovered plain text IP addresses.

(not everyone is allowed to communicate to everyone – military hierarchy is reflected in the mapping that chooses the cipher text IP addresses). Each HAIPE device (GDC) has knowledge of which IP addresses (from its plain text subnet) are associated with its unique cipher text IP address.

Each HAIPE device (GDC) can update the GDS with its address association over the network. That is, HAIPE i will have a set of plain text IP addresses discovered through its plain text IP port, which we can refer to as $R_i = \{a_1, a_2, \ldots a_n\}$ and a cipher text IP address B_i with the association ($R_i \leftrightarrow B_i$). This association is depicted in Figure 9.18.

The multicast proxy can query the GDS for the associations of plain text IP addresses to cipher text IP addresses. In this query, the proxy can run as a GDC. Regardless of the cipher text core network topology, the multicast proxy can optimize the multicast protocol based on the HAIPE tunnels and the discovered associations.

9.5   ICD X Cross Layer Signaling across the HAIPE

We have seen how MBRM and MBAC can work without assuming any information regarding the encrypted core network status. Now let us explore how the explicit congestion notification (ECN) bits, which are allowed to pass between the plain text IP layer and the cipher text IP layer (according to HAIPE standards 3.01 and later), can be utilized by the plain text resource management algorithms. The passing of these two bits can be considered as a form of cross layer signaling across the HAIPE. Other information allowed to pass through the HAIPE domain guard can also be utilized for resource management. This section focuses on utilizing of the ECN bits.

The IP header of a plain text IP layer packet, passing into HAIPE, has a one-byte type of service (TOS) field. Note that the HAIPE 3.0 and previous standards copy the left six bits of this TOS field from the plain text IP header onto the cipher text IP header TOS field (in order to allow using DiffServ QoS over the cipher text core). HAIPE 3.01 and subsequent standards allow the use of these two bits where the entire eight bits are copied from the plain text IP header to the cipher text IP header and vice versa.

Figure 9.19 shows a detailed view of the eight bits. The three leftmost bits define the CoS and we can refer to them as C1, C2, and C3. These three bits are used to define a CoS, such as voice, video, interactive data, streaming data, ftp, and so on. The next three bits are the precedence bits which we can refer to as P1, P2, and P3. These three bits are used to define

CoS (bits)			Precedence (bits)			ECN (bits)	
C1	C2	C3	P1	P2	P3	ECN1	ECN2

Figure 9.19 Breakdown of the eight bits of the TOS byte.

Table 9.1 An example of DSCP mapping

Traffic description		Markings and classifications							
Traffic type	Precedence	CoS bits			Precedence bits			COS	DSCP
Voice	Flash override	1	0	1	0	0	1	5	41
	Flash	1	0	1	0	1	0	5	42
	Immediate	1	0	1	1	0	0	5	44
	Priority	1	0	1	1	0	1	5	45
	Routine	1	0	1	1	1	0	5	46
	Routine (default)	1	1	0	0	0	0	6	48
Video	Flash override	1	0	0	0	0	1	4	33
	Flash	1	0	0	0	1	0	4	34
	Immediate	1	0	0	1	0	0	4	36
	Priority	1	0	0	1	0	1	4	37
	Routine	1	0	0	1	1	0	4	38
Data (type I)	Flash override	0	1	1	1	1	1	3	31
Data (FTP)	Flash override	0	0	1	0	0	1	1	9
	Flash	0	0	1	0	1	0	1	10
	Immediate	0	0	1	1	0	0	1	12
	Priority	0	0	1	1	0	1	1	13
	Routine	0	0	1	1	1	0	1	14
Data (BE)	Routine	0	0	0	0	0	0	0	0

the precedence within the class. This can be routine, priority, immediate, flash, and flash override. All six bits (C1, C2, C3, P1, P2, and P3) are commonly referred to as the DSCP (differentiated services code point). Marking a packet with a specific DSCP indicates the CoS and the precedence of the packet. It is essential to understand the difference between real-time requirements of traffic, which are decided by the CoS through the CoS bits, and survivability requirements, which are decided by the precedence bits. For example, a routine voice packet requires fast delivery (real-time); however, a certain packet loss ratio can be tolerable. On the other hand, a flash override data packet may not have the same real-time requirement but requires a high assurance of packet delivery. Table 9.1 demonstrates DSCP mapping for voice, video, and other data classes.

The last two bits in the TOS field are the ECN bits. ECN1 defines if the packet is an ECN capable transport (ECT), i.e., the plain text IP layer or the plain text router (see Figure 9.1) can mark this bit if it wants the core network routers or the cipher text IP layer to notify the remote plain text IP layer if the packet encountered congestion at the network core. The ECN2 bit will be toggled (set to one to mark congestion and to zero otherwise) by the cipher text router if the packet happens to pass through a cipher text IP layer where congestion is encountered.

Thus, one can see how the plain text IP layer is responsible for marking the exact DSCP (six bits) in order to define the packet's CoS and precedence. Based on the DSCP value, the packet will experience differentiated service at the encrypted core. This differentiated service is vital in ensuring that the cipher text IP layer services real-time packets (e.g., VoIP) promptly and avoids dropping packets for applications where the packet drop is sensitive, and so on.

The ECN bits were originally designed to control the TCP flow in a COTS network. However, the use of ECN bits for flow control is not possible with HAIPE encryption because the transport layer information is obscured on the cipher side. On the other hand, since it is included in the cipher text IP header, the ECN field can be a good place to communicate the encrypted core network congestion status to the plain text IP layer.

The method for using ECN bits in HAIPE encrypted networks explained in this section differs from the intended use of ECN bits within commercial networks. Here, in the cipher text IP layer, the proposed operation follows the ECN operational principles used with COTS routers. That is, an explicit congestion experienced (ECE) flag is set if the network is congested. On the plain text IP side, a bit, which is normally set by an end terminal, is set at the plain text IP layer or the tactical edge (see Figure 9.2). Upon receiving the ECE flag, the remote plain text IP layer estimates the status of the encrypted core network pass. A feedback token is needed much like the MBAC technique described above. This technique can be referred to as ECN-based MBAC.

The following further describes the operation:

- The plain text IP layer determines which categories of packets need to be configured for cross layer signaling capabilities (which packets in which the plain text IP layer can mark the ECT bit). For example, we would enable the ECN for streaming or for interactive data packets of certain precedence, assuming that these packets can give us the best indication of the cipher text IP layer congestion.
- The ECN threshold in a cipher text IP layer is set for the traffic of each class and priority and a queue is established. The threshold (i.e., the threshold queue depth above which the ECE flag is set) will be set differently for each class and priority as decided by the network management configurations.

Note that congestion estimation and admission control using ECN (ECN-based MBAC) has many advantages over the pure measurement approach (MBAC) covered early in this chapter. Consider the following advantages:

1. ECN-based MBAC gives a clear indication of whether or not the problem is at the cipher text IP layer or is a result of RF signal blockage. The estimation of $C(P_i)$ in Equation 9.7 can be very accurate. If the estimation of $T(pi)$ in Equation 9.9 is accurate, then the estimation of RF signal blockage is fairly accurate as well.
2. Since congestion estimation is more accurate, CAC and flow control at the plain text IP layer, using ECN-based MBAC, can react faster to the core network dynamics. That is, the Markov chain in Figure 9.4 can move between states faster and can adjust to the core network dynamics more gracefully.

3. There is no need to stamp the packet with GPS timing at the plain text enclaves in order to estimate end-to-end delay. With ECN-based MBAC, the cipher text IP layer decides if the packet has been delayed at the hop without the need for GPS timing at the plain text enclaves.
4. Since packet stamping is not needed with ECN-based MBAC, the overhead associated with packet stamping can be avoided. Especially for small size packets, such as those from a VoIP stream, avoiding packet stamping can save considerable bandwidth.

9.6 Concluding Remarks

Although we covered ICDs I, V, and X (from Figure 8.2) in detail, we also came across other ICDs defined in Chapter 8. For example, Sections 9.4.5.1 and 9.4.5.2 gave an example of what ICD VI can cover. The segmentation of cipher text IP packets mentioned in Section 9.3 are related to ICD II since the HAIPE cipher text side can be configured for the MTU size going to the cipher text IP layer. Examples of the capabilities of ICDs III and IV are detailed within the WNW waveform in Chapter 6. Cross layer signaling between the cipher text IP layer and the peer-to-peer relation of cipher text IP layers are well studied and can be turned into ICDs. ICD VIII is radio specific and there are ongoing efforts to define the details of this ICD (see reference [1]).

9.7 Historical Perspective

The US government relies on its National Security Agent (NSA) for ensuring the security of its communication and information systems. NSA worries about confidentiality and authentication of any messages sent over government networks. NSA algorithms are usually classified, especially those algorithms used to protect classified information. The HAIPE encryption mentioned in this chapter is sometimes referred to as Type I encryption by the NSA. In 2003, the NSA took a major step (the first in its history) of publishing the Skipjack and AES algorithms, categorized as Type I encryption.

The NSA created many security techniques for computer networking, including link encryption, key management, and access of encrypted communications. They also ensured the compatibility between defense and commercial networks, while also establishing compatibility with communications of US allies (such as NATO countries).

For over 60 years, the NSA has been developing encryption systems, starting with electromechanical systems introduced in the 1950s. These were built on the legacy of World War II systems (built by the NSA) that used rotor machines for encryption. These systems used distribution paper key lists, that changed daily, which described to the operator how to manage the rotor (changed every day).

With the use of vacuum tubes came a new generation of electronic encryption systems that used logic or algorithms based on feedback shift registers. Encryption keys were loaded into these systems using punch cards. These systems, introduced in the 1960s, stayed in use until the mid 1980s.

The third generation of the NSA's encryption systems was introduced in the 1980s. These systems leveraged the explosive use of transistors and integrated circuits, and used much more complex logic for their encryption. They were smaller and more reliable than the vacuum tube systems, and encryption keys were loaded using a connector. Similar systems, which rely on connectors, are currently in use by military radios, where a key can be loaded by a portable electronic device.

In the 1990s, the NSA rolled out its fourth generation of encryption systems. These systems relied on electronic key distribution. Public key methods appeared, such as the electronic key management system (EKMS). The military's use of these keys allowed individual commands to generate them. This was in sharp contrast with the previous system where each key was received from the NSA via a courier. The NSA introduced encryption support for commercial systems such as Ethernet IP and fiber multiplexing in the 1990s. As the Defense Department began building classified networks, (such as the secret Internet Protocol router network (SIPRNET) which used IP) it relied on the NSA to provide heavy support. The NSA developed the techniques to secure communication links between "enclaves" or subnets where the encrypted core, as mentioned in this chapter, was beginning to take root. The NSA also developed the standards for connecting these classified networks with commercial networks.

In the new millennium, the military dependence on computer networks has increased drastically, and reliance on encryption is now a vital aspect of protecting sensitive information. The NSA has recently begun providing guidance to commercial firms designing systems for government use. The HAIPE solution covered in this chapter is an example where the products of commercial companies, such as General Dynamics and ViaSat, (e.g., KG-245 and KG 250) are based on NSA standards. Since the NSA categorizes algorithms for classified use "in NSA approved systems," one can assume that in the future the NSA may use more non-classified algorithms. The KG-245A and KG-250 use both classified and unclassified algorithms. The NSA has an Information Assurance Directorate that works with the department of defense on cryptographic modernization. This partnership ensures that as military communications systems get more complex, security needs are always met.

Bibliography

1. http://www.opengroup.org/tech/direcnet-task-force/.
2. Elmasry, G.F., McCann, C.J., and Welsh, R. (2005) Partitioning QoS management for secure tactical wireless ad-hoc networks. *IEEE Communications Magazine*, **43** (11) 116–123.
3. Elmasry, G.F., Russell, B., and McCann, C.J. (2005) Enhancing TCP and CAC performance through detecting radio blockage at the plain text side. Proceedings of Milcom 2005, U306.
4. Lee, J., Elmasry, G.F., and Jain, M. (2008) Effect of security architecture on cross-layer signaling in network centric systems. Proceedings of Milcom 2008, NC9-3.
5. Elmasry, G.F., Lee, J., Jain, M., *et al.* (2009) ECN-based MBAC algorithm for use over HAIPE. Proceedings of Milcom 2009, U310.
6. Elmasry, G.F., Lee, J., Jain, M., *et al.* (2008) Achieving consistent PHB across the GIG, a QoS common denominator. Proceedings of Milcom 2008, NC7-1.
7. Elmasry, G.F., Jain, M., Lee, J., *et al.* (2009) Reservation-based quality of service (QoS) in an airborne network. Proceedings of Milcom 2009, U240.
8. Faulkner, E.A. IV (2004) Interactions between TCP and link layer protocols on mobile satellite links. Masters thesis. Department of Electrical Engineering and Computer Sciences, Massachusetts Institute of Technology, September 2004.
9. Henderson, T.R. and Katz, R.H. (1999) Transport protocols for Internet-compatible satellite networks. *IEEE Journal on Selected Areas in Communications*, **17** (2), 345–359.
10. Goode, R., Guivarch, P., and Stell, M. (2002) Quality of service in an IP crypto partitioned network. Proceedings of MILCOM 2002, U605.
11. Wydrowski, B. and Zukerman, M. (2002) QoS in best-effort network. *IEEE Communications*, **40** (12), 44–49.

12. Giordano, S., Salsano, S., Van den Berghe, S., *et al.* (2003) Advanced QoS provisioning in IP networks: the European premium project. *IEEE Communications*, **41** (1), 30–36.

13. Ahlswede, R., Cai, N., Li, S-Y.R., and Yeung, R.W. (2000) Network information flow. *IEEE Transactions on Information Theory*, **46** (4), 1204–1216.

14. http://www.mobilenetworks.org/nemo/drafts/draft-ivancic-layer3-encryptors-00.txt.

15. http://www.estoile.com/links/ipsec.htm.

16. https://ges.dod.mil/ (need security authorization).

17. United States National Security Agency (2006) High Assurance Internet Protocol Encryptor Interoperability Specification, Version 3.1.0, December 31, 2006.

18. Racz, A. (1999) How to build a robust call admission control based on on-line measurements. Proceedings of Globecom 99, pp. 1634–1640.

19. Grossglauser, M. and Tse, D. (1997) Measurement-based call admission control: a heavy traffic framework. Proceedings of the 36th Conference on Decision and Control, 1997, pp. 1792–1797.

20. Xu, Y., Westhead, M., and Baker, F. (2004) An investigation of multilevel service provision for voice over IP under catastrophic congestion. *IEEE Communications*, **42** (6), 94–100.

21. More, A.W. (2002) Measurement Based Management of Network Resources. Technical Report No. 528, University of Cambridge Computer Laboratory, April 2002.

22. Landry, R., Grace, K., and Saidi, A. (2004) On the design and management of heterogeneous networks: a predictability-based perspective. *IEEE Communications*, **42** (11), 80–87.

23. Marshall, P. (2006) Adaptation and integration across the layers of self-organizing wireless networks. MILCOM 2006.

24. Brehmer, J. and Utschick, W. (2005) Modular cross-layer optimization based on layer descriptions. WPMC 2005, Aalborg, Denmark, September 2005.

25. Christopher Ramming, J. (2005) *Control-Based Mobile Ad-Hoc Networking (CBMANET) Program Motivation and Overview*, DARPA, August 30, 2005.

26. Braden, R., Zhang, L., Berson, S. *et al.* (1997) Resource Reservation Protocol (RSVP) Version 1 Functional Specification, RFC 2205, September 1997.

27. Le Faucheur, F. (2007) Aggregation of Resource reSerVation Protocol (RSVP) reservations over MPLS TE/DS-TE tunnels, RFC 4804, February 2007.

28. Baker, F., Iturralde, C., Le Faucheur, F., *et al.* (2001) Aggregation of RSVP for IPv4 and IPv6 reservations, RFC 3175, September 2001.

29. Le Faucheur, F., Davie, B., Bose, P., *et al.* (2007) Generic Aggregate Resource reSerVation Protocol (RSVP) Reservations, RFC 4860, May 2007.

30. Berger, L., Gan, D., Swallow, G., *et al.* (2001) RSVP Refresh Overhead Reduction Extensions, RFC 2961, April 2001.

10

Bringing Commercial Cellular Capabilities to Tactical Networks

Now that we have covered the open architecture approach, we are ready to see how commercial cellular capabilities can be part of the tactical theater communications infrastructure. Commercial cellular brings many benefits to the tactical theater, such as low cost, application richness, maturity, and familiarity of use. The current generation of warfighters can greatly benefit from the use of smartphones, given the additional features they provide. Smartphones are putting never-before-seen communications capabilities into the hands of new tactical users. These include, but are not limited to, voice and command and control (C2) data, social networking, anthropological decision aids, real-time data feeds (from airborne and ground-based sensors), and remote controls of unmanned vehicular platforms. Deployment of these technologies presents a variety of challenges, such as environmental considerations, energy longevity, and spectrum management. However, security considerations, information assurance (IA) and network operations (NetOps) are the most crucial needing to be addressed since they play a major role in ensuring that the missions are not compromised. This chapter presents a comprehensive architecture that would support tactical deployment of smartphones while highlighting the critical areas needing attention. This approach is built on the open architecture principles discussed in the previous two chapters.

There are numerous pros and cons to the deployment of cellular versus WiFi services in tactical environments. These relate to security, access control, authentication, coverage, interference, and scalability. As commercially specified, 4G/LTE enables a mixed WiFi/cellular environment that could address these needs. The architecture described in this chapter is designed to instantiate a cellular core framework that can support the evolution of both WiFi and cellular services. The proposed architecture relies on a tactical cellular gateway (TCG) and a domain guard to solve interoperability and access challenges.

This chapter covers three possible use cases (scenarios) where cellular capabilities are to be integrated into tactical networks. Security and network management (NM) vulnerabilities will be addressed, as will a distributed technique for 3G/4G/LTE IP core services.

Tactical Wireless Communications and Networks: Design Concepts and Challenges, First Edition. George F. Elmasry.
© 2012 John Wiley & Sons, Ltd. Published 2012 by John Wiley & Sons, Ltd.

10.1   Tactical User Expectations

Recent technical leaps in commercial wireless products, such as the features of modern smartphones have prompted researchers in the tactical wireless networking community to pursue the exploitation of 3G/4G/LTE technologies as an integral piece of the tactical theater's communications capabilities. While cellular technology is more susceptible to intentional jamming and interference from local commercial waveforms, there are also huge benefits derived by leveraging and optimizing commercial waveforms for military use. The following are some objectives for developing tactical solutions, with commercial cellular technology, that would properly support all concepts of operations (ConOps) at all echelons from the division down:

1. **Optimizing bandwidth**: The solution must minimize the bandwidth consumption at all levels while also minimizing the number of necessary hops. Bandwidth and hop count optimization should be considered in terms of user traffic routing and reach-back to the cellular core set of services for provisioning, call signaling, and authentication.

2. **Accommodating deployment needs**: The solution should be deployable as either a stand-alone system or a part of an enterprise system. Cellular services can be made available at all military hierarchies, ranging from the dismounted soldier to the C2 or to headquarters platforms.

3. **Meeting QoS requirements**: The solution must maximize (and even take measures to enhance or guarantee) voice or video quality, and also provide reliable multicast. Addressing QoS needs is a critical issue when one cellular user tries to communicate with another user over a tactical wireless link, with limited bandwidth and high packet loss.

4. **Addressing security concerns**: The solution must minimize reach-back into cellular service and NetOps services at the network core. This will not only minimize bandwidth consumption, but will also scope and optimize security enclave isolation.

5. **Controlling call routing**: The solution must provide policy based call routing, as required by military doctrine. Call routing policies must be updateable to accommodate for future evolution of cellular technologies and changes to military doctrines. Controlling call routing is critical for keeping in line with military hierarchy and for meeting the needs of joint missions, where specific traffic flows can be permitted to cross between networks participating in a joint mission.

6. **Acquisition flexibility**: The solution needs to be vendor agnostic. Within a given commercial technology space and set of standards, it must support products from multiple vendors. Consider the deployment of specific core cellular services: one needs to be able to plug in any femtocell or base station (BS) that adheres to the deployable core service standards seamlessly.

7. **Adhering to military hierarchy**: Policy based roaming must be addressed by the solution. Roaming typically implemented in commercial wireless can easily violate the essence of military hierarchy.

8. **Simplified architecture**: The simplicity of the solution depends on the technology used for voice. Historically, the voice solutions for the US military have been based on circuit switched technology for tactical operation center (TOC) and reach-back communications. The combat net radio (CNR) has provided voice communications at the company or platoon level. As explained, with the TI (tactical Internet) IP touch

points work with the existence of circuit switched links and legacy waveforms. The tactical Global Information Grid (GIG), as the IP core network for the US military, is currently moving to VoIP. Current commercial cellular technology has VoIP as a user option on 3G cellular networks which support data. The future path for commercial cellular is what is known as LTE and provides a full VoIP solution. For the purpose of seamless IP capabilities, the cellular solution, integrated with tactical networks, must be an LTE solution of VoIP. This is consistent with the GIG vision addressed in this book, where we cover fully IP based tactical networks. Note that integration of voice technologies, other than VoIP, can complicate the architecture and would require additional gateway capabilities.

9. **Addressing vulnerabilities**: The solution needs to address several vulnerabilities relating to traffic policies and NM that arise with this integration of commercial cellular networks into the secure military infrastructure. These include:

    (a) route enforcement of certain traffic types that cannot be allowed to enter the core network
    (b) certain traffic types that must stay within a 3G/4G/LTE access point
    (c) call handovers that could violate the military hierarchy
    (d) call admission control (CAC) issues (e.g., VoIP SIP – session initiation protocol – call control from COTS has a long list of known vulnerabilities).

It is critical to understand that security, IA, and NetOps concerns must be an integrated part of any solution that addresses the introduction of commercial technologies to the tactical theater. These concerns must be mitigated to ensure that the commander's intent is fully carried out as the warfighter is enabled with this powerful technology.

The use of COTS cellular technologies in combat and non-combat communications is the focus of some recent studies. The Joint Tactical Radio System (JTRS) program studied possible architecture for the use of COTS cellular technologies in conjunction with the JTRS radios.

The introduction of 3G/4G/LTE technology into the tactical battlefield with smartphone capabilities must be enabled at the dismounted soldier level and at the command posts. This will allow 3G/4G/LTE infrastructure components to be introduced from any vendor, provided that they comply with the existing standards. It will also allow users to roam from cell to cell, within the limitations of the actual waveforms. One must consider the deployment of cellular technology with the introduction of a TCG that will attempt to provide the necessary functionality to meet the goals presented above in most situations. This chapter assesses the TCG architecture and provides several use cases. NetOps will be a key component of deploying such a system and its role is illustrated, in detail, throughout the chapter.

Figure 6.21 showed that a tactical platform can have three types of security enclaves (plain text subnets, top secret (TS), secret (S), and sensitive but unclassified (SBU).[1] One important aspect of the presented architecture is that all 3G/4G/LTE technology components must be at the SBU enclaves. If our goal is to serve different cellular technology users, including those without clearance, then the 3G/4G/LTE technology needs to be at the unclassified level.

---

[1] Some tactical platforms can have more security enclaves than these three. A coalition enclave can be dedicated for coalition specific communications. Another enclave can be dedicated for WiFi and cellular access. The separation of enclaves ensures that information assurance needs are met. This book assumes the more common three enclave architecture where the SBU enclave can be used for cellular access.

The architecture presented takes the security challenges into consideration while assuming that smartphones can be enabled for Type I or Type II encryption,[2] depending on the application or the user. Access control, smartcards, biometrics, and other approaches will ensure that the identity of the user matches the security ability of the device by reaching into the tactical network and potentially tunneling through the SBU enclave into higher enclave levels.

10.2 3G/4G/LTE Technologies within the War Theater

Figure 10.1 illustrates a potential deployment scenario for commercial 3G/4G/LTE cellular technologies in a tactical network. Essentially, the figure shows three separate mechanisms or use cases (described later) for connecting smartphone or 3G/4G/LTE technology into tactical infrastructure. In the figure, we assume that we are working with the US Army's tactical GIG. Each use case requires some sort of gateway functionality at the access point to the core network. This functionality provides additional functions and security which interfaces with the NetOps environment for configuration, monitoring, and control. The TCG, mentioned early in this chapter, will be used to encompass this collection of features that will vary depending on its deployed location in the tactical architecture and which use case is in effect.

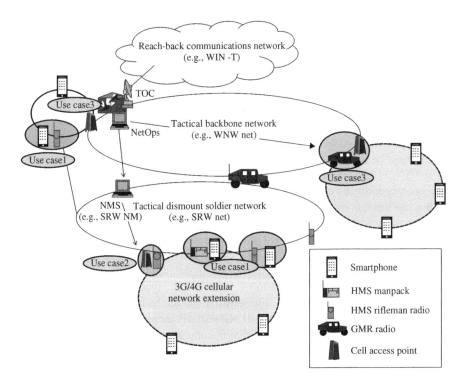

Figure 10.1 3G/4G/LTE commercial cellular technologies in a tactical network.

---

[2] Type I encryption (e.g., HAIPE) supports classified data traffic, while Type II encryption has fewer constraints.

10.3   The Tactical Cellular Gateway

The architecture proposed employs a TCG for interfacing disparate cellular technologies. The TCG is responsible for providing robust voice, video, and multicast. It also provides tactical security, policy based routing, and an efficient use of available bandwidth while distributing smartphone apps and data packages according to mission needs. The TCG is intended to meet as many of the previously mentioned goals from Section 10.1. The TCG was envisioned as both a stand-alone cellular solution and a part of a tiered, distributed deployment that enables robust cellular capabilities at all echelons. As presented, it leverages the latest technology to provide bandwidth efficient capabilities, while providing the ability to instill older (or less capable) technologies into the overall architecture. The architecture can proxy some of the cellular services into the network core, providing a distributed, disruption-tolerant architecture.

Figure 10.2 shows the overall system hierarchy envisioned for deploying the TCG. For a full deployment, each SBU enclave will have an embedded TCG. Two key roles of the TCG hierarchy will be to provide universal VoIP capabilities, while minimizing signaling and forwarding traffic reach-back. Instead of relying on a single root core network service enclave for all 3G/4G/LTE support (i.e., call control, routing, addressing, and roaming), the TCG distributes these functions hierarchically in order to keep all data and signaling traffic local wherever possible. As an oversimplified example, a smartphone, when initially

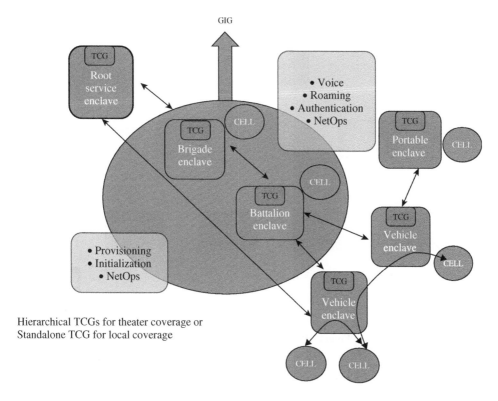

Figure 10.2   Tactical cellular gateway (TCG) hierarchy.

connecting to a cell, will reach back to the root service enclave for setup and IP address assignment. With the proposed hierarchical solution, this addressing information will be cached at each level in the TCG. GPRS (general packet radio service) is the set of protocols and devices that provide IP network capabilities to 3G networks. Thus, each TCG will have a version of a gateway GPRS support node (GGSN) that provides a view "down" in the hierarchy, in terms of end user location and routing. The hierarchical approach mirrors the current military doctrine, where much of the data and voice flow stay within the security boundaries and echelon policies.[3] This approach is similar to the domain name service (DNS) used in IP networks for name and address resolution, where proxying the core network functions reduces traffic flow to the core (reach-back).

Policy enforcement at the TCG enhances security in multiple ways. Domain guards are used in the architecture and will potentially be integrated into the TCG to manage security boundaries between incompatible security domains or levels of classification. They can be used to limit traffic types, service types, or message types from traversing the boundary, prevent/enable certain users from crossing the boundary, manage the direction of data and control traffic, block based on addresses or combinations of the above such as only allowing voice traffic to cross from specific users or hosts. TCG policies can also be established to manage routing of different traffic types (unicast, multicast, and broadcast). How the proxies are configured to support these different traffic types and how the distributed cellular infrastructure manages roaming, address allocation, and authentication are all handled by the TCG.

The address management and allocation is very critical. One possible solution is the use of a distributed GPRS support node (DGSN), within the TCG, as shown in Figure 10.3,

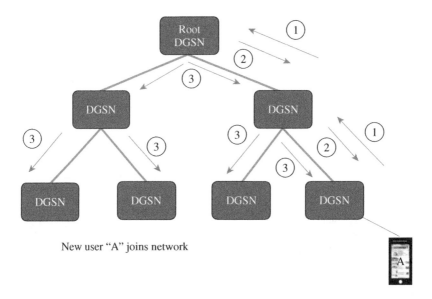

Figure 10.3   DGSN address allocation.

---

[3] The TCG could be the way that information dissemination policies are enforced on cellular traffic. Note also that the tactical world is moving to all IP (including VoIP), parallel to LTE moving to all IP based applications including VoIP.

which illustrates the steps in address allocation. In sequence 1, user "A" joins the network and during initialization, requests an address. This request goes to the root DGSN, where the address is allocated and forwarded to the user in sequence 2. Each DGSN caches the address information on the return path and uses that to prune the routing tree.

In sequence 3, the root DGSN then also floods an "active" notification concerning user A to all DGSN nodes. Thus, if another user in a different branch of the "tree" wants to send traffic to user A and user A is not active, then the packets will be dropped immediately. This prevents the traffic from traversing up to the root of the tree and also prevents potential denial-of-service attack.

The TCG would provide robust voice and video transmissions, as well as reliable multicast through the use of advanced techniques such as error correction, error recovery, admission control, and proxying. These approaches increase bandwidth savings, stabilize the performance of applications, and control signaling in high packet loss environments. This is very helpful since such environments are very common in tactical networks. A critical challenge in a tactical network is the secure distribution and revision of applications or their received data. In addition to user authentication and policies related to classification levels and user authorities, limited bandwidth significantly impacts the ability to deliver data. The TCG interacts with the authentication module and the NetOps manager to provide applications and data to the right people, with the correct configuration and classification level. This support, provided to both a user and to the smartphone is delivered at a time and fashion that does not critically impact the overall network activity.

The TCG is the key to creating secure, scalable, joint, and coalition interoperable 3G/4G/LTE functionality that can readily be distributed. The TCG must be built with a collection of modules that provide proxy capabilities such as cellular core services, GGSN and DGSN services, policy enforcement, and security functions, as well as the ability to provide applications and data to the end user, according to the defined policies. The TCG must employ edge-centric services for cell management, routing policy control, voice, video and data robustness, and platform data distribution.

10.4   Deployment Use Cases

The three use cases enable the connection of cellular based handhelds into the tactical environment. Use case I uses a wire for the interface, and the TCG is fairly simple since the smartphone becomes another source of traffic over the tactical radio network. Use cases II and III interface 3G/4G/LTE components, such as base stations, microcells, and femtocells, into the tactical network, and their TCG requires additional functionalities. All three use cases assume that the 3G/4G/LTE IP core component (a crucial part of deploying 3G/4G/LTE technology where admission control, call routing, etc. occur) is deployed at an SBU enclave at the root node network. If we take the WIN-T (Warfighter Information Network – Tactical) architecture as an example, an HQ node will have this 3G/4G/LTE IP core component at its SBU enclave. In the proposed architecture, we will use a proxy server that will proxy the 3G/4G/LTE IP core component functions through a domain guard to the local 3G/4G/LTE access point. The proxy server will be an embedded component of the TCG package, enabling greater robustness, reliability, and scalability through minimizing core network access by decentralizing or localizing the cellular infrastructure.

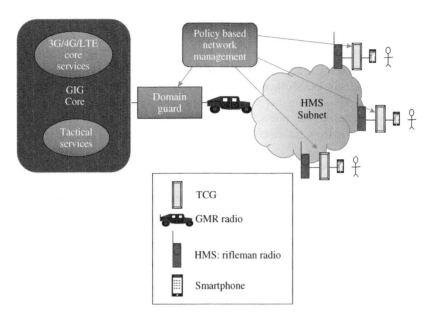

Figure 10.4   Use case I with smartphones tethered to the SRW radios.

10.4.1   Use Case I: Smartphone Tethered to a Soldier Radio Waveform (SRW) Radio

In this use case, a dismounted soldier is carrying a smartphone, tethered to a JTRS soldier radio waveform (SRW) radio, via a USB (universal serial bus) port on the phone, as shown in Figure 10.4. The phone could be tethered either to a JTRS HMS (handheld manpack and small-form-fit) rifleman radio or a JTRS HMS manpack radio.[4]

In this case, the smartphone becomes just another IP device, with the radio forwarding traffic to and from the smartphone. Additionally, the smartphone could be used as a local device to control/configure/monitor the radio, with application development on the radio and the smartphone. VoIP could be supported here, given the proper network infrastructure. When SRW supports VoIP,[5] the smartphone could talk to rifleman radios. In this example, the smartphone will likely be utilized extensively by the warfighter for collaborative, situational awareness (SA), voice, and video applications.

Here, a TCG (embedded with each smartphone), and a domain guard (at the GIG core's ingress point), both configured by NetOps, would have the onus to institute NetOps policies to ensure the following:

1. The voice traffic that is initiated at the smartphone (and now is user datagram proto-col – UDP – over IP in the SRW radio) is not allowed to propagate to the core network

---

[4] JTRS HMS program has different form factors (e.g., small-form-fit and manpack). The form factor refers to the hardware used to build the radio. The SRW waveform can be ported on either of these form factors. The manpack radio is larger in size and can carry multiple channels to work as a gateway between the squad subnet and the rest of the tactical GIG. The small-form-fit hardware is a single channel compact radio.

[5] The current implementation of SRW supports CNR voice type.

(i.e., does not leave the SRW subnet), since the dismounted soldiers should not be talking to someone at the core network.

2. The application traffic (e.g., the position locations) enters the core network. The traffic would then reach a server that processes all tracking of friendly force locations or other applications approved for this user.

3. The application traffic from the server, associated with the friendly force/enemy force tracking,[6] is allowed to reach the dismounted soldiers. This information will exist on the server in the core network.

4. The SRW network manager may also need to implement some policies to ensure the QoS of the smartphone traffic. This may include DSCP (differentiated services code point) mapping of the smartphone applications (voice, video, and data) to proper QoS mapping.

It is important that NetOps propagates its policies to the domain guard as shown in Figure 10.4. For this scenario to function, the SRW subnet gateway node (where traffic leaves the SRW subnet to a core subnet, such as the GIG network core) may need to define the traffic, generated at the 3G/4G/LTE phones, by their type and intended use. Techniques such as IP, subnet, or application discrimination would allow traffic from smartphones to be categorized. In this case, the TCG shown in Figure 10.4 will include these features, as well as other potential features to support routing, reliable multicast, and improved voice and video. This TCG can be implemented as a software solution that could be ported to the smartphone kernel.

A multifunctional proxy server will be necessary as part of the TCG. One function would be the proxy of the 3G/4G/LTE IP core to the local subnet. If present, the TCG could leverage cellular core services for authentication and routing, or it could just appear as another IP device if no cellular infrastructure is available. Another function would be to proxy any tactical core network services that can be at a core server, such as friendly force tracking. The policies to the domain guard, communicated from NetOps, would be based on the mission's needs. For example, if a mission needs position location from the GIG core server, it would get this via the proxy server, and the proxy server would then continue retrieving updated position location information from the GIG core server through the domain guard.

Note, that use case I provides the most jamming and interference resistant solution as it leverages military waveforms, designed for a hostile RF environment.

10.4.2 Use Case II: 3G/4G/LTE Services on a Dismounted Unit

Use case II, depicted in Figure 10.5, introduces the use of 3G/4G/LTE infrastructure by attaching femtocells or BSs to HMS assets (i.e., manpack radios). Using current terminology for this technology, a BS is commonly referred to as a node B. A femtocell is a specialized form of a node B with a limited range of several hundred meters and potential energy considerations. As an example of this deployment, one could have a manpack radio with a femtocell and a software based TCG. A femtocell and TCG would be smaller than a manpack radio. HMS connectivity provides reach-back from the cell to core network cellular services

---

[6] The term blue force/red force tracking is used for friendly force/enemy force position location tracking since the first implementation of this technology displayed friendly force locations as blue dots and enemy force locations as red dots.

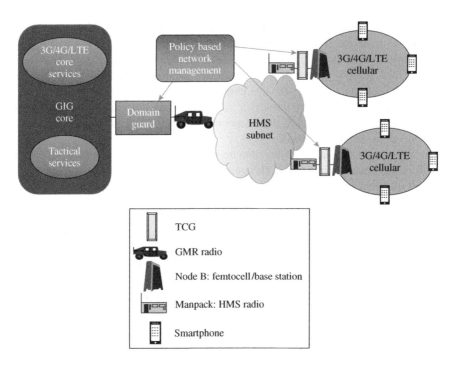

Figure 10.5 Use case II with a 3G/4G/LTE access point.

to support provisioning, mobility, addressing, IP access, and authentication. By using all VoIP, the core network is not required to support circuit switched voice. So, there is no call control reach-back required, although VoIP will require some form of support (as will be discussed later).

Figure 10.5 shows use case II where the 3G/4G/LTE access point is used by the squad leader. The remainder of the squad members will rely on smartphones as their means for communication. As described for use case I, this requires reach-back to the 3G/4G/LTE IP core which is also limited by policies and routing challenges.

In addition to the NM challenges mentioned with use case I, CAC is a growing challenge in this situation. The 3G/4G/LTE IP core component is responsible for admission control of the voice communications for the 3G/4G/LTE end terminal users. The IP core component also negotiates the rates for other non-real-time applications. Keep in mind that the SRW subnet capacity must be factored into the CAC algorithm running on the 3G/4G/LTE IP core. Since the SRW subnet bandwidth can change dynamically, it can sense if it will be hard to communicate with the given SRW subnet bandwidth resources, availability for cellular services in real time to the 3G/4G/LTE IP core services. A possible solution is to rely on the provisioning of a certain portion of the SRW subnet bandwidth for cellular services. The provisioned bandwidth can be used by the 3G/4G/LTE IP core as part of its CAC algorithm. This provisioning must be known by the SRW NM.

Note, that the spectrum emission in this use case constitutes vulnerability. Node B emits spectrum that has no military encryption. The use of a femtocell in this scenario may be preferred since its spectrum footprint is limited.

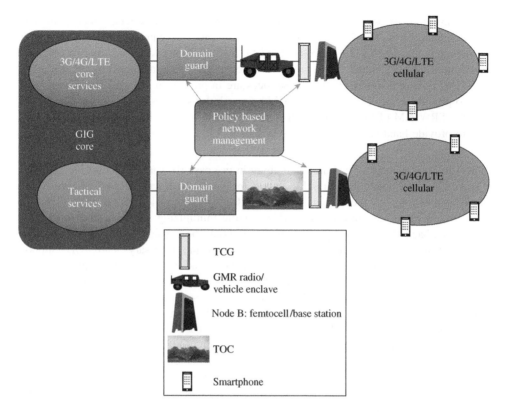

Figure 10.6 Use case III with a 3G/4G/LTE access at vehicle or TOC.

10.4.3 Use Case III: 3G/4G/LTE Access at an Enclave

Figure 10.6 illustrates use case III, where a vehicle – or upper echelons where plain text SBU enclaves (each with a COTS router) – becomes the entry point for the 3G/4G/LTE technology at the tactical network infrastructure. As Figure 10.6 demonstrates, the TCG at the TOC can be used by the network manager to ensure that all NetOps policies are enforced. It is expected that the TCG functionality can eventually be integrated into the 3G/4G/LTE node B access point, the node B itself and/or the COTS router.

As before, the NM needs to ensure the enforcement of NetOps policies such that friendly force and enemy force tracking traffic will flow properly from the client (the smartphone) to the server in the GIG network core, while the voice traffic stays within the access point subnets.

It should be noted that there are many differences between use cases II and III, including the following:

1. In use case II, we are likely to use a small footprint access point (femtocell), while in use case III one could easily use a full BS for a large volume of cellular users. This would then raise more security concerns since the 3G/4G/LTE spectrum coverage area would be much broader for use case III.

2. In use case III, we could be working with a standard NM system, which is built to communicate to the COTS routers. In such a case one can use this NM system to communicate to the TCGs through the COTS router. Congestion information collected from the COTS router can be communicated to the TCGs and used by the TCGs for CAC (where 3G/4G/LTE IP core services are in proxy). In use case II, we may need to provide some of the SRW bandwidth for cellular services and inform the TCG and the SRW NM of this provision. It may be difficult to alter how the SRW NM works to provide bandwidth for cellular services.

For a full deployment of cellular technology in use case III, each SBU enclave can have an embedded TCG and the hierarchy shown in Figure 10.2 will be part of the overall system architecture. As mentioned in Section 10.3, two key roles of the TCG hierarchy are to provide VoIP capabilities everywhere, while minimizing signaling and forwarding traffic reach-back. The hierarchy in Figure 10.2 should be considered in conjunction with Figure 10.6 to emphasize that the TCG can distribute the necessary 3G/4G/LTE support (i.e., call control, routing, addressing, and roaming) instead of relying on a single root core network service enclave. This distributed hierarchy will maintain all data and signaling traffic locally, whenever possible. The example mentioned in Section 10.3 can now be applied for use case III. When a smartphone is initially connecting to a cell, it will reach back to the TCG where the addressing information is cached. Thus each TCG will have a version of a GGSN that provides a view "down" in the hierarchy, in terms of end user location and routing. The distributed proxy of some of the cellular service in the network core, at the access points, is an essential part of the presented architecture in order to meet many of the tactical user expectations set forth at the start of this chapter. Notice that for use case I, the distributed TCG proxy could reside in the handheld instrument itself, as part of the smartphone kernel.

10.5   Concluding Remarks

As cellular technology progresses toward the LTE standards, the additional use of more bandwidth will enable even better and more reliable applications. Techniques, such as mesh networking and cellular-to-WiFi handoffs, will lead to greatly improved bandwidth utilization. Broadly integrated security architectures will then allow all users, at all echelons, to access *their* SA and *their* data wherever they are, regardless of where the information resides. Tools in the TOC today can now be pushed to the edge to empower the warfighter. Information and data generated at the edge can be processed at the edge itself and distributed, with minimal latency, to those whose lives might depend on it. The proposed TCG is architected to evolve with both LTE and tactical network evolutions.

The open architecture theme, presented in this third part of the book, could allow for utilization of the continuously evolving cellular capabilities within the tactical theater. This can be achieved by creating standards that describe the TCG, the necessary functions and interfaces at each hierarchal level, and what 3G/4G/LTE IP core services need to be in proxy at the TCG for each case.

Table 10.1 Comparison of the three use cases

Use case	Encryption	Policy based management	Management scope	End-to-end robustness	Cellular core proxy	Application proxy	Additional hardware	Jamming interference	Call admission control
I	Tactical waveform based	In guard and handheld	To be determined	Yes	In handheld	In handheld	Tether	Tactical waveform	No
II	In handheld	In guard and gateway TCG	Tactical waveform manager	Only in core	At gateway TCG	In handheld or TCG	TCG (add to router?)	Commercial waveform	In TCG
III	In handheld	In guard and gateway TCG	Core network manager	Only in core	At gateway TCG	In handheld or TCG	TCG (add to router?)	Commercial waveform	In TCG

Table 10.1 provides a comparison of the three use cases analyzed above. The table is illustrates the differences in functionality and deployment architecture for the three cases.

The TCG is responsible for many functions. One area addressed in this section is the mapping of data classes for services (TCP sessions in the core network) between the cellular network and the tactical core network. It is critical to ensure that the end-to-end operational requirements can be met in a mix of cellular and tactical core networks. Please refer to Table 9.1 to review the example that details the number of DSCP mappings that can be implemented at a tactical network.

Theoretically, there could be up to 64 different DSCPs (differentiated services) in a tactical network. When considering the use of cellular technology (which offers a limited number of service classes), meeting end-to-end delay requirements can be challenging. Achieving end-to-end target delay, for survivability information and time-sensitive data, is critical in this mix. It is essential for the TCG to map the wide range of differentiated services in the tactical network to the limited classes of service in the cellular network in such a manner that mission goals are not compromised. For example, the TCG could map the survivability and time-sensitive data of the tactical network to the cellular network streaming class and map the routine and non-time-sensitive data of the tactical network to the background class in the cellular network. The TCG should also consider assigning cellular, high data rate streaming channels to the critical time-sensitive tactical data. This would ensure that sum of the delay introduced by the cellular network and the tactical network delay does not violate the objective end-to-end message delay set forth by operational needs.

Bibliography

1. Elmasry, G.F., Welsh, R., Jain, M., *et al.* (2011) Security and network operation challenges with cellular infrastructure in tactical theater. *IEEE Communications Magazine*, 72–80.
2. Cheah, J., North, R., Wang, W., and Perkins, R. (2010) Thoughts on adjunct cell phone support within JTRS network. Proceedings of MILCOM 2010, NPP-12, pp. 1143–1148.

3. El-Damhougy, H., Yousefizadeh, H., Lofquist, D., *et al.* (2005) Hierarchical and federated network management for tactical environments. Proceedings of MILCOM 2005, U402.
4. The Latest Versions of all 3GPP Specifications, Containing the Most Recent Corrections and Additions, http://www.3gpp.org/ftp/Specs/latest/.
5. ICWG Network Management, Interface Control Document (for FCS, JTRS Cluster 1, WIN-T External Network Management Interfaces), Version 1.5 Draft, June 17, 2005.
6. Global Information Grid Net-Centric Implementation Document, Network Management (T500).
7. Global Information Grid Net-Centric Implementation Document, Enterprise Operations Management (EM000).
8. Global Information Grid Net-Centric Implementation Document, Cross Segment Enterprise Management (CrS-EM000).

11

Network Management Challenges in Tactical Networks

In Chapter 10, we saw the important role that network management (NM) and network operations (NetOps) play in bringing cellular capabilities to tactical networks. In this chapter, we will cover network management challenges in tactical networks, especially those challenges faced when creating joint NM for joint missions. Although many of the joint mission challenges can be political, there are significant technical hurdles. This chapter presents some challenges faced by the NM architecture from the tactical edge and across the GIG. At the GIG, the IP-based networks and radios must work seamlessly together on joint missions, in the presence of encryption tunnels. We will address some scenarios where unconventional NM techniques must be considered in order to meet the objectives of the joint mission. These techniques should be considered when different networks, must interface together. This chapter is divided into four main sections. The first reviews policy based network management (PBNM) and the use of gaming theory concepts in NM. The second presents some of the urgent challenges facing NM interoperability across the GIG, while the third covers an architectural approach that relies on gateway nodes (GNs) and abstraction techniques. The fourth section presents a conflict resolution case, showing how different tactical network managers need to interface to each other with techniques that are not considered in commercial networks.

If the reader is specialized in the area of NM/NetOps, references covering PBNM and the use of gaming theory in NM can offer more details than this chapter. There are many commercial and military references that cover PBNM in thorough detail. There are also numerous papers covering the use of gaming theory in NM. This chapter is meant to emphasize *some* of the NM challenges that are specific to tactical network.

11.1 Use of Policy Based Network Management and Gaming Theory in Tactical Networks

Many industry standards support PBNM, which uses policy driven automation to manage complex enterprise and service provider networks. There have been many attempts to automate the management of tactical networks. The recent maturity of PBNM, to the level

Tactical Wireless Communications and Networks: Design Concepts and Challenges, First Edition. George F. Elmasry.
© 2012 John Wiley & Sons, Ltd. Published 2012 by John Wiley & Sons, Ltd.

needed for dynamic NM, makes it a well-suited approach for use in a tactical network. In the commercial world, PBNM involves the use of policies to enable automated management of the enterprise with a business-centric view. In the tactical world, PBNM can use policies, driven by the commander's intent and the military rules, to automate the management of the war theater networks, with a military operational-centric view. PBNM is always tied to other technologies such as identity management (IdM) and service oriented architecture (SOA). These technologies support PBNM in achieving the end-to-end service delivery requirements.

The developers of PBNM for tactical networks study how the commercial world aims to distill business-oriented policy into implementable NM rules. Policy management is considered from a business perspective. A policy is defined with an understanding of why we do business, how we do business, and with whom we do business. This understanding is fundamental to a managed policy framework for the enterprise network. One can easily draw parallels between commercial networks and tactical networks; in the latter we need to understand why the military forces are deployed, how they execute the mission and with whom the mission is conducted (joint force NM is covered here in detail). This understanding is fundamental for the application of PBNM to tactical networks.

In commercial networks, delving into PBNM requires the understanding of an integrated NM model that includes different functional areas such as extensible markup language (XML), the lightweight directory access protocol (LDAP), the simple object access protocol (SOAP). Also, Java is often used to provide the needed distributed communication capabilities. This multifaceted framework facilitates the coordination of user accounts, as well as the entitlement and network service provisioning, when utilized as an "enterprise service bus." Attributes from each of these functional areas can be shared for integrated process control. The enterprise service bus can then be made to facilitate event correlation, where events can be used to trigger appropriate actions. Thus, the service bus can provide dynamic responses to critical situations.

It is very challenging for an NM utility to apply mathematical analysis to the study of tactical networks due to mobility, dynamic topology, fluctuation of traffic demand, and unpredictability of link quality that characterize tactical networks. Gaming theory offers the ability to model individual, independent decision makers whose actions potentially affect all other decision makers. By modeling these decision makers we can analyze the performance of tactical networks. The key is to model the various interactions in tactical networks as if these interactions were plays in a game. This allows the NM entity to analyze the existing protocols and resource management techniques as they pertain to the mission needs. The NM utility can then apply equilibrium-inducing mechanisms that provide incentives for individual users (node) to behave in socially constructive ways (for the good of the network and the mission at hand). The application of gaming theory, in tactical networks, can be done in many areas to include: power control, waveform adaptation, MAC layer parameters, routing, and flow control.

We discussed, in Chapter 6, how the network centric waveform (NCW) controller implements an adaptive resource management technique where nodes with more traffic demand are assigned more channels. In Section 11.4.2, we will illustrate a similar scenario where the wideband networking waveform (WNW) NM utility can flex some authority over the NCW resources allocated to NCW and WNW capable platforms. Reference [1] shows a simple implementation of gaming theory in the tactical GIG topology management, where the allocations of NCW resources are based on equilibrium between the needs of the nodes (traffic demand) and the network (full connectivity and overall throughput).

11.2 Challenges Facing Joint Forces Interoperability

Tactical networks are presently divided between the different branches of the US military. The military branches have separate, independent programs. Each has put in a great deal of effort to ensure its own success. However, these different IP-based networks and radios must interoperate seamlessly across the tactical GIG to guarantee the success of joint and coalition missions. The NM architecture, across the tactical GIG, needs to play a role in providing the joint forces with proper interoperability. This requires some level of interoperability between adjacent network managers. The challenges facing joint interoperability are numerous:

- The success of each individual network program does not automatically guarantee the success of the GIG.
- The networks need to work seamlessly to carry information to the warfighter, at the right place and time, with the right quality of service (QoS).
- The NM applications of the different services do not interoperate.
- The R&D facilities for all the services address similar issues, with minimal cross-service leveraging for developmental efficiency and interoperability.
- The development of new military systems does not include requirements or testing of interoperability with other services since there are few joint R&D programs or collaborations that address joint interoperability requirements.
- Interfaces have not been defined for NMs of different networks to pass tactical information to each other.
- Any solution must be independent of the specific NM capabilities used by any of the forces, as they were developed independently.

There have been many attempts by the US defense community to address the challenges associated with joint mission interoperability. Programs such as Joint Defense Information Infrastructure Control System (JDIICS), and its replacement, the Joint Network Management System (JNMS), were designed to provide a standardized tool for automated, joint NM.

This chapter discusses many topics related to the presence of high assurance internet protocol encryption (HAIPE) devices. As discussed throughout this book, these devices create plain text (red) subnets or enclaves and a cipher text (black) network core. The challenges of creating seamless IP flow at the encrypted core (where HAIPE tunnels are established on top of the cipher text IP layer gateways) are of specific interest.

11.3 Joint Network Management Architectural Approach

This section presents a joint NM architectural approach, focusing on the networks of different US services. While these concepts are designed for US networks, they apply equally to the various networks within the same force, as well as coalition networks. The approach passes information between the cipher text NMs of the army, navy, and air force networks, such that intelligent decisions (routing and policy) can be made. This architecture

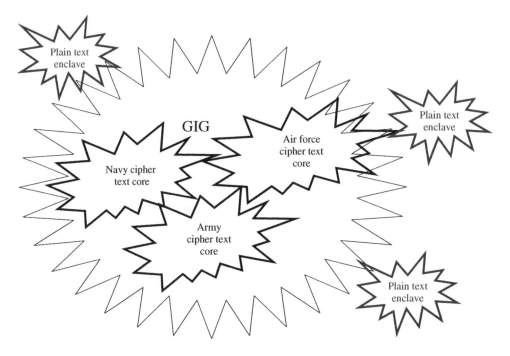

Figure 11.1   Plain text enclaves and the global information grid (GIG).

is a "strawman"[1] approach, where QoS and policy information are passed in an abstracted but sufficient manner. It can also pass along other NM information and generate a service provider interface (SPI) that will ensure joint interoperability between these otherwise independent peer networks.

One of the most important constraints of the GIG is that each network must have a cipher text core that follows National Security Agent (NSA) standards for HAIPE encryption. The cipher text cores of different networks are envisioned to merge and create one seamless cipher text core, where packets can flow from one network to another, helping the warfighter carry out joint missions.

Figure 11.1 depicts how the plain text enclaves (subnets) communicate over a cipher text core network, with different cipher text networks aggregated to create the tactical GIG. Packets are encrypted as they enter the cipher text network and decrypted as they reach their destination subnet. As discussed in earlier chapters of this book, NSA standards require a complete separation between the plain text (red) enclaves and the cipher text (black) core. This separation also implies that the NMs of the plain text enclaves are not allowed to communicate with the NM(s) of the cipher text core, unless they go through a domain guard approved by the NSA. Note that the core network can have many wireless links, especially at the lower echelons. The focus here is on the tactical level of the cipher text core, where interoperability with the other military branches is needed. An interoperability approach will be presented for the NMs associated with the GNs of different forces' networks. The

---

[1] The term "strawman" means that only a skeleton architecture for a plausible solution is presented, as a starting point.

approach will allow these networks to share minimum, but sufficient information, in line with the GIG requirements.[2]

One would need to create an SPI definition to pass information among the different NMs that will support these networks with resource management. SPIs must be abstracted, but carry sufficient information to manage resource allocation without significant control traffic. It is critical that the addition of the joint management component be independent of the specific NM capabilities used by any of the joint forces at the cipher text core. The end result is a well-defined SPI for the NM systems at each of the forces' core networks (Warfighter Information Network-Tactical (WIN-T) for the Army, Automated Digital Network System (ADNS) for the Navy, and Airborne Network (AN) for the Air Force), implemented as an add-on module. The goal to **not** affect any of the existing NM designs that support interoperability. We aim to have the SPI enable every NM capability.

11.3.1 Assumptions and Concepts for Operations (ConOps)

For tactical networks, we cannot create an SPI similar to that of commercial networks, since tactical networks have the following characteristics that generate different requirements:

- **Mobility**: Bandwidth fluctuations, due to mobility, can make it impossible to meet a specific Service Level Agreement (SLA) at all times between all access points.
- **ConOps**: Tactical networks are controlled by operational boundaries that define a specific command structure. This structure defines which nodes in peer networks can communicate with each other during defined periods of a maneuver or operation. Building NM capabilities based on ConOps is a must.
- **Red/black separation**: This is the hardest challenge for tactical networks. To generate the SLA for the red-side NMs, one must overcome the hurdle of having little knowledge of the path at the black side. This leads to a simplistic SLA that is not dependent on path, egress points, dynamics of the black core, and topology changes.

The cipher text NM interoperability (black–black interface) augments the plain text NM capabilities, with a cipher text peer NM interface relying on the GNs. This interface would contain an SPI that is path specific, tied to operational needs, quickly reacts to the dynamics of the network, and meets the Information Assurance (IA) requirements.

The rationale behind this black–black interface draws from the need to adhere to red–black separation, complement plain text NM capabilities, consider specific path use in a peer network, and create a dynamic SPI at the cipher text side that is tied to operational requirements.

As explained in Chapter 8, the plain text IP layer can use measurement based resource management (MBRM) and measurement based admission control (MBAC) which relies on a set of thresholds (delay, jitter, packet loss, etc.). If these thresholds are violated, the plain text IP layer raises its estimation of the cipher text core congestion severity level and executes admission control and flow control policies based on the estimated severity level. The plain text IP layer thresholds are based on some service level expectations and, as these

---

[2] References [2–4] (accessible only to authorized individuals) define some GIG architectural approaches meant to create commonality between the networks of the different forces.

are violated, congestion is assumed. The plain text IP layer blocks and preempts the lower precedence traffic to allow higher precedence traffic to use the limited bandwidth. The plain text IP layer thresholds are not path aware, and apply to all egress points of the encrypted core. One can identify this as a shortcoming, seeing how the plain text implementation of MBRM and MBAC is not path aware. If this implementation is to be used on paths to peer networks (e.g., from WIN-T across to the AN), it will be reactive in nature and will not create the dynamic demand for a specific service level for a specific path in a specific time, based on some operational needs.

The black–black NM interfacing between peer networks creates this path-dependent, dynamic SPI to augment the plain text side NMs' interfacing capacities. While the red-to-red interfacing can define some level of service expectation between peer networks, the black-to-black interfacing can define a dynamic, path dependent SPI that adheres to the red–black boundaries.

From an operational perspective, the cipher text side NM of the AN would have a set of addresses (i.e., HAIPE cipher text addresses as explained in Chapter 9) from the Army's WIN-T network (e.g., Points-of-Presence (POPs) to specific commanders on the ground who should be allowed to communicate to the AN). These sets of cipher text address pairings will define the specific paths to be considered for implementing dynamic SLAs between the different networks, as explained later in this chapter.

Notice here that the presence of HAIPE tunnels presents many new challenges that would not exist in non-HAIPE or IPSec encrypted networks. For multihomed gateways, the existence of multiple HAIPE versions (where older versions create static tunnels while newer versions implement dynamic discovery) increases the complexity of multihomed routing. As Figure 11.2 shows, the presence of multiple tunnels for multihoming cases results in the

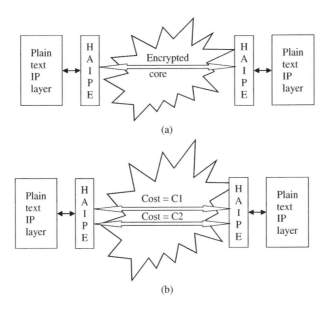

(a)

(b)

Figure 11.2 Different possibilities for multihoming with red/black separation. a) Plain text IP layer peer with single gateway address and HAIPEs form single tunnel. b) Plain text IP layers peer with multiple gateway addresses and HAIPEs form multiple tunnels to the same destination enclave.

plain text IP layer needing to make a routing decision in case (b) to reach the destination enclave. Technically, the plain text IP layers see each other, over the encrypted core, as the next hop (the encrypted core is just a black box). However, the existence of many possible paths at the encrypted core means that the encrypted core path with the lowest metric (i.e., cost) must be communicated to the plain text IP layer. The lowest metric here can mean security metrics or it can have a traffic engineering basis, where terrestrial links are used for real-time and satellite links for non-real time traffic. The plain text IP layer can be given metrics based on the traffic type to assign a tunnel end-point to each traffic type, or a single tunnel to all traffic types based on the security constraints. Due to the complexity of multihoming (which increases when protocols such as SCTP – Stream Control Transmission Protocol – are used at the red enclaves), the plain text IP layer can be configured manually, creating static paths. The gateway techniques described below assume the existence of static route configurations. Please refer to Figure 8.2 to see how interface "X" can be used to avoid the static configuration and synchronize the plain text NM with the dynamics of the encrypted core known by the cipher text side NM. However, keep in mind that interface "X" constitutes a violation of the red/black separation and any information passing through the domain guard has to be approved by the NSA.

Chapter 9 covered some of the multicast challenges faced by the plain text IP layer and explained how multicast is tied to the cipher text IP layer topology. Multicast transport over HAIPE networks presents another challenge for the NMs in joint missions. For example, reconnaissance information could be multicast to multiple destinations in different joint and coalition networks. In order to optimize multicasting, the NMs for the different joint mission's networks could face a slew of different scenarios as follows:

1. HAIPE supports multicast tunneling over the cipher text core.
2. HAIPE does not support multicast tunneling over the cipher text core.
3. HAIPE does support multicast tunneling, but the cipher text IP layer topology may not be formed considering the HAIPE multicast tunnels.
4. There is some knowledge about the cipher text side topology, that is, the layover of multicast tunnels is known by the plain text IP layer. Multicast over HAIPE could utilize this knowledge to perform multicast pruning.
5. There are multiple access waveforms at the cipher text core, with the multicast tunnels utilizing the multiple access capabilities of these waveforms.

11.3.2 The Role of Gateway Nodes

It is difficult or near impossible to suggest any changes in the joint forces' networks. US defense programs tend to be independent and focus on independent service needs. The most successful interoperability efforts create add-on capabilities at the specific interface points. The focus of the interoperability efforts is on GNs since each force has a form of GN with some NM capabilities. Examples of such efforts are the WIN-T Joint gateway Node (JGN) for the Army,[3] the ADNS GN for the Navy, and AN GN for the Air Force. The architecture focuses on the NMs resident at these gateways. Although in this book we focus on the Army, Navy, and Air Force, this approach can be easily extended to include the Marines, National Guard, Special Operations, and coalition forces.

---

[3] The Army's Gateway Node is usually referred to as Joint Gateway Node.

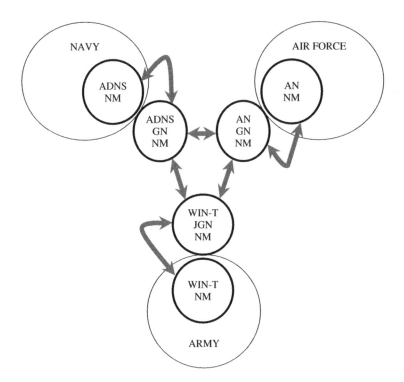

Figure 11.3 Gateway nodes' roles for interoperability.

Figure 11.3 presents a logical diagram for the interoperability solution, showing the services' gateways nodes: WIN-T JGN, ADNS GN, and AN GN. As shown, a NM at each GN coordinates the passing of information between the three gateways. The thick gray lines in the figure define the interfaces between these entities. It is crucial to leave the internals of each of the forces' networks untouched. In the following sections, we will discuss interoperability information sharing between the different NMs.

The GNs need more capabilities for the SPI, than discussed in this chapter, in order to handle issues such as multihoming and multicast. Also, topics such as the use of Border Gateway Protocol (BGP), the creation of rendezvous points at the gateways for multicast traffic and handling of address resolution, and so on, are beyond the scope of this book. The chapter discusses how unconventional NM techniques are needed and focuses on how to make a SPI works over HAIPE tunnels.

11.3.3 Abstracting Information

Joint mission networks can be quite large. For example, the Army's network(s) can contain thousands of nodes, each with multiple plain text enclaves and a gateway router for the cipher text side. The size of these networks suggests that information shared should be minimized but sufficient to meet the mission interoperability requirements. One does not expect peer NMs to need detailed information about each other's network topology. The information supplied to a peer network should be abstracted (to avoid consuming excess bandwidth or increase security vulnerabilities) but at the same time, should be sufficient. A specific path could be defined based on a HAIPE tunnel carrying the external network traffic.

Recall the discussion about HAIPE tunnel association in Section 9.4.5. We showed that the planed deployment dictates the cipher text side tunnel formation. The same principle applies in planning joint missions. As an example, consider other services utilizing the Army's network. Certain hosts (i.e., commanders or the joint missions operational commands) in the Army's network will send traffic to, and receive traffic from, external networks (e.g., AN or ADNS). These hosts belong to plain text enclaves (all have HAIPE tunnels). The cipher text side HAIPE addresses of these hosts should be known across the joint forces as part of the plan. The AF or the Navy NMs will focus on requesting specific information from the Army NM, indicating the HAIPE tunnel address that the commander on the ground is using to receive traffic from these external networks. The same concept applies to the Navy and the AF as well. Thus, the NM for each force will have a set of HAIPE tunnel addresses across the joint forces where there is a need to achieve a specific service level.

11.3.4 Creating Path Information

Figure 11.4 shows an example of how a path from the JGN to a WIN-T POP (assumed destination of traffic from the AN or the ADNS) can be defined by multiple hops. The WIN-T NM can derive the path information from the hop information, since it already has all or most of the hop information needed.

Let us consider that a WIN-T node has an agent collecting information about the radio link, and feeding this information to the WIN-T NM. Also, the WIN-T NM can query the cipher text IP layer about service policy information such as minimum, maximum, and average delays per packet per DSCP (Differentiated Services Code Point). One can divide the hop information into radio link information (coming from the radio to the agent) such as signal-to-noise ratio (SNR) and Bit Error Rate (BER), and router information (coming from the black router) such as delays (minimum, maximum, and average per DSCP) and packet drop data.

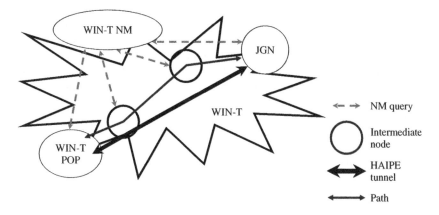

Figure 11.4 A multihop path from JGN to WIN-T POP.

Let us go back to Figure 11.3, consider the scenario where the AN NM queries the AN GN about WIN-T network status for a given destination HAIPE tunnel address. The AN GN will pass this query to JGN, which queries the WIN-T NM about the path status of a specific destination tunnel address. The WIN-T NM defines the path, as shown in Figure 11.4, where the first hop is the JGN node, the second is the first intermediate node defined by the top right circle and the next is the second intermediate hop defined by the other circle. The WIN-T NM will create the path information from the hop information to reply to the query from the JGN NM, as follows:

The radio with the lowest SNR for the hop can be chosen to reflect the path status, that is, the reply to the query can have an object with an attribute defining the lowest SNR for the hop in the path. The BER can be collective (the overall effect of the BER in the path); for instance the reply to the query can contain an attribute defining the collective BER of the path.

Using the cipher text IP parameters, one can create a Probability Distribution Function (pdf) from the minimum, average, and maximum delays collected from the cipher text IP layer at each node. The collective effect of queuing delays over the path can be calculated by convoluting the pdfs of all the hops in the path. The reply to the query can contain attributes describing the minimum, average, and maximum delays of the path. The packet drop data from the cipher text IP layer can also be used to create the cumulative packet drop of the path.

Thus, the objects describing the path information can possibly have the following attributes:

- *Lowest SNR*
- *Cumulative BER*
- *Collective Maximum Delay*
- *Collective Average Delay*
- *Collective Minimum Delay*
- *Cumulative Packet Drop.*

The objects describing the WIN-T network obtained by the JGN NM have attributes that abstract the WIN-T network, focusing on single path information instead of the entire network. The information from multiple hops is processed and abstracted to simplify the information passed by the GN. The question we must ask here is can it meet the service level requested? In case it cannot, the service provider network (WIN-T in this example) will pass the abstracted set of information to the peer network (AN in this example). The peer network will then make an intelligent decision to either request a lower level of service or proceed anyway with the demanded service with no guarantee or shape traffic at the gateway.

The general rule is that any of the three networks in Figure 11.3 can query its GN about specific path information. This GN will pass the query to the external GN with a specific HAIPE tunnel address for information about the external network; the external GN will define a path in its network and collect the path information from hop information. The reply to the query will follow the same format, with the objects describing the attributes as defined above.

Creating the path information from the hop information can require many aggregation methods, depending on how the NM monitors and analyzes the network. One possible approach is to rely on the cipher text IP service policy metrics (which are defined as COTS

(Commercial-Off-The-Shelf) standards) which indicate the packet loss, delays (maximum, average, and minimum), and radio metrics in terms of the SNR and the BER, to create the hop information. Aggregating the hop information to pass this path specific information can follow probabilistic models that create pdfs using the hop metrics. The convolution of these pdfs to create path pdfs is used to create the path metrics.

11.3.5 Sequence Diagram

In the section above, the interoperability concept was demonstrated with an example. This section makes this concept generic where Network A requests a service from Network B, and thus the NM of Network A queries the NM of Network B to obtain the necessary information about Network B. The information needed by Network A about Network B should be abstracted but sufficient. Network A can be any of the three networks in Figure 11.3, with B being either of the other two.

Figure 11.5 shows the sequence diagram where we have a GN NM for each service per the setup in Figure 11.3. The following steps detail the sequence:

a. Network A requests a level of service from Network B (A will utilize B and requests a specific service level from B). Network A NM issues a query (arrow with Step a) to its GN NM. This query contains Object 1, whose attributes include the destination HAIPE tunnel address, the DSCP (path reliability is DSCP dependent), and the time interval for service.

b. Network A's GN NM passes the query (containing Object 1) to Network B's GN NM.

c. Network B's GN NM adds an attribute to Object 1, thus creating Object 2. The added attribute is the source address (a path is defined by a source and destination address). Here, the source address is specific to Network B's GN and could be defined by the black router of the GN in Network B.

d. Network B's NM collects path information and generates Object 3. This object contains the source address specific to Network B's GN.

e. Network B's GN NM, based on the path information, decides if the requested level of service can be met and passes Object 3 to Network A's GN NM. This object indicates if the requested level of service can be met. If it cannot be met, the available level of service is provided.

f. Network A's GN NM removes the source address, specific to Network B, from Object 3, thus creating Object 4. Object 4 is the reply to the requested level of service and contains the abstracted information about Network B and is specific to a destination HAIPE tunnel address in Network B.

g. Network A's NM sends out Object 5 to its GN NM to stop this query (end the requested service). Object 5 will have the destination HAIPE tunnel address, the DSCP and a STOP as its attributes.

h. Network A's GN NM will pass Object 5 to Network B's GN NM.

i. Network B's GN NM adds an attribute to Object 5, creating Object 6 and sends it to its own NM. The added attribute is the source address.

The above sequence diagram can be implemented with any NM tool. It is an add-on to whatever solution(s) are already being used by the services and does not require a redesign of NM capabilities to create interoperability.

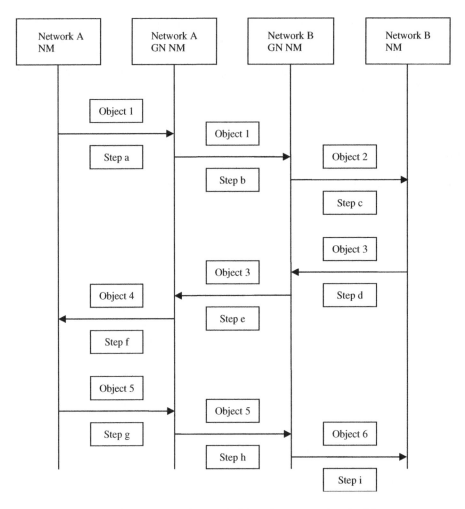

Figure 11.5 Sequence diagram defining the proposed methodology.

11.4   Conflict Resolution for Shared Resources

This section presents an approach for resolving conflicts between two NMs where one owns a specific set of resources meant to be shared with another peer network. A known case is the resource management of NCW channels (explained in Chapter 6), which is part of WIN-T and is managed by the WIN-T network manager, but shared with the Joint Tactical Radio System (JTRS) WNW. The JTRS WNW NM (known as JWNM) can optimize the WNW network throughput more efficiently if it can influence how NCW channels are allocated. Specifically, which nodes with both (WNW and NCW) capabilities can use NCW channels, and at what times, in order to optimize the WNW subnet throughput efficiency? The presented approach, while illustrated for the NCW case, can easily be adapted for other shared resources.

11.4.1 Tactical Network Hierarchy

Please refer back to Figure 6.1, which shows the US arm tactical GIG network deployment with the hierarchy of waveforms (SRW – Soldier Radio Waveform, WNW, HNW – Highband Networking Waveform, and NCW). NCW reach-back capabilities are used at the WNW local subnets level, with each WNW subnet having some WNW/NCW-capable platforms for reach-back. JWNM plays a major role in managing the WNW local and global subnets. The Brigade Combat Team Modernization (BCTM) program utilizes the concept of a Manager of Mangers (MoM) to manage everything between the tactical GIG lower hierarchies, presented in Figure 6.1, all the way up to the WNW global subnet. HNW and NCW resources are considered WIN-T resources, and like JWNM, the MoM has no authority over these resources. The split of NCW resources between WNW subnets and WIN-T nodes may be decided based on the mission planning. One could allocate some NCW resources to the WNW/NCW-capable platforms in the WNW subnets, with the remaining resources left for the WIN-T nodes.

As explained in Chapter 6, the NCW waveform uses a dynamic bandwidth-on-demand mechanism to allocate more channels to the GNs, where there is a higher demand. During a mission planning phase, the commander could consider certain strategic resources, such as WNW/NCW-capable nodes as lifelines, and designate them exempt from the dynamic resource allocation.

The NCW resources allocated to WIN-T nodes are beyond the scope of this text. We will instead focus on the portion of the NCW resource that is pre-planned for WNW/NCW-capable nodes. One can create a simple resource allocation technique, where JWNM can request NCW resources. If these resources are granted, then they are allocated. However, if they are denied, no action is taken. The JWNM may have special techniques for making a specific allocation of the NCW resources, within its nodes, that will have a greater impact on the WNW network throughput efficiency. For this, the JWNM would need to have access to some form of resource management of a subset of the NCW resources. If, for example, the WIN-T network management (WIN-NM) can allocate 30 NCW channels to the JWNM, then the JWNM must be allowed to use this subset dynamically to optimize its own network throughput. The JWNM should be allowed to choose when and which NCW/WNW-capable nodes can be NCW-enabled. If the brigade has 100 nodes with NCW/WNW capabilities, not all 100 nodes can be NCW-enabled at all times. The JWNM would be in the best position to decide how the NCW channel subset (e.g., 30 channels) should be used and at what times.

11.4.2 Dynamic Activation of NCW in WNW/NCW-Capable Nodes

To further illustrate this conflict resolution for shared resources, consider Figure 11.6a, showing a sample subnet with WNW nodes managed by the JWNM. A subset of X nodes, that are also NCW capable, are depicted with a $+$ inside the oval. In this subnet, the available NCW channels are Y, where $Y < X$. Here $Y = 2$, the two WNW/NCW-capable highlighted nodes. The JWNM can have an optimization technique for minimizing the WNW traffic hops, for the traffic leaving the WNW network before it is routed to NCW. This would be based on both the number of hops and the traffic load.

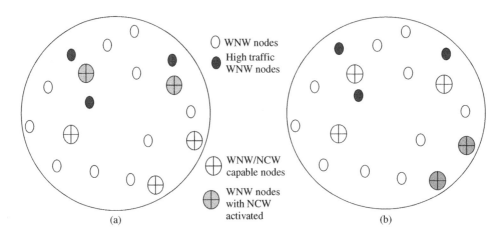

Figure 11.6 WNW subnet with: (a) desirable configuration and (b) inefficient configuration.

In Figure 11.6a, the solid oval nodes, that have high traffic loads leaving the WNW subnet, are just a single hop away from the two activated, NCW-enabled, nodes. In Figure 11.6b, these high traffic nodes are far from the two NCW activated nodes, and the high traffic load will have to traverse WNW, over multiple hops, before reaching an NCW capable node. The configuration presented in Figure 11.6a is evidently more desirable than that presented in Figure 11.6b. The configuration in Figure 11.6b would result in a much less efficient throughput.

Through this simple example, one sees that if NCW node activation is left to the WIN-T NM, a random enabling of NCW capable nodes can result in poor throughput efficiency. The JWNM will likely have a more thorough knowledge of its own network and can manage it more efficiently if it has a say in controlling the NCW resource allocation. Thus, the JWNM should be allowed to co-manage its share of the NCW resources, even though the WIN-NM continues to own them.

11.4.3   Interfacing between the WIN-NM and the JWNM for NCW Resources

In order to negotiate, the JWNM and the WIN-NM can use a defined interface for NCW resource management. With this interface, the two NM systems would interchange messages, allowing the JWNM to have a say in the NCW resource management. The message exchange for this interface can be as follows:

- **MSG 1**: the JWNM requests NCW resources for node 1.
- **MSG 2**: the WIN-NM grants the resources.
- **MSG 3**: the JWNM requests NCW resources for node N.
- **MSG 4**: the WIN-NM denies resources.
- **MSG 5**: the JWNM requests NCW resources swap between nodes 1 and N.
- **MSG 6**: the WIN-NM grants resources swap.

This is a simple example of the possible exchanges between the two NMs. Once this interface is fully defined and implemented, further enhancements can be easily added. The protocol gives the JWNM the right to choose which node(s) will receive the NCW resources. The JWNM will also determine when it is necessary to swap the resources in order to optimize the WNW throughput efficiency.

11.4.4   NCW Resource Attributes

The attributes of the message exchange, mentioned above, can be outlined as below. This outline is simple, and more attributes and messages can be added if there is a need for more efficient NCW resource management. The attributes of the messages are as follows:

- **MSG 1: Request resources**. The JWNM requests NCW resources for node 1, with these attributes:
  - mode = request
  - amount = NCW BW (or # of channels)
  - name and location of node 1
  - node 1 WNW interface IP address.
- **MSG 2: Grant resources**. The WIN-NM grants the resources to node 1, with these attributes:
  - mode = grant
  - amount = NCW BW (or # of channels)
  - name and location of node 1
  - node 1 WNW interface IP address.
- **MSG 3: Request resources**. The JWNM requests that the WIN-NM allocate NCW resources for node N, with these attributes:
  - mode = request
  - amount = NCW BW (or # of channels)
  - name and location of node N
  - node N WNW interface IP address.
- **MSG 4: Deny resources**. The WIN-NM denies the allocation since the NCW resource pool is already allocated, with these attributes:
  - mode = deny
  - name and location of node N
  - node N WNW interface IP address.
- **MSG 5: Request resources swap**. For throughput efficiency, the JWNM really needs NCW activation for node N, and determines that node 1 need not be NCW activated. Thus, the JWNM requests a NCW resource swap between nodes 1 and N, with these attributes:
  - mode = swap
  - amount = NCW BW (or # of channels)
  - name and location of node 1
  - node 1 WNW interface IP address
  - name and location of node N
  - node N WNW interface IP address.

- **MSG 6: Grant resources swap**. The WIN-NM grants resources swap, with these attributes:
  - mode = grant swap
  - amount = NCW BW (or # of channels)
  - name and location of node 1
  - node 1 WNW interface IP address
  - name and location of node N
  - node N WNW interface IP address.

11.5   Concluding Remarks

This chapter has focused on the NM architecture across the tactical GIG and described the need for the cipher text NM interoperability for the joint forces. It is important to consider the dynamic nature of the tactical GIG, which requires NM techniques that pass abstracted but sufficient information among the joint forces' networks. This in turn enables each force's NM to make intelligent resource management decisions. The abstraction methodology is based on HAIPE tunnel addresses. A query structure minimizes the control signaling, while making the networks interoperate seamlessly. While this chapter focused on the SPI, QoS metrics, and real-time resource management information, the interfaces presented here can be expanded to cover other NM areas (such as fault information, spectrum information, etc.) to create a comprehensive SPI definition of the GIG interoperability.

The presented case of sharing of NCW resources with the JWNM is simply a case study; similar situations exist in other DoD networks. For example, the AN is expected to be used by different networks as a reach-back in joint missions. If the NM of the AN were designated to provide certain resources for the ground networks, a similar interface could be used to allow the ground networks to have a say in managing the provided resources for the optimization of their own resources.

In concluding this third part of the book, it is necessary to emphasize that the concept of an open architecture in tactical networks, although better than using proprietary protocol, has many challenges. We have seen how the protocol stack layers need extra capabilities that are not needed for commercial networks. We have already seen and discussed the need for cross layer signaling, the consequences of using HAIPE encryption, and the challenges associated with interfacing tactical networks to commercial cellular wireless technologies. In this chapter, we scratched the surface of the challenges facing NM and NetOps, associated with creating the seamless tactical GIG from the networks of the different military forces.

As we conclude this book, I hope the reader has found this presentation of tactical wireless communications and networking helpful. The challenges we face in this field are enormous and one cannot specialize in one area of tactical networks without having a general idea about many of the other areas. I hope the theoretical concepts, covered in the first part, the different generations of tactical radios presented in the second part, and the open architecture message delivered in the final part help the reader gain the necessary general understanding of the field. I also hope that engineers, in the different areas of tactical communications, can find this book useful in creating design concepts for their specific area of expertise, based on the awareness of the dependencies of the different areas of tactical wireless communications and networking.

Bibliography

1. D'Amour, C., Life, R., Elmasry, G.F., and Welsh, R. (2004) Determining network topology using governing dynamics based on nodal and network requirements. Proceedings of MILCOM 2004, U076.
2. Global Information Grid Net-Centric Implementation Document, Network Management (T500).
3. Global Information Grid Net-Centric Implementation Document, Enterprise Operations Management (EM000).
4. Global Information Grid Net-Centric Implementation Document, Cross Segment Enterprise Management (CrS-EM000).
5. Srivastava, V., Neel, J., Mackenzie, A.B., *et al.* (2005) Using game theory to analyze wireless ad hoc networks. *IEEE Communications Surveys*, **7** (4), 46–56. (Fourth Quarter).
6. Zaikiuddin, I., Hawkins, T., and Moffat, N. (2004) Towards a game-theoretic understanding of Ad-Hoc routing. Proceedings of Workshop on Games in Design and Verification, July 2004, Electronic Notes in Theoretical Computer Science, Vol. 119 (1), February 2005, pp. 67–92.
7. Neel, J., Reed, J., and Gilles, R. (2002) The role of game theory in the analysis of software radio networks. SDR Forum Technical Conference, November 2002.
8. Popescu, D.C. and Rose, C. (2002) Interference avoidance applied to multiaccess dispersive channels. Proceedings of the IEEE International Symposium on Information Theory, July 2002.
9. Kosiur, D. (2001) *Understanding Policy-Based Networking*, John Wiley & Sons, Inc., ISBN: 0-471-38804-1.
10. Elmasry, G.F., Jain, M., Welsh, R., *et al.* (2011) Network management challenges for joint forces interoperability. *IEEE Communications Magazine*, **49** (10), 81–89.
11. Elmasry, G.F., Jain, M., Jakubowski, K., and Whittaker, K. *et al.* (2010) Conflict resolution for shared resources between network managers. Proceedings of Milcom 2010, CSNM-10.5.
12. Elmasry, G.F., Jain, M., Orlando, R., *et al.* (2009) Joint network management across future force tactical networks. Proceedings of Milcom 2009, U308.
13. http://www.globalsecurity.org/intell/library/reports/2001/compendium/JDIICS_D.htm.
14. ICWG Network Management (2005) Interface Control Document (for FCS, JTRS Cluster 1, WIN-T External Network Management Interfaces), Version 1.5 Draft, June 17, 2005.
15. Strassner, J. (2004) *Policy Based Network Management*, Morgan Kaufmann Publishers, Copyright, Elsevier.
16. http://www.tmforum.org.
17. http://www.dmtf.org/standards.
18. Braz, K.A., Payne, A.T., and Julasak, C. (2005) Mission-oriented NETOPS: an operational framework for coordinated planning, analysis, activation, monitoring, and response for current and transformational networks. Proceedings of MILCOM 2005, U212.
19. Hershey, P., Runyon, D., and Yangwei, W. (2007) Metrics for end-to-end monitoring and management of enterprise systems. Proceedings of MILCOM 2007, NAPM-2.
20. El-Damhougy, H., Yousefizadeh, H., Lofquist, D., *et al.* (2005) Hierarchical and federated network management for tactical environments. Proceedings of MILCOM 2005, U402.

Index

3G/4G/LTE 207, 209, 261–5, 267,
  269–72
3G/4G/LTE access at an enclave 271
3G/4G/LTE services in dismount unit
  269

Adaptive rate codec 244
Additive white Gaussian noise 14, 22–5,
  28–30, 38, 40, 57–8, 67, 109, 125,
  195, 197
ADC see Analog-to-digital converter
ADNS see Automated digital network
  system
Advanced InfoSec machine 188
AFH see Spread spectrum/Frequency
  hopping/Adaptive frequency hopping
AIM see Advanced InfoSec machine
Airborne network 141, 279–84, 290
AN see Airborne network
Analog-to-digital converter 184, 189, 193,
  201
Application layer see OSI model/
  Application layer
ARQ see Automatic repeat request
Arrival rate 88–91, 93–4, 98–102, 106,
  148–9, 168, 243
ASCII 46, 62, 69
Automated digital network system 141,
  279, 281–3

Automatic repeat request 69, 72–3, 75,
  77–8, 234–6, 239
  Hybrid automatic repeat request 69, 72,
  75, 77–8, 239
AWGN see Additive white Gaussian noise

BER see Error rate/Bit error rate
BFSK see Binary frequency shift keying
BGP see Border gateway protocol
Binary frequency shift keying see Shift
  keying/Binary frequency shift keying
Binary phase shift keying see Shift
  keying/Phase shift keying/Binary
  phase shift keying
Binary symmetric channel 53–6
Block codes 58
  Linear block codes 58, 60–62
  Systematic linear block codes 62
Blockage detection 234–6
Bluetooth 198, 200
Bootstrap slots 151–3
Border gateway protocol 134, 282
BPSK see Shift keying/Phase shift
  keying/Binary phase shift keying
BSC see Binary symmetric channel

CAC see Call admission control
Call admission control 209, 213–14, 217,
  227, 234, 257, 263, 270, 272

Tactical Wireless Communications and Networks: Design Concepts and Challenges, First Edition. George F. Elmasry.
© 2012 John Wiley & Sons, Ltd. Published 2012 by John Wiley & Sons, Ltd.

Printed and bound by CPI Group (UK) Ltd, Croydon, CR0 4YY

16/04/2025

14658388-0001